Praise for *Reliable Machine Learning*

I don't care how much data science work you've done in the past, or how expert you are on the statistical foundations of machine learning. I don't care if you have read every line of the Tensorflow source code, or implemented your own distributed ML training from scratch. Before you ever put a real system based on machine learning into deployment, you will benefit from reading this book. This is what is needed for the thousands of upcoming ML deployments where their usefulness is a double-edged sword. The more useful, the higher the stakes around safety, security, paying customers who are counting on you, fairness, or policy decisions that will be made on the basis of your system. This book thoroughly surveys the operations you need to be running if you have this level of responsibility, and you can rest assured that it comes from combined decades of hard-won experience.

—*Andrew Moore, VP and General Manager Google Cloud AI*

MLOps wouldn't be nearly as painful if we, the people who do machine learning, applied software engineering best practices. This is a well-written and comprehensive book on these engineering best practices from some of the world's top experts.

—*Chip Huyen, author of* Designing Machine Learning Systems

Reliable Machine Learning is a must-read for people building real-world machine learning systems. It provides a blueprint for thinking about the complex and nuanced issues of developing machine learning enabled products.

—*Brian Spiering, Data Science Instructor*

可靠的机器学习（影印版）

Reliable Machine Learning

[美国] 凯茜・陈（Cathy Chen）
[爱尔兰] 尼尔・理查德・墨菲（Niall Richard Murphy）
[美国] 克拉蒂・帕里萨（Kranti Parisa）
[美国]D. 斯卡利（D. Sculley）
[美国] 托德・安德伍德（Todd Underwood）著

Beijing • Boston • Farnham • Sebastopol • Tokyo

O'Reilly Media, Inc.授权东南大学出版社出版

南京 东南大学出版社

图书在版编目(CIP)数据

可靠的机器学习 = Reliable Machine Learning：影印版：英文 /（美）凯茜·陈(Cathy Chen)等著.—南京：东南大学出版社，2023.3

ISBN 978-7-5766-0552-5

Ⅰ.①可… Ⅱ.①凯… Ⅲ.①机器学习-英文 Ⅳ.①TP181

中国版本图书馆 CIP 数据核字(2022)第 253329 号

图字:10-2022-476 号

可靠的机器学习（影印版）

著　　者：[美国]凯茜·陈(Cathy Chen),[爱尔兰]尼尔·理查德·墨菲(Niall Richard Murphy),[美国]克拉蒂·帕里萨(Kranti Parisa),[美国]D.斯卡利(D. Sculley),[美国]托德·安德伍德(Todd Underwood)

责任编辑：张　烨　　封面设计：Karen Montgomery,张　健　　责任印制：周荣虎

出版发行：东南大学出版社

社　　址：南京四牌楼 2 号　　邮编:210096　　电话:025-83793330

网　　址：http://www.seupress.com

电子邮件：press@ seupress.com

经　　销：全国各地新华书店

印　　刷：常州市武进第三印刷有限公司

开　　本：787mm×1000mm　1/16

印　　张：25.75

字　　数：504 千

版　　次：2023 年 3 月第 1 版

印　　次：2023 年 3 月第 1 次印刷

书　　号：ISBN 978-7-5766-0552-5

定　　价：119.00 元

本社图书若有印装质量问题,请直接与营销部联系。电话(传真)：025-83791830

Table of Contents

Foreword

Machine learning (ML) is at the heart of a tremendous wave of technological innovation that has only just begun. Picking up where the "data-driven" wave of the 2000s left off, ML enables a new era of *model-driven* decision making that promises to improve organizational performance and enhance customer experiences by allowing machines to make near-instantaneous, high-fidelity decisions, at the point of interaction, based on the most current information available.

To support the productive use of ML models, the practice of machine learning has had to evolve rapidly from a primarily academic pursuit to a fully fledged engineering discipline. What was once the sole domain of researchers, research scientists, and data scientists is now, at least equally, the responsibility of ML engineers, MLOps engineers, software engineers, data engineers, and more.

Part of what we see in the evolution of machine learning roles is a healthy shift in focus from simply trying to get models to work to ensuring that they work in a way that meets the needs of the organization. This means building systems that allow the organization to produce and deliver them efficiently, hardening them against failure, enabling recovery from any failures that do happen, and most importantly doing all this in the context of a learning loop that helps the organization improve from one project to the next.

Fortunately, the machine learning community hasn't had to bootstrap the knowledge required to accomplish all this from scratch. Practitioners of what has come to be called MLOps have had the benefit of a vast array of knowledge that was developed through the practice of DevOps for traditional software projects.

The first wave of MLOps focused on the application of technology and process discipline to the *development* and *deployment* of models, resulting in a greater ability for organizations to move models from "the lab" to "the factory," as well as an explosion of tools and platforms for supporting those stages of the ML lifecycle.

But what about the ops in MLOps? Here again we stand to benefit from progress made operating traditional software systems. A significant contributor to maturing the operational side of DevOps was that community's broader awareness and application of site reliability engineering (SRE), a set of principles and practices developed at Google and many other organizations that sought to apply engineering discipline to the challenges of operating large-scale, mission-critical software systems.

The application of methodologies from software engineering to machine learning is not a simple lift and shift, however. While one has much to learn from the other, the concerns, challenges, and solutions can differ quite significantly in practice. That is where this book comes in. Rather than leaving it to each individual or team to identify how to apply SRE principles to their machine learning workflow, the authors of this book aim to give you a head start by sharing what has worked for them at Google, Apple, Microsoft, and other organizations.

To say that the authors are well qualified for their task is an understatement. My work has been deeply informed and influenced by several of them over the years.

In the fall of 2019, I organized the first TWIMLcon: AI Platforms conference to provide a venue for the then-nascent MLOps community to share experiences and advance the practice of building processes, tooling, and platforms for supporting the end-to-end machine learning workflow. Among us insiders it became a bit of a running joke just how many of the presentations at the event included a rendition of the "real-world ML systems" diagram from D. Sculley's seminal paper, "Hidden Technical Debt in Machine Learning Systems."[1]

At our second conference, in 2021, Todd Underwood joined us to present "When Good Models Go Bad: The Damage Caused by Wayward Models and How to Prevent It."[2] The talk shared the results of a hand analysis of approximately 100 incidents tracked over 10 years in which bad ML models made it, or nearly made it, into production.

I've since had the pleasure of interviewing D. for *The TWIML AI Podcast* for an episode titled "Data Debt in Machine Learning."[3] The depth of experience D. and Todd shared in these interactions comes through clearly in this book.

---

1 D. Sculley et al. "Hidden Technical Debt in Machine Learning Systems," *Advances in Neural Information Processing Systems* (January 2015): 2494-2502. *https://oreil.ly/lKOWR*.

2 Todd Underwood, "When Good Models Go Bad: The Damage Caused by Wayward Models and How to Prevent It," TWIMLcon, 2021, *https://oreil.ly/7pspJ*.

3 D. Sculley, "Data Debt in Machine Learning," interview by Sam Charrington, *The TWIML AI Podcast*, May 19, 2022, *https://oreil.ly/887p4*.

And, if you're coming from the SRE perspective, Niall needs no introduction. His books *Site Reliability Engineering* and *The Site Reliability Workbook* helped popularize SRE among DevOps practitioners in 2016 and beyond.

(Though I've not previously come across Cathy and Kranti's work, it is clear that their experience structuring SRE organizations and driving large-scale consumer-facing applications of ML informs many aspects of the book, particularly the chapters on implementing ML organizations and integrating ML into products.)

This book provides a valuable lens into the authors' experiences building, operating, and scaling some of the largest machine learning systems around.

The authors avoid falling into the trap of attempting to document a static set of architectures, tools, or recommendations, and in so doing succeed at offering so much more: a survey of the vast complexity and myriad considerations that teams must navigate to build and operate—and to build operable—machine learning systems, along with the principles and best practices the authors have collected through their own extensive navigation of the terrain.

Their goal is stated early on in the text: to "enumerate enough of the complexity to dissuade any readers from simply thinking… 'this stuff is easy.'"

If we've learned anything as a community over the past several years it's that the ability to create, deliver, and operate ML models in an efficient, repeatable, and scalable manner is far from easy. We've also learned, though, that because of its willingness to openly share experiences and build on the learnings of others, the machine learning community is able to advance rapidly, and what's hard today becomes easier tomorrow. I'm grateful to Cathy, Niall, Kranti, D., and Todd for allowing us all to benefit from their hard won lessons and for helping to advance the state of machine learning in production in the process.

— Sam Charrington
Founder of TWIML, host of The TWIML AI Podcast

Preface

This is not a book about how machine learning works. This is a book about how to make machine learning work—for you.

The way that machine learning (ML) works is fascinating. The math, algorithms, and statistical insights that surround and support ML are themselves of interest, and what they can achieve when applied to the right data can be nothing short of magical. But we do something a little different in this book. We are not *algorithm* oriented—we are *whole-system* oriented. In short, we talk about everything *other* than the algorithms. Plenty of other works cover the algorithmic component of ML in great detail, but this one is deliberately focused on the whole lifecycle of ML, giving it the time and attention it doesn't really get elsewhere.

This means that we talk about the messy, complicated, and occasionally frustrating work involved in shepherding data correctly and responsibly; reliable model building; ensuring a smooth (and reversible) path to production; safety in updating; and concerns about cost, performance, business goals, and organizational structure. We attempt to cover everything involved in having ML happen reliably in your organization.

Why We Wrote This Book

We firmly believe at least some of the hype: ML and AI techniques are currently reshaping computing and society at an accelerating rate. To that extent, the public hype has not caught up with the private reality in some respects.[1] But we are also grounded and experienced enough to understand just how laughably unreliable and

1 See "27 Incredible Examples of AI and Machine Learning in Practice" (*https://oreil.ly/ITkPX*) by Bernard Marr for examples that might surprise you. This Forbes article is from 2018, now ancient history in ML terms, and the industry has continued to expand across many industries and applications. There is simultaneously both completely unjustified ML/AI hype and way more real, working applications to more industries than most people are aware of.

problematic many real-world ML systems actually are. The technology press writes about space flight, while most organizations still have trouble staying upright on their bicycles; these are the early days still. Now is the perfect time to actively pay attention to what ML can do and how your organization might benefit from it.

Having said this, though, we recognize that many organizations are worried about "missing out" on ML, and everything it could do for (and to) their organization. The good news is, there's no need to panic—it is possible to get started now and to be sensible and disciplined about how you work with ML, in a way that successfully balances both obligation and reward. The bad news, and the reason many organizations are worried, is that the curve of complexity is quite steep. Once you get past the simpler aspects, many of the techniques and technologies are just being invented, and it's hard to find a solid, paved path.

This book should help you navigate that complexity. We believe that, despite the immaturity of the industry, there is much to be gained by focusing on simplicity and standardization, an approach that has the beneficial side effect of making it easier to get started. Ultimately, organizations that deeply integrate ML into their business will benefit—some substantially[2]—but they will, of course, need a degree of sophistication about how that is done. A simpler, standardized foundation will facilitate developing that capability better than ad hoc experiments, or even worse, a system that works but no one knows how or why.

SRE as the Lens on ML

A plethora of ML books exist already, many of which promise to make your ML journey better in some way, often focusing on making it easier, faster, or more productive. Few, however, talk about how to make ML more *reliable*, an attribute often overlooked or undervalued.[3] That is what we focus on, since looking at how to do ML well through that lens has specific benefits you don't get in other ways. The reality is that current development best practices don't map straightforwardly onto the challenges of doing ML well end to end. Instead, seeing these questions through a site reliability engineering (SRE) lens—holistically, sustainably, and with the customer experience in mind—is a much better framework for understanding how to meet those challenges.

You can find a parallel argument in *Building Secure and Reliable Systems* (*https://oreil.ly/bpErx*) by Heather Adkins et al. (O'Reilly, 2020). An unreliable system can

2 See the Google Cloud report "Business Impacts of Machine Learning" (*https://oreil.ly/eWgDg*), this excerpt being perhaps the biggest reason: "Standard ML projects tend to have an ROI of between two to five times the investment in the first year of implementation."

3 Or to put it another way, there is plenty of material on how to build an ML model, but not much on how to build an ML *system*. A model can also be unreliable in different ways than a system.

often be parlayed into system access for an attacker—security and reliability are intimately connected. Doing one well is not easily separated from doing the other. Similarly, ML systems, with their surprising behaviors and indirect yet profound interconnections, motivate a more holistic approach toward deciding how to integrate development, deployment, production operations, and long-term care.

We believe, in short, that ML systems being *reliable* captures the essence of what customers, business owners, and staff really want from them.

Intended Audience

We are writing for anyone who wants to take ML into the real world and make a difference in their organization. Accordingly, this book is for data scientists and ML engineers, for software engineers and site reliability engineers, and for organizational decision makers—even nontechnical ones, although parts of the book are quite technical:

Data scientists and ML engineers
: We'll explore how the data, features, and model architecture you use change the way your model works, and how manageable it is in the long run, all with an eye to model velocity.

Software engineering building ML infrastructure or integrating ML into existing products
: We address both how to integrate ML into systems and how to write ML infrastructure. An improved understanding of how the ML lifecycle works helps with developing functionality, designing application programming interfaces (APIs), and supporting customers.

Site reliability engineers
: We'll show how ML systems typically break and how best to build (and manage) them to avoid those failure modes. We'll also explore the implications of ML model quality not being something a reliability engineer can entirely ignore.

Organizational leaders who want to add ML to their existing products or services
: We will help you understand how best to integrate ML into your existing products and services, and the structures and organizational patterns required. Having a sensible way of assessing risks and advantages when making ML-related decisions is important.

Everyone who is rightfully concerned about the ethical, legal, and privacy implications of developing and deploying ML
: We will lay out the issues clearly and point to practical steps you can take to address these concerns before they cause damage to your users or your organization.

One, perhaps counterintuitive, thing to note: many of the chapters are potentially most valuable to the people whose work is *not* the topic of that chapter. For example, Chapter 2 can certainly be read by data scientists and ML engineers. But it's potentially even more useful to infrastructure/production engineers and organizational leaders. For the former groups, of course, fine-tuning what you're already working on is useful, but for the latter, it can provide a fresh and complete introduction to a topic area that may be entirely new.

How This Book Is Organized

Before we talk about the structure of the book in detail, let's provide broader context about how we selected the topics and their organization. It might not be what you were expecting.

Our Approach

Engineers need to employ specific approaches and techniques to make ML systems work well. But each of these approaches is subject to an enormous number of decisions once put into place in a particular organization and for a particular purpose. It is not feasible for this book to cover all, or even most, of the implementation choices that readers will generally face. Similarly, we will de-emphasize concrete recommendations for specific pieces of software. We hope that this separation from the day-to-day will allow us to express ideas more clearly, but for this kind of book, remaining platform agnostic is beneficial in and of itself.

Let's Knit!

Though from time to time we use other examples, our main method of illustrating the book's content is a hypothetical online store—a purveyor of textile supplies via a website called *yarnit.ai*. This concept is worked through in some detail throughout the book in order to demonstrate how choices about one stage (say, data acquisition or normalization) have consequences for the rest of the stack, business, and so on.

This store is a single, relatively simple business (buy knitting and crocheting products, put them on a website, and market and sell them to customers). This explicitly does not capture the full complexities of the sectors you see using ML in the real world, such as manufacturing, self-driving cars, real-estate marketing, and medical technology. However, our example provides enough insight (and manages the scope) such that we feel the implementation complexities we deal with here offer lessons applicable to other domains. (In other words, the limitations of our example are, we believe, worth it.)

To explore the example in more depth, let's consider the case of a retailer sourcing products from a huge variety of suppliers, and selling them globally to the public on a website. The business is to attract and maintain a customer base that purchases products at a reasonable margin to cover the costs of operating the site and produce a profit. In some ways, it's quite a simple business, but the complexity will show up almost immediately as we try to add ML. At a high level, we are interested in improving sales, improving customer experience, lowering costs, improving margins, and making our whole business operate more efficiently.

While adding ML technologies to business operations at first is usually driven by narrow and concrete goals, such as efficiency, that is not the endgame. ML has the potential to positively or negatively transform the business *fundamentally*, changing the way products are created and selected, customers are identified and served, and commercial opportunities are uncovered. ML-adopting businesses that successfully deploy these technologies will outperform their competitors in the long run; a recent survey (*https://oreil.ly/FETOy*) of over 2,000 executives conducted by McKinsey indicated that 63% of execs had ML/AI projects that improved the bottom line, though organizations are often cagey about precisely how much. But in the short run, adoption must start with concrete, intelligible goals like "increased conversions"—i.e., more sales.

Our website, *yarnit.ai*, has many sources of data to implement these initial, concrete improvements and many potential applications for ML. To begin, we will consider customer-facing examples as simple as recommendation (other products a customer may like) and prediction (which products are most likely to sell to a given customer, for example). Then we will move on to more behind-the-scenes applications where ML might be used to optimize entire business processes. Concretely, we'll work through several examples of areas where ML can improve our operations, although, of course, not limited to these requirements:

Website search results
: Customers should get the best rankings for their search queries, showing them the products that most match those terms and that they are most interested in purchasing.

Discovery and suggestion
: Customers should be presented the opportunity to consider products related to the products they are looking at or purchasing. We can find products that might have a high likelihood of being useful and therefore purchased.

Dynamic pricing
: We should identify products that are not selling as quickly as we want, and reduce their prices to clear up space in the warehouse. Likewise, if we are running low on extremely popular products, we may want to temporarily increase their price in order to slow sales and make more money, while also possibly ordering more inventory.

Cart abandonment
: What is driving customers to add items to their cart but not complete purchases? Can we predict this? Are there ways for us to learn how to intervene with a reminder, a well-timed discount, or other features that improve checkout completion?

Inventory and ordering automation
: We might want to use ML to predict how to order replacement products from our suppliers based on a prediction of future sales and the predicted delivery delays.

Trust and safety
: This is an industry-wide term for detecting likely fraudulent behavior—in this case, attempted purchases—and increasing verification steps as necessary. Manual review and heuristics won't scale, so it makes sense to turn to ML for help.

Margin improvement
: We can use various techniques to try to improve the profit that we make on each sale—ranging from suggesting additional higher-margin products to customers while they shop, to running marketing campaigns to generate more demand for the highest-margin products.

These are just a few obvious places that ML might be trialed to determine whether it adds value worth the added complexity and cost. To be clear: ML is not always successful, and even when it *is* successful, it's not always worth the trouble. Standing up a complex pipeline with exacting data formats, significant engineering, and professional production operations requirements is a costly undertaking that needs to produce noticeable and definite value for the customers and the business. The maintenance costs are also considerable. All of this effort might not be worth it for your organization. This book will help you to understand whether it is.

Organizations like our yarn store should approach ML with an open mind but a willingness to experiment, measure, and possibly cancel the applications if they do not work out. This is why planning to assess the probability of success before making unalterable changes to the store's website infrastructure is critical.

For completeness, we need to say that obviously this is only a single example of the kind of organization and application that might find uses for ML.

Navigating This Book

We begin with an introduction and general principles. Here, relative newcomers to ML (or ML in production) can orient themselves to the problem space as a whole. It is also where we cover critical topics that impact all of the rest of the chapters, such as data management, what an ML model is, how to evaluate its quality, what a feature is (and why you would care), and fairness and privacy.

Next we focus on ML models and their lifecycle. We explain how models are created (known as *training*) and how to use them in production (known as *serving*). We outline how incredibly important it is to know what your models are doing, and how to do that at model development time and while the model is running in production. Last in this section, we cover the question of continuous ML: models that are *continually* updated with reference to a constantly adjusting reality.

Finally, we touch on the complicated question of how ML is introduced into organizations, and what happens to them when this takes place. We start with a concrete illustration of the complexities of ML in the incident response domain—something we expect basically every engineer with production responsibilities can relate to. We look at some of the critical questions about how organizations could integrate ML into existing (and emerging) products, often a process best done with some thought put into it in advance. Then we look at how ML could be implemented organizationally: centralized, distributed, and every point in between, followed by concrete guidelines and recommendations in the next chapter. Finally, we end with real case studies of ML implemented in organizations around the world.

After all of that, well, first of all, you deserve a break—but second of all, you should be well equipped to understand everything you're likely to come across when doing ML for the first time, or even when refining how it works inside your organization when you already have experience.

A Note on Chapter Credits

In general, each chapter had significant input from all authors, whether it be in the conceptual, review, first-draft, edit, or finalization stage, so by default we do not distinguish individual chapter authorship. However, when a chapter has had a significant contribution outside of the core team, we note it here as well as in the chapter itself:

- Chapter 4, written by Robbie Sedgewick and Todd Underwood
- Chapter 6, written by Aileen Nielsen
- Chapter 9, written by Niall Murphy and Aparna Dhinakaran

The individual stories in Chapter 15 are credited to each individual author (Cheng Chen, Daniel Papasian, Todd Philips, Harsh Saini, Riqiang Wang, and Ivan Zhou) and have been edited for clarity and consistency.

About the Authors

The authors of this book collectively have decades of experience building and running various kinds of ML systems in production. We have helped productionize large ad-targeting systems; built large search-and-discovery systems; published groundbreaking research on ML in production; and constructed and run the critical data ingestion, processing, and storage systems wrapped around them all. We have collectively had the (unfortunate) opportunity to witness firsthand most of the spectacular and fascinating ways these systems can break. But the good news is, we learned how to build systems that are resilient to the most common failure modes that ML systems suffer—technically and organizationally.

Conventions Used in This Book

The following typographical conventions are used in this book:

Italic
: Indicates new terms, URLs, email addresses, filenames, and file extensions.

`Constant width`
: Used for program listings, as well as within paragraphs to refer to program elements such as variable or function names, databases, data types, environment variables, statements, and keywords.

`Constant width bold`
: Shows commands or other text that should be typed literally by the user.

`Constant width italic`
: Shows text that should be replaced with user-supplied values or by values determined by context.

This element signifies a tip or suggestion.

This element signifies a general note.

This element indicates a warning or caution.

O'Reilly Online Learning

For more than 40 years, *O'Reilly Media* has provided technology and business training, knowledge, and insight to help companies succeed.

Our unique network of experts and innovators share their knowledge and expertise through books, articles, and our online learning platform. O'Reilly's online learning platform gives you on-demand access to live training courses, in-depth learning paths, interactive coding environments, and a vast collection of text and video from O'Reilly and 200+ other publishers. For more information, visit *https://oreilly.com*.

How to Contact Us

Please address comments and questions concerning this book to the publisher:

O'Reilly Media, Inc.
1005 Gravenstein Highway North
Sebastopol, CA 95472
800-998-9938 (in the United States or Canada)
707-829-0515 (international or local)
707-829-0104 (fax)

We have a web page for this book, where we list errata, examples, and any additional information. You can access this page at *https://oreil.ly/reliable-machine-learning-1e*.

Email *bookquestions@oreilly.com* to comment or ask technical questions about this book.

For news and information about our books and courses, visit *https://oreilly.com*.

Find us on LinkedIn: *https://linkedin.com/company/oreilly-media*

Follow us on Twitter: *https://twitter.com/oreillymedia*

Watch us on YouTube: *https://www.youtube.com/oreillymedia*

Acknowledgments

The authors would like to take the opportunity to thank the following people.

We would like to collectively thank the following contributors to the book: Ely M. Spears, who provided invaluable and detailed technical and structural feedback on many chapters in this work—thank you for helping this be better. Robbie Sedgewick, who provided similar feedback, as well as encouragement when we felt we were writing things that would not resonate with the general public. James Blessing, who reviewed a number of chapters quickly, and whose useful feedback helped us improve. We benefited additionally from thoughtful and careful reviews from Andrew Ferlitsch, Ben Hutchinson, Benjamin Sloss, Brian Spiering, Chenyu Zhao, Christina Greer, Christopher Heiser, Daniel H. Papasian, David J. Groom, Diego M. Oppenheimer, Goku Mohandas, Herve Quiroz, Jeremy Kubica, Julian Grady, Konstantinos (Gus) Katsiapis, Lynn He, Michael O'Reilly, Parker Barnes, Robert Crowe, Salem Haykal, Shreya Shankar, Tina H. Wong, Todd Phillips, and Vinsensius B. Vega S. Naryanto. Finally, thank you as well to our team at O'Reilly: John Devins, Mike Loukides, Angela Rufino, Ashley Stussy, Kristen Brown, and Sharon Wilkey.

Cathy Chen

Thank you to my partner, Morgan, who has been lovingly supportive when I've needed to work weekends or evenings on the book. I blame Todd for pulling me into this project but thank him not only for being a great leader on this project but also for giving me a great place to work. Thank you to all the coauthors and our amazing group of volunteers who have reviewed, edited, commented, and generally helped us improve this book.

Niall Richard Murphy

I dedicate this book to those who died during the period of its writing. While this book was being written—and I still have trouble believing it, even as I put it down here—we all experienced a global epidemic, a war of choice in Europe, and numerous political, personal, and professional events, and though I have often wished otherwise, this blizzard does not seem to be slowing down. Accordingly, I'd like to remember here both my paternal step-grandmother Winifred, a victim of COVID-19 in 2020, and my aunt Esther Gray, who passed away in Belfast in 2021 and is much missed. Of those still with us, I'd like to thank my mother, Kay Murphy, the best mother I ever could have had, and a shining moral light in dark times, as well as my wife, Léan, and my children, Oisín and Fiachra. Since my previous thanks section, Oisín has been national U19 chess champion, and Fiachra has had his art featured in another book, and their dad will never stop being proud of them. Finally, I would be remiss if I did not thank Todd Underwood himself, for exemplifying constant good-humored leadership in similarly trying times. This project owes its existence to him, above all others. I've learned a lot and discovered I knew more than I thought, but a lot of what I learned, I learned from him.

Kranti Parisa

I dedicate this book to all the frontline and essential workers, our heroes, in the COVID-19 crisis. Their immense dedication toward their job and humanity is truly inspirational. I'd like to thank all the coauthors and contributors for their extraordinary commitment and patience. Special thanks to Todd Underwood—you're a great inspiration and the force behind this book. I'd like to thank my friend, Dave Rensin, for looping me into this amazing opportunity and motivating me to share my knowledge and experiences building ML systems at scale with the rest of the world. Finally, to my parents, especially my mother, Nagarani, my wife, Pallavi, my little princess, Sree, and my friends for their constant love and support.

D. Sculley

I'd like to thank my remarkable coauthors and the many colleagues who have taught me so much—and lived with me through all the production fires and hairy issues along the way. My wife, Jessica, and daughter, Sofia, remain the inspiration for everything I do.

Todd Underwood

I dedicate this book to my family, who either benefited from or paid for my absence for many hours while writing. Beth, Ágatha, and Beatrix: I don't believe that you'll read this but I do hope you're happy that it's finally done. I dedicate it also to my brother, Adam, who died in the middle of a pandemic and who I didn't get to see nearly as often as I would have liked. And finally, to my coauthors who stuck with this project when it seemed likely to never be done. I hope we've done something worthwhile here—something useful and something we can feel proud of.

CHAPTER 1

Introduction

We begin with a model, or framework, for adding machine learning (ML) to a website, widely applicable across a number of domains—not just this example. This model we call the *ML loop*.

The ML Lifecycle

ML applications are never really done. They also don't start or stop in any one place, either technically or organizationally. ML model developers often *hope* their lives will be simple, and they'll have to collect data and train a model only once, but it rarely happens that way.

A simple thought experiment can help us understand why. Suppose we have an ML model, and we are investigating whether the model works well enough (according to a certain threshold) or doesn't. If it doesn't work well enough, data scientists, business analysts, and ML engineers will typically collaborate on how to understand the failures and improve upon them. This involves, as you might expect, a lot of work: perhaps modifying the existing training pipeline to change some features, adding or removing some data, and restructuring the model in order to iterate on what has already been done.

Conversely, if the model is working well, what usually happens is that organizations get excited. The natural thought is that if we can make so much progress with one, naïve attempt, imagine how much better we can do if we work harder on it and get more sophisticated. This typically involves—you guessed it—modifying the existing training pipeline, changing features, adding or removing data, and possibly even restructuring the model. Either way, more or less the same work is done, and the first model we make is simply a starting point for what we do next.

Let's look at the ML lifecycle, or loop, in more detail (Figure 1-1).

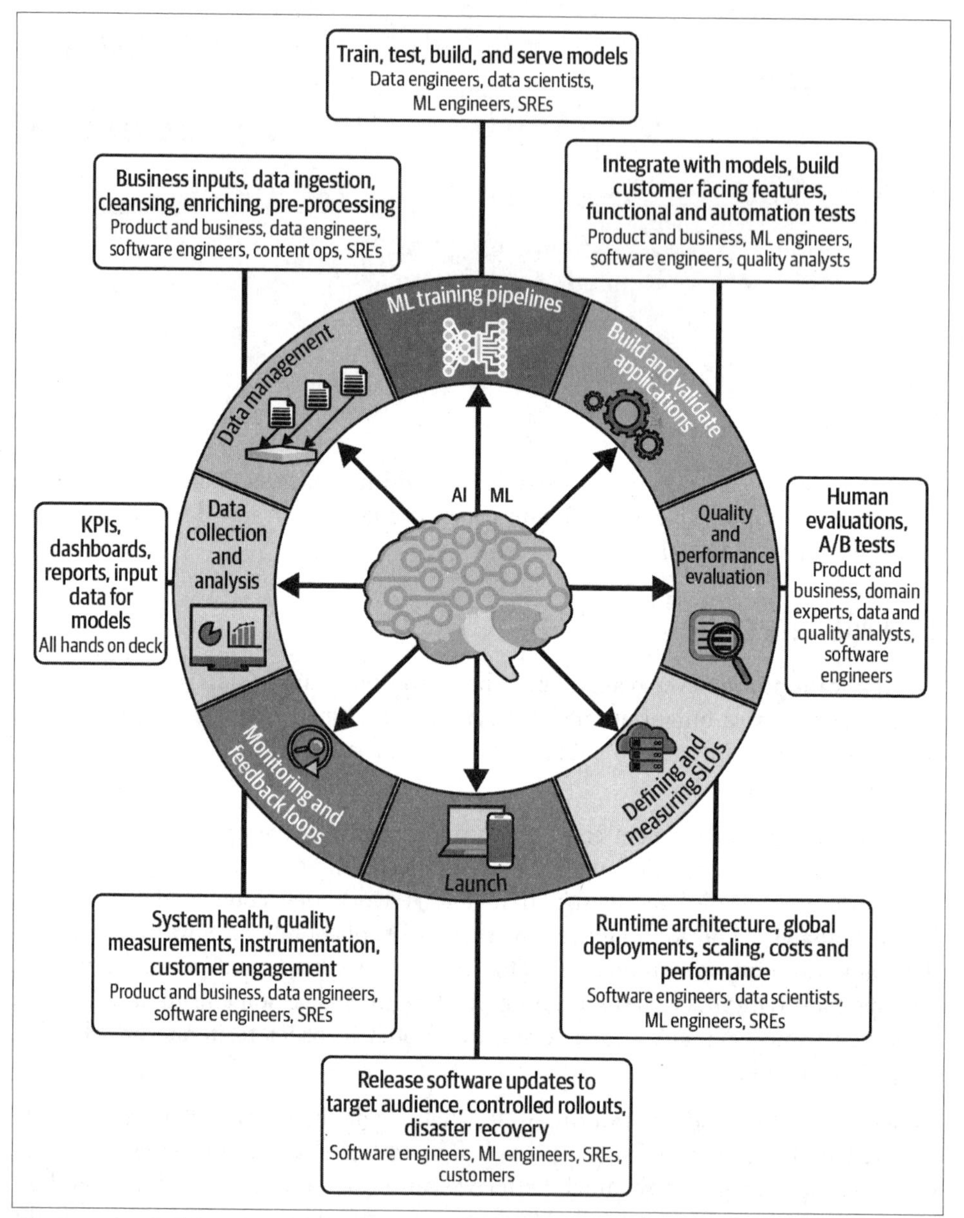

Figure 1-1. ML lifecycle

ML systems start with data, so let's start on the left side of the diagram and go through this loop in more detail. We will specifically look at each stage and explain, in the context of our shopping site, who in the organization is involved in each stage and the key activities they will carry out.

Data Collection and Analysis

First, the team takes stock of the data it has and starts to assess that data. The team members need to decide whether they have all the data they require, and then prioritize the business or organizational uses to which they can put the data. They must then collect and process the data.

The *work* associated with data collection and analysis touches almost everyone in the company, though how precisely it touches them often varies a lot among firms. For example, business analysts could live in the finance, accounting, or product teams, and use platform-provided data every day. Or data and platform engineers might build reusable tools for ingesting, cleaning, and processing data, though they might not be involved in business decisions. (In a smaller company, perhaps they're all just software or product engineers.) Some places have formal data engineering roles. Others have data scientists, product analysts, and user experience (UX) researchers all consuming the output of work from this phase.

For YarnIt, our web shop operator, most of the organization is involved in this step. This includes the business and product teams, which will know best the highest-impact areas of the business for optimization. For example, they can determine whether a small increase in profit for every sale is more important to the business, or whether instead it makes more sense to slightly increase order frequency. They can point to problems or opportunities with low- and high-margin products, and talk about segmentation of the customers into more and less profitable customers. Product and ML engineers will also be involved, thinking about what to do with all of this data, and site reliability engineers (SREs) will make recommendations and decisions about the overall pipeline in order to make it more monitorable, manageable, and reliable.

Managing data for ML is a sufficiently involved topic that we've devoted Chapter 2 to data management principles and later discuss training data in Chapters 4 and 10. For now, it is useful to assume that the proper design and management of a data collection and processing system is at the core of any good ML system. Once we have the data in a suitable place and format, we will begin to train a model.

ML Training Pipelines

ML training pipelines are specified, designed, built, and used by data engineers, data scientists, ML engineers, and SREs. They are the special-purpose extract, transform, load (ETL) data processing pipelines that read the unprocessed data and apply the

ML algorithm and structure of our model to the data.[1] Their job is to consume training data and produce completed models, ready for evaluation and use. These models are either produced complete at once or incrementally in a variety of ways—some models are incomplete in that they cover only some of the available data, and others are incomplete in scope as they are designed to cover only part of the ML learning as a whole.

Training pipelines are one of the only parts of our ML system that directly and explicitly use ML-specific algorithms, although even here these are most commonly packaged up in relatively mature platforms and frameworks such as TensorFlow and PyTorch.

Training pipelines also are one of the few parts of our ML system in which wrestling with those algorithmic details is initially unavoidable. After ML engineers have built and validated a training pipeline, probably by relying on relatively mature libraries, the pipeline is safe to reuse and operate by others without as much need for direct statistical expertise.[2]

Training pipelines have all the reliability challenges of any other data transformation pipeline, plus a few ML-specific ones. The most common ML training pipeline failures are as follows:

- Lack of data
- Lack of correctly formatted data
- Software bugs or errors implementing the data parsing or ML algorithm
- Pipeline or model misconfiguration
- Shortage of resources
- Hardware failures (somewhat common because ML computations are so large and so long-running)
- Distributed system failures (which often arise because you moved to using a distributed system for training in order to avoid hardware failures)

All of these failures are also characteristic of the failure modes for a regular (non-ML) ETL data pipeline. But ML models can fail silently for reasons related to data distribution, missing data, undersampling, or a whole host of problems unknown in the

1 ETL is one common abstraction to represent this kind of data processing. Wikipedia's "Extract, transform, load" page (*https://oreil.ly/XqcQs*) has a reasonable overview.

2 Which mature libraries and systems we use depends mostly on application. These days, TensorFlow, JAX, and PyTorch are all widely used for deep learning, but there are many other systems if your application benefits from a different style of learning (XGBoost is common, for example). Selecting a model architecture is mostly beyond the scope of this book, although small pieces of it are covered in Chapters 3 and 7.

regular ETL world.[3] One concrete example, covered in more detail in Chapter 2, hinges on the idea that missing, misprocessing, or otherwise not being able to use subsets of data is a common cause of failure for ML training pipelines. We'll talk about ways to monitor training pipelines and detect these kinds of problems (generally known as *shifts in distribution*) in Chapters 7 and 9. For now, let's just remember that ML pipelines really are somewhat more difficult to operate reliably than other data pipelines, because of these kinds of subtle failure modes.

In case it's not already clear, ML training pipelines are absolutely and completely a production system, worthy of the same care and attention as serving binaries or data analysis. (If you happen to be in an environment where no one except you believes this, it is small comfort to know there will be enough examples to the contrary to persuade anyone—eventually.) As an example of what can happen if you don't pay sufficient attention to production, we're aware of stories told about companies built on models generated by interns who have now left the company, and no one knows how to regenerate them. It is probably facile to say so, but we recommend you never end up in that situation. Making a habit of writing down what you've done and turning that into something automated is a huge part of avoiding the outcomes we allude to. The good news is that it's eminently possible to start small, with manual operations and no particular reproducibility required. However, becoming successful will require automation and auditing, and our view is that the sooner you can move to your model training being automated, gated by some simple checks for correctness and model preservation, the better.

In any event, assuming we can successfully build a model, we will need to integrate it into the customer-facing environment.

Build and Validate Applications

An ML model is fundamentally a set of software capabilities that need to be unlocked to provide value. You cannot just stare at the model; you need to *interrogate* it—ask it questions. The simplest way to do this is to provide a direct mechanism to look up predictions (or report on another aspect of the model). Most commonly, though, we have to integrate with something more complicated: whatever purpose the model has is generally best fulfilled by integrating the model with another system. The integration into our applications will be specified by staff in our product and business functions, accomplished by ML engineers and software engineers, and overseen by quality analysts. For much more detail on this, see Chapter 12.

Consider *yarnit.ai*, our online shopping site where people from all walks of life and all over the world can find the best yarn for knitting or crocheting, all with

3 Consider reading Andrej Karpathy's excellent 2019 blog post, "A Recipe for Training Neural Networks," (*http://karpathy.github.io/2019/04/25/recipe*) for more.

AI-based recommendations! Let's examine, as an example, a model that recommends additional purchases to a shopper. This model could take the shopping history of a user as well as the list of products currently in their cart, along with other factors like the country they normally ship to, the price ranges they normally purchase, and so on. The model could use those features to produce a ranked list of products that shoppers might contemplate purchasing.

To provide value to the company and to the user, we have to integrate this model with the site itself. We need to decide where we will query the model and what we'll do with the results. One simple answer might be to show some results on a horizontal list just below the shopping cart when a user is thinking about checking out. This seems like a reasonable first pass, providing some utility to shoppers and possibly some extra revenue for YarnIt.

To establish how well we are doing with our integration, the system should log what it decides to show and whether users take any actions—do they add items to their cart and ultimately buy them? By logging such events, this integration will provide new feedback for our model, so that it can train on the quality of its own recommendations and begin improving.[4] At this stage, though, we will simply validate that it works at all: in other words, that the model loads into our serving system, the queries are issued by our web server application, the results are shown to users, the predictions are logged, and the logs are stored for future model training. Next up is the process of evaluating the model quality and performance.

Quality and Performance Evaluation

ML models are useful only if they work, of course. It turns out that surprisingly detailed work is required to actually answer that question—beginning with the almost amusing, but absolutely true point that we have to decide what will count as *working*, and how we will evaluate model performance against that target. This usually involves identifying the effect that we're trying to create, and measuring it across various subsets (or slices) of representative queries or use cases. This is covered in much more detail in Chapter 5.

Once we have decided what to evaluate, we should begin the process by doing it offline. The simplest way to think about this is that we issue what we believe to be a representative set of queries and analyze the results, comparing the answers to a believed set of "correct" or "true" responses. This should help us determine how well the model *should* work in production. Once we have some confidence in the basic performance of the model, we can do an initial integration, either live or dark

4 If you are familiar with the concept of A/B testing from ecommerce generally, this is also an appropriate place to make sure that the plumbing for such testing is correctly working as part of the integration testing. A great use case here is to be able to distinguish user behavior in the presence and absence of ML suggestions.

launching the system. In a *live launch*, the model takes live production traffic, affects the website and dependent systems, and so on. If we are careful or lucky, this is a reasonable step to take, as long as we are monitoring key metrics to make sure we're not damaging the user experience.

A *dark launch* involves consulting the model and logging the result, but not using it actively in the website as users see it. This can give us confidence in the technical integration of the model into our web application but will probably not give us much confidence about the quality of the model.

Finally, there's a middle ground: we might build the capability in our application to only *sometimes* use the model for a *fraction* of users. While the selection of this fraction is a surprisingly advanced topic beyond the scope of this book,[5] the general idea is simple: try out the model on some queries and gain confidence in not only the integration but also the model quality.

Once we gain confidence that the model is not causing harm and is helping our users (and our revenue, hopefully!), we are almost ready to launch. But first, we need to focus on monitoring, measurement, and continuous improvement.

Defining and Measuring SLOs

Service-level objectives (*SLOs*) are predefined thresholds for specific measurements, often known as *service-level indicators* (*SLIs*), that define whether the system is performing according to requirements. A concrete example is "99.99% of HTTP requests completing successfully (with a 20x code) within 150 ms." SLOs are the natural domain of SREs, but they are also critical for product managers who specify what the product needs to do, and how it treats its users, as well as data scientists, ML engineers, and software engineers. Specifying SLOs in general is challenging (*https://www.alex-hidalgo.com/the-slo-book*), but specifying them for ML systems is doubly so because of the way that subtle changes in data, or even in the world around us, can significantly degrade the performance of the system.

Having said that, we can use obvious separations of concern to get started when thinking about SLOs for ML systems. First of all, we can use the divisions between

5 Naively, we might just generate a random number and select 1% of them to get the model. But this would mean that the same user would, even in the same web session, sometimes get model-generated recommendations and sometimes not. This is unlikely to help us figure out all aspects of whether the model works and might generate genuinely bad user experiences. So then, for a web application, we might select 1% of all logged-in users to get the model-generated results or perhaps 1% of all cookies. In that case, we will not easily be able to tell the impact of model-generated results on users, and there might be bias in the selection of current users versus new users. We might want the same user to sometimes get model-generated results and sometimes not, or we might want some users to always do so, but others only on particular sessions or days. The main point is that *how* to randomize access to ML results here is a somewhat statistically complicated question.

serving, training, and the application itself. Second, we have the divisions between the traditional golden four signals (*https://oreil.ly/hl4Vd*) (latency, traffic, errors, saturation) and the internals of ML operations, themselves substantially less generic than the golden signals, but still not completely domain specific. Third, we have SLOs related to the working of the ML-enhanced application itself.

Let's look more concretely at some very simple suggestions of how these ideas about SLOs might apply directly to *yarnit.ai*. We should have individual SLOs for each system: serving, training, and the application. For serving the model, we could simply look at error rates, just as we would any other system. For training, we should probably look at throughput (examples per second trained or perhaps bytes of data trained if our models are all of comparable complexity). We might establish an overall SLO for model training completion as well (95% of training runs finish within a certain number of seconds, for example). And in the application, we should probably monitor metrics such as number of shown recommendations, and successful calls to the model servers (from the perspective of the application, which may or may not match the error rate reported by the model serving system).

Notice, however, that none of these examples is about the ML performance of the models. For that, we'll want to set SLOs related to the business purpose of the applications themselves, and the measurement might be over considerably longer periods of time. Good starting places for our website would probably be click-through rate on model-generated suggestions and model-ranked search results. We should probably also establish an end-to-end SLO for revenue attributable to the model and measure that not just in aggregate but also in reasonable subslices of our customers (by geography or possibly by customer type).

We examine this in more detail in Chapter 9, but for the moment we ask you to accept there are reasonable ways to arrive at SLOs for an ML context, and they involve many of the same techniques that are used in non-ML SLO conversations elsewhere (though the details of how ML works are likely to make such conversations longer). But don't let the complexities get in the way of the basics. Ultimately, it is critical that product and business leads specify which SLOs they can tolerate, and which they cannot, so the production engineering resources of the organization are all focused on accomplishing the right goals.

Once we have gathered the data, built the model, integrated it into our application, measured its quality, and specified the SLOs, we're ready for the exciting phase of launching!

Launch

We will now get direct input from customers for the first time! Here product software engineers, ML engineers, and SREs all work together to ship an updated version of our application to our end users. If we were working with a computer-based or mobile-based application, this would involve a software release and all of the quality testing that those kinds of releases entail. In our case, though, we're releasing a new version of the website that will include the recommendations and results driven by our ML models.

Launching an ML pipeline has factors in common with launching any other online system, but also has very much its own set of concerns specific to ML systems. For general online system launch recommendations, see Chapter 32 of *Site Reliability Engineering: How Google Runs Production Systems* (*https://oreil.ly/OsNL3*), edited by Betsy Beyer et al. (O'Reilly, 2016). You'll definitely want the basics of monitoring/observability, control of releases, and rollback to be covered—going forward with a launch that doesn't have a defined rollback plan is dangerous. If your infrastructure doesn't allow you to roll back easily, or at all, we strongly recommend you solve that first before launching. For ML-specific concerns, we outline a few of them in detail next.

Models as code

Remember that models are code every bit as much as your training system binaries, serving path, and data processing code are. Deploying a new model can most definitely crash your serving system and ruin your online recommendations. Deploying new models can even impact training in some systems (for example, if you are using transfer learning to start training with another model). It is important to treat code and model launches similarly: even though some organizations ship new models over (say) the holiday season, it's entirely possible for the models to go wrong, and we've seen this happen in a way that required code fixes shortly thereafter. In our view, they have equivalent risk and should use equivalent mitigation.

Launch slowly

When deploying a new version of an online system, we are often able to do so progressively, starting with a fraction of all servers or users and scaling up over time only as we gain confidence in our system behaving correctly and the quality of our ML improvements. Explicitly here, we are trying to limit damage and gain confidence in two dimensions: users and servers. We do not want to expose all users to a terrible system or model if we happen to have produced one; instead, we show it to a small collection of end users first and incrementally grow thereafter. Analogously, for our server fleet, we do not want to risk all of our computing footprint at once if we happen to have built a system that doesn't run or doesn't run well.

The trickiest aspect of this is ensuring that the new system cannot interfere with the old system during the rollout. The most common way this would happen for ML systems is via intermediate storage artifacts. Specifically, changes in *format* and changes in *semantics* cause errors in the interpretation of data. These are covered in Chapter 2.

Release, not refactor

The general precept of changing as little as possible at one time applies in many systems, but is particularly acute in ML systems. Behavior of the overall system is so prone to change (by changes in underlying data, etc.) that a refactoring that would be trivial in any other context could make it impossible to deduce what is going wrong.

Isolate rollouts at the data layer

When doing a progressive rollout, remember that the isolation *must be at the data layer* as well as at the code/request/serving layer. Specifically, if a new model or serving system logs output that is consumed by older versions of the code or model, diagnosing problems can be long and tricky.

This is not just an ML problem, and failure to isolate the data of a new code path from an older code path has provided some exciting outages over the years. This can happen to any system that processes data produced by a different element of the system, although the failures in ML systems tend to be subtler and harder to detect.

Progressive Rollouts in a Stateful System

Story time: one of the authors worked on a payments system that experienced errors during a new feature rollout. While it was unexpected, it's not exactly unprecedented in the world of running systems, and so it was an easy fix to just roll back the update. This was especially true given errors were rising in proportion to the rollout.

Or so the team thought! Unfortunately, as the rollback completed, errors shot to 100%, and the team was in a panic. After much debugging, it turned out the update had actually changed some data formats in anticipation of the new feature, and they were incompatible with the format the old system expected. The errors occurred when a newer component wrote a log that happened to be picked up by an older binary, as happened during the rollout. A rollback removed the ability to correctly process the new format logs that had already been written. In fact, if the team had just let the rollout complete (or done it all at once), the errors would have gone away. The main lesson here is that you need to think holistically about all the system components that might need to participate in a rollback—and particularly the data layer.[6]

6 For completeness, it is also true that there's a safe way to roll out a new data format: specifically, by adding support for reading the format in a progressive rollout that completes before the system starts writing the format. This was not what was done in this case, obviously.

Measure SLOs during launch

Ensure that you have at least one dashboard that shows the freshest and most sensitive metrics, and keep track of those during the launch. As you figure out which metrics you care most about and which are most likely to indicate some kind of a launch failure, you can encode these in a service that can automatically stop your launch if things are going badly in the future.

Review the rollout

Either manually or automatically, make sure that someone or something is watching during a launch of any kind. Smaller organizations or bigger (or more unusual) launches should probably be watched by humans. As you get confidence, as mentioned previously, you can start to rely on automated systems to do this and can significantly increase the rate of launching!

Monitoring and Feedback Loops

Just as for any other distributed system, information about the correct, or incorrect, functioning of our ML system is key to operating it effectively and reliably. Identifying the primary objectives of "correct" functioning is still clearly the role for product and business staff. Data engineers will identify signals, and software engineers and SREs will help implement the data collection, monitoring, and alerting.

This is closely related to the SLO discussion earlier, since monitoring signals often feed directly into selection or construction of SLOs. Here we explore the categories in slightly more depth:

System health, or golden signals
: These are no different from any non-ML signal. Treat the end-to-end system as a data ingestion, processing, and serving system and monitor it accordingly. Are the processes running? Are they making progress? Is new data arriving? And so on (you'll see more detail in Chapter 9). It is easy to be distracted by the complexity of ML. It is important to remember, however, that ML systems are just that: systems. They have all of the same failure modes as other distributed systems, plus some novel ones. Don't forget the basics, which is the idea behind the *golden signal* approach to monitoring: find generic, high-level metrics that are representative of system behavior overall.

Basic model health, or generic ML signals
: Checking on basic model health metrics is the ML equivalent of systems health: it is not particularly sophisticated, or tightly coupled to the domain, yet includes basic and representative facts about the modeling system. Are new models of the expected size? Can they be loaded into our system without errors? The key criterion in this case is whether you need any understanding of the model's

contents in order to do the monitoring; if you don't, the monitoring you are doing is a matter of basic model health. There is substantial value to be had in this context-free approach.

Model quality, or domain-specific signals
: The most difficult thing to monitor and instrument is model quality. There is no hard line between an operationally relevant model quality problem and an opportunity for model quality improvement. For example, if our model has poor recommendations for people shopping for needles but not yarn on our site, that could be an opportunity to improve our model (if we chose to launch with this level of quality), or it could be an urgent incident that requires immediate response (if this is a recent regression).[7] The difference is context. This is also the most difficult aspect of ML systems for most SREs to come to terms with: there is no objective measure of "good enough" for model quality, and, worse yet, it's a multidimensional space that is hard to measure. Ultimately, product and business leaders will have to establish real-world metrics that indicate whether models are performing according to their requirements, and the ML engineers and SREs will need to work together to determine which quality measures are most directly correlated with those outcomes.

As a final step in the loop, we need to ensure that the ways that our end users interact with the models make it back into the next round of data collection and are ready to travel the loop again. ML serving systems should log anything they think will be useful so they can improve in the future. Typically, this is at the very least the queries they received, the answers they provided, and something about why they provided those answers. "Why" can be as simple as a single-dimensional relevance score, or it can be a more complex set of factors that went into a decision.

We've completed our first trip around the loop and are ready to start all over. By this point, *yarnit.ai* should have at least minimal ML functionality added, and we should be in a position to start continuously improving it, either by making the first models better or by identifying other aspects of the site that could be improved with ML.

Lessons from the Loop

It should be clear now that *ML begins and ends with data.* Successfully, reliably integrating ML into any business or application is not possible without understanding the data that you have and the information you can extract from it. To make any of this work, we have to tame the data.

7 Don't forget data drift either: a model from 2019 would have a very different idea about the importance and meaning of face masks in most parts of the world than a model from 2020.

It should also be clear that there is no single order to implementing ML for any given environment. It usually makes sense to start with the data, but from there, you will need to visit each of these functional stages and even potentially revisit them. The problems we want to solve inform the data we need. The serving infrastructure tells us about the models we can build. The training environment constrains the kind of data we will use and how much of it we can process. Privacy and ethics principles shape each of these requirements as well. The model construction process requires a holistic view of the entire loop, but also of the entire organization itself. In the ML domain, a strict separation of concerns is not feasible or useful.

Beneath all of this is the question of organizational sophistication, and risk tolerance with respect to ML. Not all organizations are ready to make massive investments in these technologies, and to risk their critical business functions on unproven algorithms—and they shouldn't! Even for organizations with a lot of experience with ML and the ability to evaluate the quality and value of models, most new ML ideas should be trialed first, because most new ML ideas don't work out. In many ways, ML engineering is best approached as a continual experiment, deploying incremental changes and optimizations and seeing what sticks by evaluating success criteria with the help of product management. It's not possible to treat ML as a deterministic development process, as much of software engineering attempts to do today. Yet even given the baseline chaos of today's world, you can significantly improve the chances of your ML experiments eventually working out by being disciplined about how you do the first one.[8]

As the implementation is cyclical, this book can absolutely be read in almost any order. Pick a chapter that is closest to what you care most about right now and start there. Then, figure out your most pressing questions and head to that chapter next. All of the chapters have extensive cross-references into the other chapters.

If you are an in-order sort of reader, that works fine too, and you'll start with the data. People who are curious about the way that fairness and ethics concerns have to be incorporated into every part of the infrastructure should skip ahead to Chapter 6.

By the end of the book, you should have a concrete understanding of where to start the journey of incorporating ML into your organization's services. You will also have a roadmap of changes that will need to take place for that process to be successful.

8 "Taming the Tail: Adventures in Improving AI Economics" (*https://oreil.ly/474mq*) by Martin Casado and Matt Bornstein is an article that's useful to consider in this context.

CHAPTER 2

Data Management Principles

In this book, we are rarely concerned with the algorithmic details of how models are constructed or how they're structured. The most exciting algorithmic development of last year is the mundane executable of next year. Instead, we are overwhelmingly interested in two things: the data used to construct the models, and the processing pipeline that takes the data and transforms it into models.

Ultimately, ML systems are data processing pipelines, and their purpose is to extract usable and repeatable insights from data. There are some key differences between ML pipelines and conventional log processing or analysis pipelines, however. ML pipelines have some very different and specific constraints and fail in different ways. Their success is hard to measure, and many failures are difficult to detect. (We cover these topics at length in Chapter 9.) Fundamentally, they consume data, and output a processed representation of that data (though vastly different forms of both). As such, ML systems depend thoroughly and completely on the structure, performance, accuracy, and reliability of their underlying data systems. This is the most useful way to think about ML systems from the reliability point of view.

In this chapter, we will start with a deep dive on data itself:

- Where data comes from
- How to interpret data
- Data quality
- Updating data sources (which we use and how we use them)
- Assembling data into an appropriate form for use

We'll cover the production requirements of data and show that, just like models, *data in production has a lifecycle*:

- Ingestion
- Cleaning and data consistency
- Enrichment and extension
- Storage and replication
- Use in training
- Deletion

The stability of data and metadata definitions as well as version control of those definitions are crucial, and we'll explain how to achieve them.[1] We'll also cover data access constraints, privacy, and auditability concerns and show some approaches to ensuring *data provenance* (where the data comes from) and *data lineage* (who has been responsible for it since we got it). At the end of this chapter, we expect you to have a complete but superficial understanding of the primary issues involved in making the data processing chain reliable and manageable.

Data as Liability

Writing about ML almost universally suggests that data is an important *asset* in ML systems. This perspective is sound: it's certainly impossible to have an ML system without data. As shown in Figure 2-1, it is often true that a simple (or even simplistic) ML system with more (and higher-quality) training data can outperform a more sophisticated system with less, or less representative, data.[2]

Organizations continue to scramble to collect as much data as possible, hoping to find ways to turn that data into value. Indeed, many organizations have made this into a profoundly successful business model. Think of Netflix, whose ability to recommend high-quality shows and movies to customers was an early differentiator. Netflix also reportedly used this data, once it got into the content production side of the business, to figure out what shows to make for which audiences, based on a detailed understanding of what people want to watch.

---

1 Version control for different versions of the data itself might also be warranted if the data is mutable and updated.

2 For the data to be useful, it has to be of high quality (accurate, sufficiently detailed, representative of things in the world that our model cares about). And for supervised learning, the data has to be consistently and accurately labeled—that is, if we have pictures of yarn and pictures of needles, we need to know which ones are which so that we can use that fact to train a model to recognize these kinds of pictures. Without high-quality data, we cannot expect high-quality results.

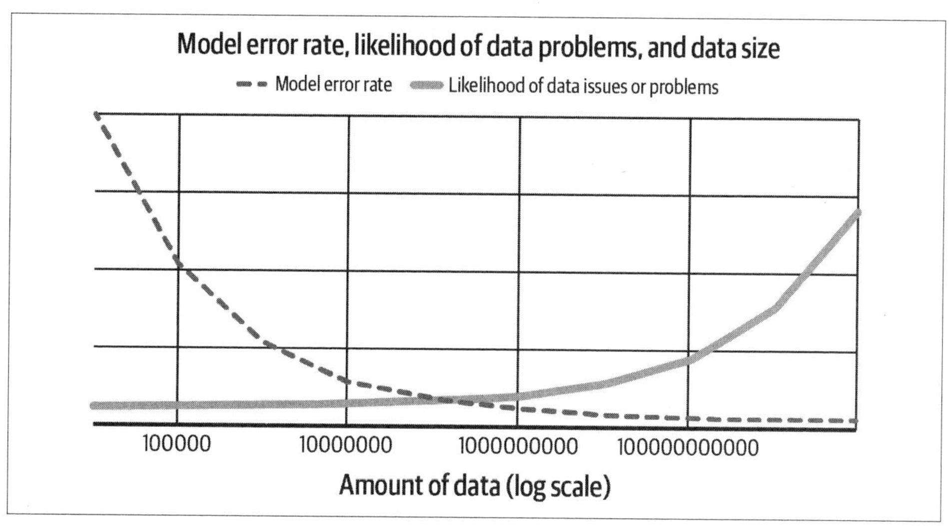

Figure 2-1. Illustrative trade-offs of data size, model error rates, and risk of problems or issues associated with the data

Of course, just like anything could be an asset, under the right (wrong) circumstances, it can also be a liability. In the case of data, the most important thing to say is that the acquisition, collection, and curation of data can expose areas of unexpected nuance and complexity in the data. Not accounting for these can lead to potential harm for us and for our users. All of these methods must be scoped appropriately for the type of data—medical records probably require different treatment from job history, for example. Of course, the best ways to curate data are high cost, so there's no free lunch here.

The intent of this short section is not to be the authoritative work on data collection, storage, reporting, and deletion practices. That is well beyond the scope of this section and even of this book. The intent here is to enumerate enough of the complexity to dissuade any readers from simply thinking "more data == better" or thinking "this stuff is easy." Let's go through the lifecycle of data and see where some of the challenges come from.

First, the data must be collected in compliance with applicable laws, which might be based on where our organization is located, where the data originates, and on organizational policies. We'll definitely need to think this through (and talk to lawyers for all of the jurisdictions we might be operating in). There are significant restrictions on what counts as data about people, how to get permission to store the data, how to store and retrieve the permission that was granted, whether we need to provide access to the data to the people who provided it, and under what circumstances. These restrictions may come from laws, industry practices, insurance regulations, corporate governance policies, or any range of other sources. Examples of common restrictions in some jurisdictions include prohibition against collecting personally identifiable information (PII) about an individual without their explicit written consent, along with the requirement to delete that data upon request by the data subject. Whether and how to collect data is not a technical question. It is a question of policy and governance. (Parts of this topic are covered much more thoroughly in Chapter 6.)

If we are allowed to collect and store the data, it must be secured from external access. Very few good things happen to organizations as a result of revealing their users' private data. Moreover, access must be restricted, even to employees. Employees should not be able to view or change private user data without restriction and without detailed logging of that.

Another approach to reducing data access and reducing the auditing surface is to anonymize the data. One relatively simple and valuable option is to use *pseudonymization*. Here, private identifiers are replaced with others in a reversible fashion, and reversing the pseudonymization requires access to an additional data or system. This protects the data from casual inspection by engineers working on the pipeline but permits discovery if we find that we need to reverse the anonymization. Pseudonymization also hopefully preserves the properties of data that are relevant to our model. In other words, if it is important that a data field be similar in a certain way under particular circumstances (think of postal codes, for example, which are prefix-identical when they are in the same town or the same part of the same city), then our pseudonymization might need to preserve that. While this level of protection has value against casual inspection, it is important to treat pseudonymized data as potentially just as risky as completely nonanonymized data. History is full of cases of this data being used to expose the very real and private information of users (see the following sidebar).

Example Applications for Different Levels of Anonymization

Different controls are required for data, depending on the sensitivity of the data that is being accessed:

Raw data
: The data contains no PII *or* we have extremely strict access controls in place and the PII itself is critical (e.g., medical data).

Pseudonymized data
: Some strict access controls are in place. Occasional, controlled manual inspection is required, but the PII data itself is not relevant to the inspection (e.g., credit card data with the prefix blanked or transformed, leaving only the final four digits: XXXX-XX-1234—still a lot of data, but some reasonable protection from casual inspection).

Anonymized data
: All other cases for which we collect PII but do not require it for troubleshooting or model performance.

A better approach is permanently removing any direct connection between private data about a person and the data that we use to train on. If we are able to permanently remove any connection between the private data and the person, we substantially reduce the risk of the data and increase the flexibility we have to handle it. This is, of course, harder than it seems.[3] Doing this in a way that's not trivially reversible but still valuable can be difficult. Although many good techniques exist, one common basic idea is to combine collections of data to ensure that no reported piece of data is tied to a unique identifier for any fewer than a certain number of real people. This is the approach taken by many serious population research organizations, including, notably, the United States Census Bureau. There are, however, many subtleties in getting this right. Correct anonymization is a topic that is mostly beyond the scope of this book, although we'll refer to it.

Don't Anonymize Yourself

Several very high-profile failures over the past several years should have made the difficulties of anonymization clear. The work, in particular, of Dr. Latanya Sweeney (*http://latanyasweeney.org*) in Harvard's government department is worth reading. Sweeney demonstrated with ease that she could identify the health records of William Weld, then governor of Massachusetts, although they were "anonymized," simply by knowing his gender, age, and zip code (which were freely available from voter records). Sweeney further demonstrated that 87% of all people in the US could be uniquely identified via only three pieces of information: gender, age, and zip code; see "Anonymized Data Really Isn't—and Here's Why Not" (*https://oreil.ly/V39HT*) by Nate Anderson for more details.

Similar work has shown that a browser user agent string combined with zip code or several other kinds of information uniquely identifies humans. The moral of the story is that anonymization is hard, and you will almost certainly have to develop expertise in it yourself, or outsource it to people or platforms that do; it is a significant specialization in and of itself. The topic of correct anonymization is closely aligned with other mathematically complex topics like cryptography that are properly a separate specialization.

3 The case of AOL search logs is the most famous such debacle; see "A Face Is Exposed for AOL Searcher No. 4417749" (*https://oreil.ly/WALx5*) by Michael Barbaro and Tom Zeller Jr. This incident is also explained at Wikipedia's "AOL search log release" page (*https://oreil.ly/cBpOve*).

Finally, we will ultimately need to be able to delete data. We might do this at the request of individual users, local laws, regulations like the European Union's General Data Protection Regulation, or GDPR (*https://gdpr-info.eu*), or in other cases where we no longer have permission to store the data. It turns out that deleting data and having it *actually* be deleted is surprisingly hard.

This has been true since at least the early MS-DOS days, when deleting a file just removed the reference to it and not the actual data itself, meaning you could reconstruct the file with sufficient determination and luck. Everything about today's computing environment makes deletion even harder than it was in those days, from having to track down multiple copies of the data to metadata management. In most distributed storage systems, data is divided into many pieces and stored across a collection of physical machines. Depending on the implementation, it may be practically impossible to determine every durable storage device (hard disk drive or solid state drive) that might have the data written on it.

It's important to be certain that people want their data deleted without putting up arbitrary barriers. One way to balance this is to impose a short delay before really deleting the data. A user might be granted a few hours or even days after requesting data be deleted to cancel that request. But at some point, once we've confirmed the request is intended and legitimate, we will need to track down every copy of the data and eliminate it. Depending on how it is stored, data structures, indices, and backups may be reconstructed to make accessing nearby data as efficient and reliable as it had been.

As with most things, the task of deleting data is made better by putting explicit thought into it. If your system hasn't had that much thought put in, you can use a couple of workarounds. Here are two common optimizations:

Periodically rewrite the data
: When there is a process that regenerates the data, we can take advantage of the "don't delete data immediately" recommendation noted previously to simply schedule the "deleted" data to not be included the next time the data is rewritten. This assumes that the period of data regeneration matches the expected and acceptable delay in deletion. This also assumes that the rewrite of the data is effective at actually deleting the data, which may well not be the case at all.

Encrypt all of the data and throw away some keys
: A system designed in this way has some significant advantages. In particular, it protects private data "at rest" (written to persistent storage). Deleting data is also trivial: if we lose the key to a user's data, we can no longer read that data. The downsides are mostly avoidable but worth considering seriously: anyone employing this strategy will need very, very reliable key-handling systems because if the keys are lost, *all* of the data is lost. This can also make it difficult to reliably delete a single key from every backup of the key system.

The Data Sensitivity of ML Pipelines

The primary difference between ML pipelines and most data processing pipelines is that ML pipelines are unusually sensitive to their input data compared to most other data processing pipelines. All data processing pipelines are, in some sense, subject to the correctness and volume of their input data, but ML pipelines are furthermore sensitive to subtle changes in *distribution* of the data. A pipeline can easily go from mostly right to significantly wrong simply by omitting a small fraction of the data, provided that small fraction is not random, or is somehow not evenly sampled in the range of characteristics our model is sensitive to.

An easy thought experiment here is to consider a real-world system like *yarnit.ai* that somehow loses all of the data from a particular country, region, or language. For example, if we drop all of the data from December 31 of a given year, we lose the ability to detect New Year's Eve shopping trends, which may be substantially different from the surrounding days in December and January. In many of these cases, losing a small amount of data that turns out to be systematically biased results in significant confusion in the understanding and predictions of our models.

As a result of this sensitivity, the ability to aggregate, process, and monitor *data*, rather than only the live systems, is critical to successfully managing ML data pipelines. We discuss monitoring data in some depth in Chapter 9, but here is a preview. A key insight to monitoring data is the slicing, or division, of the data along various axes (determining *which* axes are the best to slice the data is an important activity in exploratory data analysis, or EDA, and is beyond the scope of this book). In a system that is trying to track real-time activities, we might divide data in buckets of data age: recent, one to two hours old, three to six hours old, etc. We might track which buckets we are currently processing data from so that we can understand how far behind we have gotten. But we can, and should, track various other histograms that are relevant to our application. The ability to detect when *all* or *almost all* of a subset of the data is gone will matter enormously.

For example, in our shopping site *yarnit.ai*, we may train on searches to try to predict the best results for any given search (where "best" is "most likely to be purchased"—we are in the selling business, after all!). We operate our site in multiple languages in multiple markets. Let's say that a widespread payments outage affects only our Spanish language site, resulting in a drastically lower number of completed orders from people searching in Spanish. A model whose job is to recommend products for customers to buy will learn that Spanish-language results are significantly less likely to result in purchases than results in other languages. The model will show fewer results in Spanish and may even start showing results in other languages to Spanish-speaking users. Of course, the model will not be able to "know" the reasons for this change in behavior.

This might result in a small decline in total purchases if our site is predominantly a North American or European site, but a massive decline in the total number of searches and purchases in Spanish. If we train on this data, our model will probably have terrible results for Spanish-language searches. It might learn that Spanish-language queries never buy anything (since this will essentially be true). The model might start exhibiting exploratory behavior—if all of the Spanish language results are equally terrible, then any result *might* be good, so the model will try to find anything that Spanish-language searchers might actually buy. This will result in terrible search results and lower sales once our Spanish site's payment outage is over. This is a pretty bad outcome. (For a more complete treatment of a similar problem, see Chapter 11.)

The change in query volume might not be easily detectable at the gross query volume level either, as the total queries in Spanish might be small compared to those in other languages in total. We'll talk about ways to monitor training pipelines and detect these kinds of shifts in distribution in Chapter 9. This example is given just to motivate the thought that ML pipelines really are somewhat more difficult to operate reliably because of these kinds of subtle failure modes.

With these constraints in mind, let's review the lifecycle of data in our system. We'll have to think about the data as under our care from the moment of its creation until we delete it.

Phases of Data

Most teams rely on the platforms they use, including their data storage and processing platforms, to provide most of what they need. YarnIt is not a large organization, but we'll still involve staff responsible for business management, data engineering, and operations to help us understand and meet the requirements here. We are lucky enough to have SREs who will address reliability concerns related to the storage and processing of data.

The data management stage is fundamentally about transforming the data we have into a format and storage organization suitable for using it for later stages of the process. During this process, we will also probably apply a range of data transformations that are model specific (or at least model-domain specific) in order to prepare the data for training. Our next operations on the data will be to train ML models, anonymize certain sensitive items of data, and delete data when we no longer need it or are asked to do so. To prepare for these operations on the data, we will continue to take input from the business leaders to answer questions about our primary use cases for the data, along with possible areas for future exploration.

As with most of the sections of this book, deep familiarity with ML is not required or sometimes even desirable in order to design and run reliable ML systems. However, a basic understanding of model training does directly inform what we do with data as we get it ready. Data management in modern ML environments consists of multiple phases before feeding the data into model-training pipelines, as illustrated in Figure 2-2:

- Creation
- Ingestion
- Processing (which includes validation, cleaning, and enrichment)
- Post-processing (which includes data management, storage, and analysis)

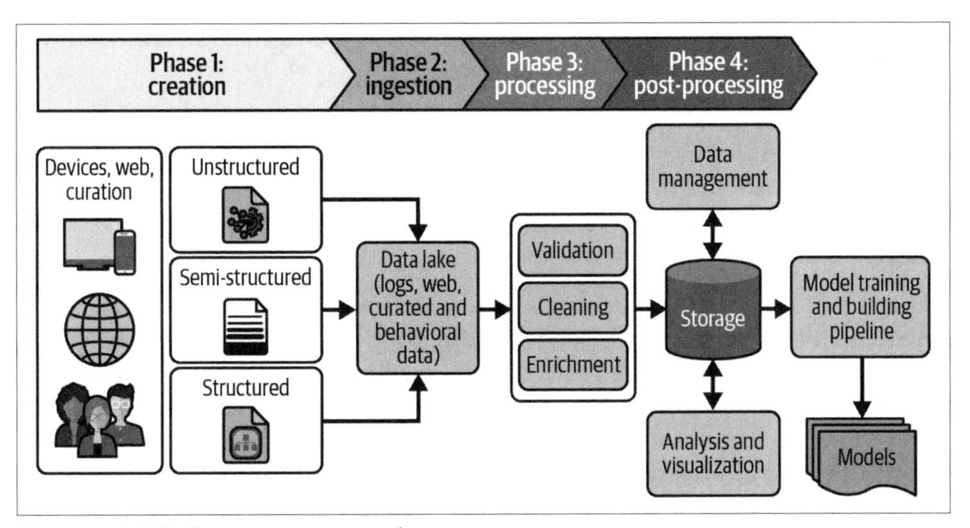

Figure 2-2. ML data management phases

Creation

It may seem either odd or obvious to state, but ML training data comes from *somewhere*. Perhaps your dataset comes from elsewhere, such as a colleague in another department or from an academic project, and was created there. But datasets are all created at some point via some process. Here we are referring to data creation as the process of generating or capturing the data in *some* data storage system but not *our* data storage system. Examples here include logs from serving systems, large collections of images captured from an event, diagnostic data from medical procedures, and so on. Implicit in this process is that we will want to design new systems and adapt existing systems to generate more data than we might otherwise, so that our ML systems have something to work with.

Some datasets work well when they are static (or at least don't change quickly), and others are useful only when frequently updated. A photo-recognition dataset, for example, may be usable for many months, as long as it is suitably representative of the types of photos we would like to recognize with our model. On the other hand, if the outside photos represent only winter environments from a temperate climate, the distribution of the photo set will be significantly different from the expected set of images we need to recognize. It will therefore not be useful as the environment warms during spring in those places. Likewise, if we're trying to recognize fraud in transactions at *yarnit.ai* automatically, we will want to be training our model continuously on recent transactions along with information about whether those transactions were fraudulent. Otherwise, someone might come up with a neat idea of how to steal all of our knitting supplies that is hard to detect, and we might never teach our model how to detect it.

The kinds of data we collect and the data artifacts we create can be unstructured, semi-structured, or very well structured (Figure 2-3).

Structured data is quantitative with a predefined data model/schema, highly organized, and stored in tabular formats like spreadsheets or relational databases. Names, addresses, geolocation, dates, and payment information are all common examples of structured data. Since it is well formatted, structured data can be easily processed by relatively simple code using obvious heuristics.

On the other hand, *unstructured data* is qualitative without a standard data model/schema, so it cannot be processed and analyzed using conventional data methods and tools. Examples of unstructured data include email body text, product descriptions, web text, and video and audio files. Semi-structured data doesn't have a specific data model/schema but includes tags and semantic markers and so is a type of structured data that lies between structured and unstructured data. Examples of semi-structured data include email, which can be searched by Sender, Receiver, Inbox, Sent, Drafts, etc., and social media content that may be categorized as Public, Private, and Friends, as well as by user-maintained labels such as hashtags. The character of the internal structure of the data will have significant implications for the way we process, store, and use it.

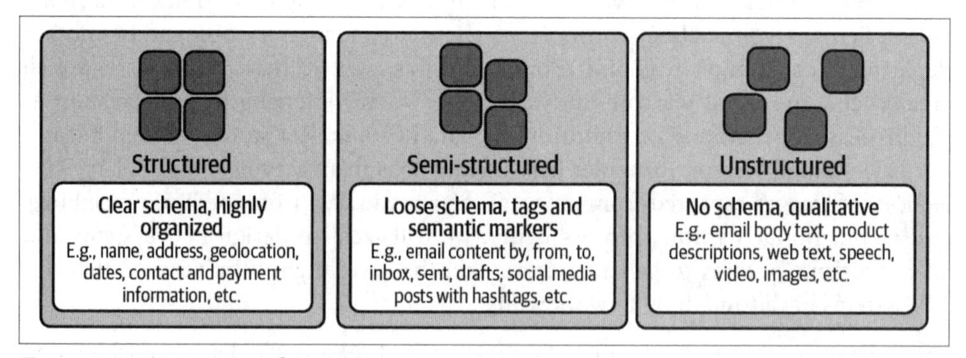

Figure 2-3. ML training data categories

Although bias in models comes from the structure of the model as well as the data, the circumstances of data creation have profound implications for correctness, fairness, and ethics. Though we treat this in considerably more detail in Chapters 5 and 6, the primary recommendation we can make here is that you have some kind of process to establish whether your model is biased. We can do this in numerous ways and the simplest is probably the Model Cards (*https://oreil.ly/h7E8h*) approach, but having any process that does this and having it be organizationally accepted is much better than not having such a process.[4] It's definitely the first thing to do as we begin to address ethics and fairness considerations in our ML system. You could relatively easily combine the effort to detect bias into a data provenance or data lifecycle meeting or tracking process, for example. But every organization doing ML should establish some kind of process, and review it as part of continuous improvement.

Recall, though, that bias comes from many sources and shows up at many stages through the process. There is no guaranteed way to ensure data fairness. One useful precursor to success here is to have an inclusive company culture full of people from all kinds of backgrounds with differing and creative points of view. There's good evidence that people with very different backgrounds and perspectives, while working in an environment of trust and respect, produce much better and more useful ideas than teams that are all similar. This can be part of a robust defense against the kinds of bias that slips through the checks described previously. It will not, however, prevent all bad outcomes without process and tooling, just as no human effort prevents all bad outcomes without systems help.

One final note on dataset creation, or rather, dataset augmentation: if we have a small amount of training data but not enough to train a high-quality model on, we may need to augment that data. Tools are available that can do so. For example, Snorkel (*https://www.snorkel.org/features*) provides a programmatic interface for taking a small number of data points and permuting them into a larger number of more varied data points, essentially making up imaginary but statistically valid training data. This is one good place to start, as it allows us to easily expand a small dataset into a large one. Although it might appear that these programmatically created datasets are somehow less valuable or less useful, there is good evidence that this approach can yield extremely good results at low cost, although it does need to be used with caution.[5]

4 A complementary approach to Model Cards for datasets is known as Data Cards; see the Data Cards Playbook site (*https://oreil.ly/aaSMr*).

5 See "Learning to Compose Domain-Specific Transformations for Data Augmentation" (*https://oreil.ly/uxLdr*) by Alexander J. Ratner et al. for more details.

Ingestion

The data needs to be received into the system and written to storage for further processing. At this stage, a filtering and selection step necessarily occurs. Not all data created may be ingested or collected, because we don't want or need to.

We may filter data by type at this stage (data fields or elements that we do not believe will be useful for our model). We may also simply sample at this stage if we have so much data we do not believe we can afford to process all of it, as ML training and other data processing is often extremely computationally expensive. Sampling data can be an effective way to save money on intermediate processing and training costs, but it is important to measure the quality cost of sampling and compare that to the savings. Data should also be sampled proportional to the volume/rate per time period or per other slice in the data we care about. This will avoid missing detail during bursty periods. However, sampling is likely to occasionally lose some detail for some events. This is unavoidable.

In general, ML training systems perform better with more data. Pedants will immediately think of many exceptions, but this is a useful starting point. Therefore, any reduction in data may well impact quality at the same time as reducing costs.

Depending on the volume of data and the complexity of our service, the ingestion phase may be as simple as "dump some files in that directory over there" or as sophisticated as a remote procedure call (RPC) endpoint that receives specifically formatted files and confirms a reference to the data bundle that was received so that its progress through the system can be tracked. In most cases, we will want to provide data ingestion via at least a simple API because this provides an obvious place to acknowledge receipt/storage of the data, log the ingestion, and apply any governance policies about the data.

Reliability concerns during the ingestion phase typically focus on correctness and throughput. *Correctness* is the general property that the data is properly read and written in the correct place without being unnecessarily skipped or misplaced. While the idea of misplacing data sounds amusing, it absolutely happens, and it's easy to see how it could. A date- or time-oriented bucketing system in storage combined with an off-by-one error in the ingestion process could end up with every day's data stored in the previous day's directory. Monitoring the existence and condition of data before and during ingestion is the most difficult part of the data pipeline.

Processing

Once we have successfully loaded (or ingested) the data into a reasonable feature storage system, most data scientists or modelers will go through a set of common operations to make the data ready for training. These operations—validation, cleaning and ensuring data consistency, and enriching and extending—are detailed next.

Validation

No matter how efficient and powerful our ML models are, they can never do what we want them to do with bad data. In production, a common reason for errors in the data is bugs in the code that is collecting the data in the first place. Data ingested from external sources might have a lot of errors even though a well-defined schema exists for each source (for example, having a float value for an integer field). So, it is extremely important to validate the incoming data, especially when there is a well-defined schema and/or an ability to compare against the last-known valid feed.

Validation is performed against a common definition of the field—i.e., is it what we expect it to be? To perform this validation, we need to both store and be able to reference those standard definitions. Using a comprehensive metadata system to manage the consistency and track the definition of fields is critical to maintaining an accurate representation of the data. This is covered in much more detail in Chapter 4.

Cleaning and ensuring data consistency

Even with a decent validation framework in place, most data is still messy. It may have missing fields, duplicates, misclassifications, or even encoding errors. The more data we have, the more likely that cleaning and data consistency will be its own stage of processing.

This might seem frustrating and unnecessary: building an entire system simply to check the data sounds a bit overblown to many people doing so for the first time. The reality is that our ML pipeline will assuredly have code whose job is to clean the data. We can either put this code in a single place, where it can be reviewed and improved, or put aspects of it throughout the training pipeline. That second strategy makes for extremely fragile pipelines as the assumptions about data correctness grow but our ability to ensure that they are met does not. Moreover, as we improve some of the code for validating and correcting data, we might neglect to implement those improvements in all of the many places where we are performing that work. Or worse, we can be counterproductive. For example, we can "correct" the same data multiple times in ways that eliminate the original information in the data. We can also have potential race conditions whereby different parts of the process are cleaning or making consistent the same data differently.

Another set of data consistency tasks during this portion is the normalization of data. *Normalization* generally refers to a set of techniques used to transform the input data into a similar scale, which is useful for methods like deep learning that rely on gradient descent or similar numerical optimization methods for training. Some standard techniques here, depicted in Figure 2-4, include the following:

Scaling to a range
: Mapping all of the X values of the data to a fixed range, often 0 to 1, but sometimes (for something like height or age) other values that represent the common min and max values.

Clipping
: Cutting off the maximum values of the data. This is useful when the dataset has a small number of extreme outliers.

Log scaling
: $x' = \log(x)$. This is useful when the data has a power law distribution, with a small number of very large values and a large number of very small values.

Z-score normalization
: Maps the variable to the number of standard deviations from the mean.

It is important to note that each of these techniques can be dangerous if any of the range or distribution or means are calculated on a set of test data with different properties than the dataset where it is later applied.

A final related and common technique is that of putting the data into *buckets*: we map a range of data into a much smaller set of groups that represent the same range. For example, we could measure age in years, but when training we could bucket on decades so that everyone who is 30 to 39 gets put into the "30s" bucket. Bucketing can be the source of many difficult-to-detect errors. Imagine, for example, that one system buckets age on decade boundaries and another on five-year boundaries. When we bucket the data, we have to give serious consideration to preserving the existing data and writing out a new, correctly formatted field for each record. If (when?) we change the bucketing strategy, we'll be glad we did; otherwise, it will be impossible to switch.

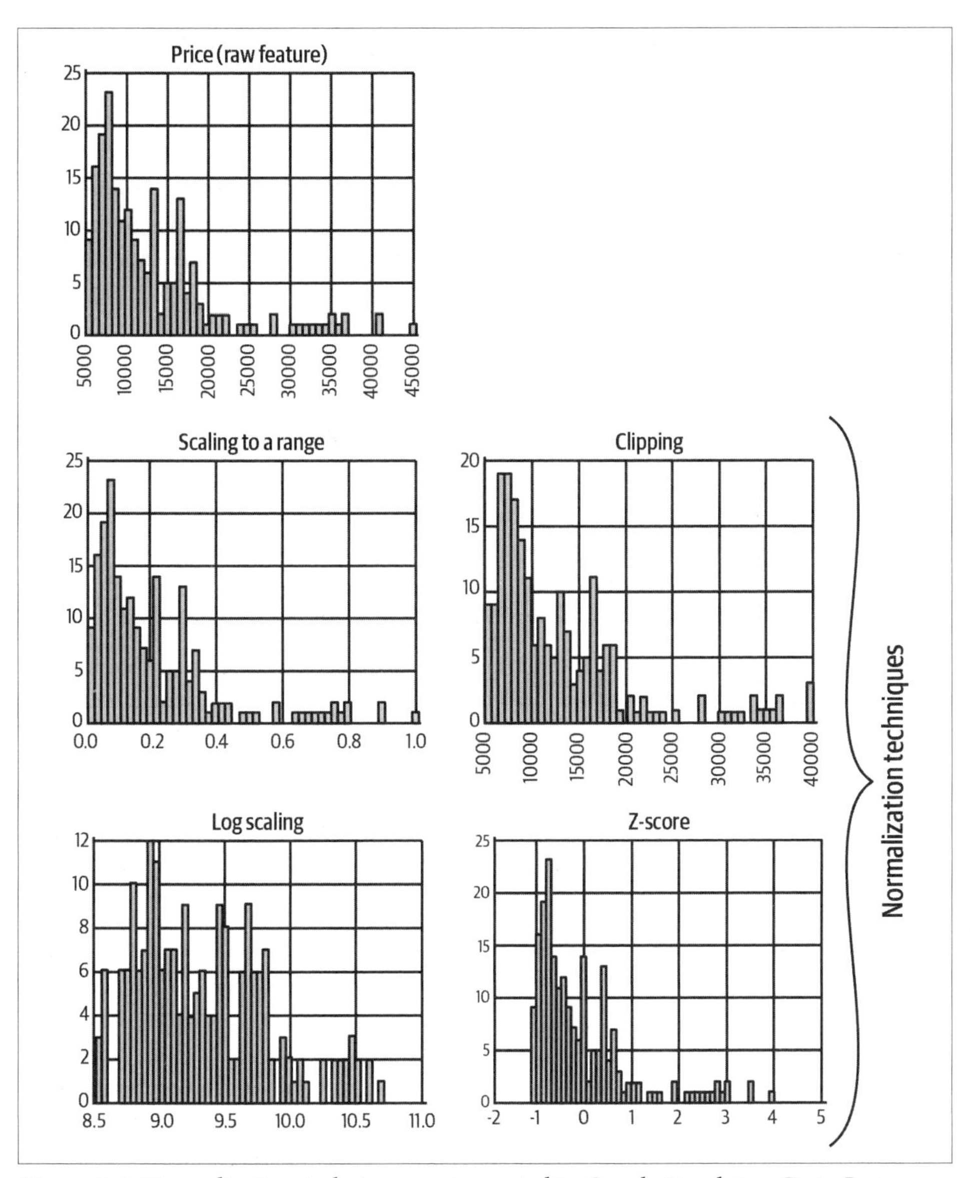

Figure 2-4. Normalization techniques as presented in Google Developers Data Prep course (https://oreil.ly/0cgBm)

Enriching and extending

In this stage, we combine our data with data from other sources. The most common and fundamental way to extend the data is through *labeling*. This process identifies a given event or record by bringing in confirmation from an outside source of data (sometimes a human). Labeled data is the key driver of all supervised ML and is often one of the most challenging and expensive parts of the whole ML process. Without a sufficient volume of high-quality labeled data, supervised learning won't work. (Labeling and labeling systems are covered extensively in Chapter 4.)

But labeling is only one way to extend data. We might use many external data sources to extend our training data. Let's say we believe for some reason that the temperature at a person's location will predict what they will buy.[6] We can take our logs of searches on the *yarnit.ai* site and add the temperature at the approximate geolocation of the user at the time they were visiting the web page. We could do this by finding or creating a temperature history service or dataset. This will allow us to train a model based on `source temperature` as a feature and see what kind of predictions we can make with it. This is probably a terrible idea, but it is not completely impractical.

Storage

Finally, we need to store the data somewhere. How and where we store the data is mostly driven by how we tend to use it, which is really a set of questions about training and serving systems. We go into this considerably more in Chapter 4, but there are two predominant concerns here: efficiency of storage and metadata.

The efficiency of a storage system is driven by access patterns that are influenced by the model structure, team structure, and training process. To make sensible choices about our storage system, here are some basic questions we need to answer:

- Are we training models once over this data or many times?
- Will each model read *all* of the data or only parts of it? If only some of the data will be read, is the subset being read selected by the type of data (some fields and not others) or by randomly sampling the data (30% of all records)?
- In particular, do related teams read slightly different subsets of the fields in the data?
- Do we need to read the data in any particular order?

On the subject of reuse of data, it turns out that almost all data is read multiple times and the storage system should be built for that, even if model owners assert that they will train only one model on the data once. Why? Model development is inherently an iterative process. An ML engineer makes a model (reading the data necessary to do so), measures how well the model performs at its designed task, and then deploys it. Then they get another idea: an idea of how to improve the model in some way. Before you know it, they're back rereading the same data to try out their new idea.

6 There is very little reason to believe this is true, but it's vaguely plausible and moderately amusing. If anyone implements "source temperature" as a feature and finds it to be valuable, please contact the authors for an autographed book.

Every system we build, from data to training all the way to serving, should be built with the assumption that model developers will semi-continuously retrain the same models in order to improve them. Indeed, as they do so, they might read different subsets of the data each time.

Given this, a column-oriented storage scheme with one column per feature is a common design architecture, especially for models training on structured data.[7] Most readers will be familiar with row-oriented storage, in which every fetch of data from the database retrieves all of the fields of a matching row. This is an appropriate architecture for collections of applications that all use most or all of the data—in other words, a collection of very similar applications. Column-oriented data facilitates the retrieval of only a subset of fields. This is much more useful for a collection of applications (ML training pipelines in this case) that each use a given subset of the data. In other words, column-oriented data storage allows different models that efficiently read different subsets of features and do so without reading in the whole row of data every time. The more data we collect in one place, the more likely it is that we will have different models using very different subsets of that data.

This approach is way too complicated for some training systems, however. As an example, if we're training on a lot of images that aren't preprocessed, we really don't need to have a column-oriented storage system—we can just read image files from a directory or a bucket. However, say we are reading something more structured, such as transaction data logs with fields like timestamp, source, referrer, amount, item, shipping method, and payment mechanism. Then, assuming that some models will use some of those features and others will use others will motivate us to use a column-oriented structure.

Metadata helps humans interact with the storage. When multiple people work on building models on the same data (or when the same person works on this over time), metadata about the stored features provides huge value. It is the roadmap to understanding how the last model was put together and how we might want to put together the model. Metadata for storage systems is one of the more commonly undervalued parts of an ML system.

This section should clarify that our data management system is primarily motivated by two factors:

The business purpose to which we intend to put the data
: What problem are we trying to solve? What is the value of that problem to our organization or our customers?

---

7 Most common, large data storage and analysis services from large cloud providers are column oriented. Google Cloud BigQuery, Amazon RedShift, and Microsoft Azure SQL Data Warehouse are all column oriented, as is the main data store for data services provider Snowflake. Both PostgreSQL and MariaDB Server have column-oriented configuration options as well.

The model structure and strategy
: What models do we plan to build? How are they put together? How often are they refreshed? How many of them are there? How similar to one another are they?

Every choice we make about our data management system is constrained by, and constrains, these two factors. And if data management is about how and why we write data, ML training pipelines are about how and why we read it.

Management

Typically, data storage systems implement credential-based access controls to restrict unauthorized users from accessing the data. Such simple techniques can serve only a basic ML implementation. In more advanced scenarios, especially when the data contains confidential information, we need to have more granular data access management methodologies in place. For example, we might want to allow only model developers to access the features that they directly work on or restrict their access to a subset of the data in some other way (perhaps only recent data). Alternatively, we might anonymize or pseudononymize the data either at rest or when it is being accessed. Finally, we might allow production engineers to access all of the data, but only after demonstrating that they need to during an incident and having their access carefully logged and monitored by a separate team. (Some interesting aspects of this are discussed in Chapter 11.)

SREs can configure data access restrictions on the storage system in production to allow data scientists to read data securely via authorized networks like virtual private networks (VPNs), implement audit logging to track which users and training jobs are accessing what data, generate reports, and monitor usage patterns. Business owners and/or product managers can define user permissions based on the use cases. We may need to generate and use different kinds of metadata for these dimensions of heterogeneity to maximize our ability to access and modify the data for subsequent phases.

Analysis and Visualization

Data analysis and visualization is the process of transforming large amounts of data into an easy-to-navigate representation using statistical and/or graphical techniques and tools.[8] It is an essential task of ML architectures and knowledge-discovery techniques to make data less confusing and more accessible. It is not enough just to show a pie chart or bar graph. We need to provide the human reader an explanation of what

8 Data scientists, data analysts, research scientists, and applied scientists use various data visualization tools and techniques including infographics, heatmaps, fever charts, area charts, and histograms. For more insights, refer to Wikipedia's "Data and information visualization" page (*https://oreil.ly/DL2B2*).

each record in the dataset means, how it is linked with the records in other datasets, and whether it is clean and safe to use for training models. It is practically impossible for data scientists to look at large datasets without well-defined and high-performant data visualization tools and processes.

Data Reliability

Because our data processing system needs to work, the data must have several characteristics as it traverses the system. Articulating these characteristics can be controversial. To some people, they seem obvious, and to others, they seem impossible to really guarantee. Both of these perspectives miss the point.

The intent of articulating a set of invariants that should always be true about our data is that it permits us all to notice when they are not true, or when a system cannot properly guarantee that they will be true. This permits us to take action to do better in the future. Note that the topic of reliability, even of data management systems, is extremely broad and cannot be covered completely here; for much more detail, see *Site Reliability Engineering: How Google Runs Production Systems* (*https://oreil.ly/ZzIkN*), edited by Betsy Beyer et al. (O'Reilly, 2016).

This section covers just the basics of making sure that the data is not lost (durability), is the same for all copies (consistency), and is tracked carefully as it changes over time (version control). We also cover how to think about how fast the data can be read (performance) and how often it's not ready to be read (availability). A quick overview of these concepts should prepare us to focus on the right areas.

Durability

When spelling out the requirements for storage systems, durability is the most often overlooked because it is assumed. *Durability* is the property of the storage system having your data, and having not lost, deleted, overwritten, or corrupted it. We definitely want that property with a very high probability.

Durability is normally expressed as an annual percentage of bytes or blocks that are not lost irretrievably. Common values for good storage systems here are 11 or 12 nines, which can also be expressed as "99.999999999% or more of bytes stored are not lost." While this may be the offering of the raw underlying storage system, our guarantees might be quite a bit more modest because we're writing software that interacts with the storage system.

One important note is that some systems have extremely durable data, in the sense that nothing is lost, but do have failure modes leaving some data inaccessible for extremely long periods of time. This might include cases where the data needs to be recovered from another slower storage system (say, a tape drive) or copied from off-site over a slow network connection. If this is raw data and is important to the model, you may have to recover it. But for data that is somehow derived from existing raw data, reliability engineers will want to consider whether it might be easier to simply re-create the data rather than restore it.

For an ML storage system with many data transformations, we need to be careful about how those transformations are written and monitored. We should log data transformations and, if we can afford to, store copies of the data pre- and post-transformation. The hardest place to keep track of the data is on ingestion, when the data goes from an unmanaged to a managed state. Since we recommend using an API for ingestion, this provides a clear place to ensure that the data is stored, to log the transaction, and to acknowledge the receipt of the data. If the data is not cleanly and durably received, the sending system can, of course, retry the send operation while the data is still available.

At each stage of data transformation, if we can afford it, we should store pre- and post-transformation copies of the data. We should monitor throughput of the transformations as well and expected data size changes. For example, if we sample the data by 30%, the post-transformation data should obviously be 30% the size of the pre-transformation data, unless an error occurs. On the other hand, if we're transforming a float into an integer by bucketing, depending on the data representation, we could expect the resulting data size to remain unchanged. If it's much bigger or much smaller, we almost certainly have a problem.

Consistency

We may want to guarantee that, as we access data from multiple computers, the data is the same with every read; this is the property of *consistency*. ML systems of any scale are usually distributed. Most of the processing that we are doing is fundamentally parallelizable, and assuming that we'll be using clusters of machines from the beginning is generally worthwhile. This means that the storage system will be available via a networking protocol from other computers, and introduces challenges for reliability. Significantly, it introduces the fact that different versions of the same data might be available at the same time. It is difficult to guarantee that data is replicated, available, and consistent everywhere.

Whether the model-training system cares about consistency is actually a property of the model and the data. Not all training systems are sensitive to inconsistency in the data. For that matter, not all *data* is sensitive to inconsistencies. One way to think about this is to consider how dense or sparse the data is. Data is *sparse* when the information that each piece represents is rare. Data is *dense* when the information that each piece of data represents is common. Data is sparse when the dataset has a lot of zero values. So if *yarnit.ai* has 10 popular yarns that represent almost all of what we sell, the data for any given purchase of one of those yarns is dense—it's unlikely to teach us very much that's new. If a single purchase of a popular yarn is readable in one copy of our storage system but not another, the model will be essentially unaffected. On the other hand, if 90% of our purchases are for different yarns, every single purchase is important. If one part of our training system sees a purchase of a particular yarn and another does not, we might produce an incoherent model with respect to that particular yarn, or yarns that our model represents as similar to that yarn. Under some circumstances, consistency is difficult to guarantee, but often if we can wait somewhat longer for the data to arrive and sync, we can easily guarantee this property.

We can eliminate consistency concerns in the data layer in two straightforward ways. The first is to build models that are resilient to inconsistent data. Just as with other data processing systems, ML systems offer trade-offs. If we can tolerate inconsistent data, especially when the data is recently written, we might be able to train our models significantly faster and operate our storage system more cheaply. The cost, in this case, is flexibility and guarantees. If we go down this path, we limit ourselves to indefinitely operating the storage system with these guarantees and we can train only models that are satisfied with this property. That's one choice.

The second choice is to operate a training system that provides consistency guarantees. The most common way to do that for a replicated storage system is for the system itself to provide information about what data is completely and consistently replicated. Systems that read the data can consume this field and choose to train only on the data that is fully replicated. This is often more complicated for the storage system since we need to provide an API to replication status. It may also be significantly more expensive or slower. If we want to use the data quickly after ingestion and transformation, we may need to have lots of resources provisioned for networking (to copy the data) and storage I/O capacity (to write the copies).

Thinking through consistency requirements is a strategic decision. It has long-term implications for balancing costs and capabilities and should be made with both ML engineers and organizational decision makers involved.

Version Control

ML dataset versioning is, in many ways, similar to traditional data and/or source code versioning used to bookmark the state of the data so that we can apply a specific version of the dataset for future experiments. *Versioning* becomes important when new data is available for retraining and when we're planning to implement different data preparation or feature engineering techniques. In production environments, ML experts deal with a large volume of datasets, files, and metrics to carry out day-to-day operations. The varying versions of these artifacts need to be tracked and managed as experiments are performed on them in multiple iterations. Version control is a great practice for managing numerous datasets, ML models, and files in addition to keeping a record of multiple iterations—i.e., when, why, and what was altered.[9]

Performance

The storage system needs fast-enough write throughput to rapidly ingest the data and not slow transformations. The system needs fast-enough read bandwidth to enable us to rapidly train models using the access patterns that fit our modeling behavior. It is worth noting that the cost of slow read performance can be quite significant for the simple reason that ML training often uses relatively expensive compute resources (GPUs or high-end CPUs). When these processors are stalled waiting on data to train on, they are simply burning cycles and time with no useful work done. Many organizations believe they cannot afford to invest in their storage system, when they actually cannot afford not to.

Availability

The data we write needs to be there when we read it. *Availability* is, in some ways, the product of durability, consistency, and performance. If the data is in our storage system and is consistently replicated, and we are able to read it with reasonable performance, then the data will count as available.

Data Integrity

Data that is valuable should be treated as such. This means respecting provenance, security, and integrity.[10] Our data management system will need to be designed for these properties from the beginning to be able to make appropriate guarantees about the kind of access controls and other data integrity we can offer.

Data integrity has three other big themes in addition to security and integrity: privacy, policy compliance, and fairness. It is worth taking a moment to consider these topics from a general point of view. We need to ensure that we understand the requirements presented by these areas so that we can ensure that the storage system and API that we build offer the kinds of guarantees we require.

9 Some readers might read "version control" and think "Git." A content-indexed software version control system like Git is not really appropriate or necessary to track versions of ML data. We are not making thousands of small and structured changes, but rather generally adding and deleting whole files or sections. The version control we need tracks what the data refers to, who created/updated it, and when it was created. Many ML modeling and training systems include some version control. MLflow is one example.

10 Data durability, discussed previously in "Durability" on page 33, is often included as a key concept in data integrity. Since this entire chapter is about data management, durability has been grouped with reliability concepts, and *integrity* here refers to properties we can assert about the data, beyond its mere existence and accessibility.

Security

Valuable ML data often begins its life as private data. Some organizations choose to build processes to simply exclude all PII from their datastore. This is a good idea for several reasons. It simplifies access control problems. It eliminates the operational burden of data deletion requests.[11] Finally, it removes the risk of storing private information. As we've discussed, data should properly be considered a liability as much as an asset.

We may have successfully excluded PII from the ML datastore. But we probably shouldn't count on that for two reasons. On the one hand, we may not have excluded PII as effectively as we think we have. As previously mentioned, it is notoriously difficult to identify PII without a thoughtful analysis so unless there is careful, time-consuming, human review of *all* data that is added to the feature store, it is extremely likely that some of the data in combination with other data does contain PII. On the other hand, for many organizations, it may simply not be feasible to reasonably exclude all PII from the datastore. Such organizations are obliged to strongly guard their datastore as a result.

Beyond concerns about PII, teams will likely develop particular uses for particular kinds of data. Reasonable use of the datastore will restrict access to certain data to the team most likely to need and use that data. Thoughtful restriction of access will actually increase productivity if model developers can easily access (and only access) the data they are most likely to use to build models.

In all cases, systems engineers should track metadata about which development teams build which models and which models depend on which features in the feature store —effectively an audit trail. This metadata is useful, if not required, for operational and security-related purposes.

Privacy

When ML data is about individuals, the storage system will need to have privacy-preserving characteristics. One of the fastest ways to transform data from an asset to a liability is by leaking private information about customers or partners.

We can make two architecturally different choices about dealing with private data: eliminate it or lock it down. To the extent that we can still get excellent results, eliminating private data is an extremely sound strategy. If we prevent PII data from ever being stored in the data storage system, we eliminate most of the risk of holding private data. This may be difficult—not only because it is not always easy to recognize

11 One useful summary can be found in the Databricks article "Best Practices: GDPR and CCPA Compliance Using Delta Lake," especially the section on pseudonymization (*https://oreil.ly/I5hPt*).

private data, but also because it's not always possible to get great results without private data.

Privacy Choices for Recommendations at YarnIt

Let's think about a recommendation or discovery system for YarnIt. The general idea is that we would like to show customers items that they might be interested in buying at various stages of their visit to the *yarnit.ai* website. That might include when they land on the page, when they search for a type of yarn or a brand of knitting needle, when they put something into their cart, or when they check out. Ideally, we will present them with suggestions that they find appealing. So, what pieces of information would we need as inputs to our system to identify products they might also consider?

One of the time-honored approaches is "people who bought *X* also bought *Y*." It makes sense and allows us to make reasonable recommendations for a wide set of products for which there is commonality or homogeneity among the customers. If everyone who buys a particular type of mohair yarn also buys a specific type of needles, we should be able to recommend them with no private information at all about the individual user right now. But if there is some variety among our customers, things get more interesting.

For example, what if one customer is considerably more or less price sensitive than other customers? If they have a much smaller budget than the typical mohair yarn purchaser, they may choose not to purchase additional needles, or may buy them only below a certain price. Or what if the system knows that this customer had already purchased those needles in a previous transaction? In that case, recommending more of them is probably a waste of screen space and precious attention. We should recommend things we have reason to predict the customer will actually be interested in buying.

To make those kinds of recommendations, however, we need private data. Specifically, we need individual users' purchase histories. With that data, we can easily determine things like estimated budget and types of items previously purchased, including specific items already purchased. If we decide that our models can achieve their goals only by having access to private data, we will need to have a serious conversation about the architecture of storing, using, and eventually deleting that private data. The most thorough structural approach generally requires creating per user datastores that are encrypted at rest and unlocked by a key under control of only the customer. This is most common when handling other organizations' data and less common for individuals running their own training system. Additionally, using data like this, combined with general datastores with data from multiple users, requires federated learning—an advanced topic that's beyond the scope of this book.[12] (See Figure 2-5 for a list of the types of data and access control implications.)

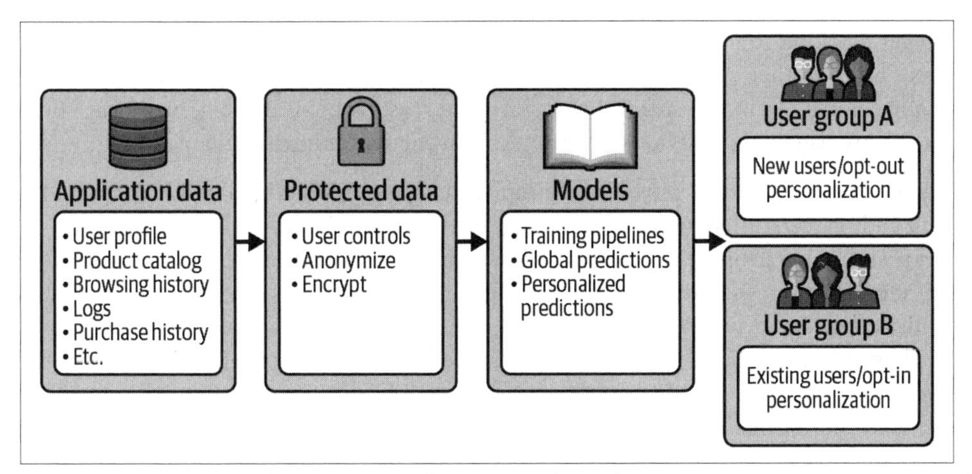

Figure 2-5. Choices and processing as data moves through an ML system

Given all of this complexity, it's substantially better and easier to anonymize data as we ingest it. As previously mentioned, the topic of anonymization is technically complex, but everyone building an ML system needs to know two key facts:

Anonymization is hard
: It's a topic people study and develop expertise on. Don't try to just muddle through. Take it seriously and do it right.

Anonymization is context dependent
: There is no guaranteed way to anonymize data without knowing what other data exists and how the two bits of data relate to each other.

Anonymization is difficult but not at all impossible, and when done properly, it avoids a host of other problems. Note that doing so durably will require periodic review to ensure that current anonymization still matches the assumptions made about the data and access permissions when it was implemented, as well as a review every time new data sources are added to ensure that connections among data sources do not undermine anonymization. This topic is covered more extensively in Chapter 6.

12 See "Federated Learning: Collaborative Machine Learning Without Centralized Training Data" (*https://oreil.ly/ptj9h*) by Brendan McMahan and Daniel Ramage for a general overview and reasonable links to the topic as it was in 2017. Federated learning has, of course, continued to evolve since then.

Policy and Compliance

Policy and compliance typically flow from requirements originating outside your organization. In some cases "outside the organization" actually means a boss or a lawyer working for YarnIt but implementing an external legal requirement of some kind, and in other cases it means a national government directly intervening. These requirements often have painful and powerful reasons behind them, but these backstories are usually not obvious when looking at the requirements themselves.

Here's an annoying but powerful example: European regulations about cookie consent in browsers often seem overbearing, intrusive, or silly to web users, both European and non-European. The idea that websites should get explicit consent to store an identifier on the user's machine might appear unnecessary. But anyone who understands the privacy-violating power of the third-party ad cookie can attest that a really strong rationale lies behind at least some restrictions on cookies. While the "ask every user for every website" approach probably isn't the most elegant and scalable, it is much more understandable when we know more about how these cookies can be used and how difficult it is to protect against their bad usage.

Policy and compliance requirements for data storage should be taken seriously. But it's a mistake to read the letter of a requirement or standard without understanding the intent behind it. Often, whole industries of consultants have developed difficult compliance practices when a simple approach might also be in compliance.

Anonymization, as mentioned previously, is one potential compliance shortcut. If private data requires special treatment, there may be a way to avoid those requirements simply by determining (and *documenting*) that we are not storing any private data.

There are two other things to note about policy and governance requirements: jurisdictions and reporting.

Jurisdictional rules

The world is increasingly filled with governments that assert control over the handling of data stored in or sourced from their geographic location. While this seems reasonable in principle, it does not map at all cleanly onto the way that the world has been building networked computer systems for the past few decades. It may not even be possible for some cloud providers to ensure that data generated in one country is processed in that country. YarnIt plans to sell globally, even though we may launch with only a few supported countries to start. So we will have to think carefully about what data storage and processing requirements we need to comply with.

For larger organizations, one choice of jurisdiction is more important than any other: the host country of the corporate headquarters. This matters because the government of that country is able to exert authority over data residing in any other country. Picking the country of incorporation can have profound implications to our data management system, but it's not a factor most people carefully consider. They should.

Reporting requirements

Remember that compliance work entails reporting. In many cases, reporting can be integrated into the way we monitor the service. Compliance requirements are SLOs, and reporting includes the SLIs that establish the status of our implementation with respect to those compliance SLOs. Thinking of it this way normalizes this work alongside the rest of the implementation and reliability work we need to do.

Conclusion

This has been a rapid and superficial introduction (though it probably doesn't feel like that) to thinking about data systems for ML. At this point, you may not be comfortable building a complete system for data ingestion and processing, but the basic elements of such a system should be clear. More importantly, you should be able to identify where some of the biggest risks and pitfalls lie. To start making concrete progress on this, most readers will want to divide their efforts into the following areas:

Policy and governance
: Many organizations start work on ML from the product or engineering groups. As we have highlighted, however, having a consistent set of policy decisions and a consistent approach to governance will be critical in the long run. Organizations that haven't yet started on this effort should do so immediately. Chapter 6 is a good place to start.

  To have the most impact in this area, you should identify the likely biggest problems or gaps and address them first. Perfection is not possible, given the current state of our understanding of the risks of using ML incorrectly and the tools available. But reducing instances of egregious violations absolutely is possible and is a reasonable goal.

Data sciences and software infrastructure
: If we have started using ML at all, it is likely that our data science teams are already building bespoke data transformation pipelines in various places around the organization. Cleaning, normalizing, and transforming the data are normal operations that are required to do ML. To avoid future technical ML debt, we

should start building software infrastructure to centralize these transformation pipelines as soon as is practical.[13]

Teams that have already solved their own problems may be resistant to this centralization. However, by operating data transformation as a service, it is sometimes possible to entice all new users and even some existing users to transfer allegiance to the centralized system. Over time, we should try to consolidate data transformations in a single, well managed place.

Infrastructure
: We obviously need significant data storage and processing infrastructure to manage ML data well. The biggest element here is the feature storage system (often simply referred to as the *feature store*). We discuss the useful elements of a feature store in detail in Chapter 4.

13 The technical debt in ML systems is often fairly different from what we see in other software systems. One paper that explains this in detail is "Hidden Technical Debt in Machine Learning Systems" (*https://oreil.ly/3SV7Q*) by D. Sculley (a coauthor of this book) et al.

CHAPTER 3
Basic Introduction to Models

Most of this book is about managing ML systems and production-level ML pipelines. This involves work that is quite different from the work often performed by many data scientists and ML researchers, who try to spend their days developing new predictive models and methods that can squeeze out another percentage point of accuracy. Instead, in this book, we focus on ensuring that a system that includes an ML model exhibits consistent, robust, and reliable *system-level* behavior. In some ways, this system-level behavior is independent of the actual model type, how good the model is, or other solely model-related considerations. Still, in certain key situations, it is *not* independent of these considerations. Our goal in this chapter is to give you enough background to understand which situation you are in when the alarms start to go off or the pagers start to fire for your production system.

We will say at the outset that our goal here is *not* to teach you everything about how to build ML models, which models might be good for what problems, or how to become a data scientist. That would be a book (or more) all to itself, and many excellent texts and online courses cover these aspects.

Instead of going too deep into the minutia, in this chapter our goal is to give a quick reminder about what ML models are and how they work. We'll also provide some key questions that machine learning operations (MLOps) folks should ask about the models in their system so they can understand the kinds of problems to plan for appropriately.

What Is a Model?

In mathematics or science, the word *model* refers to a rule or guideline, often expressible in math or code, that helps take inputs and give predictions about the way that the world might work in the future. For example, here is a famous model you might recognize:

$$E = mc^2$$

This is a lovely little model that tells you how much energy (E) you are likely to get if you convert a given input amount of mass (m) into something super hot and explosive, and the constant c^2 tells you that you get really quite a lot of E out for even a little bit of m. This model was created by a smart person who thought carefully for a long time and holds up well in various settings. It doesn't require a lot of maintenance and generalizes beautifully to a huge range of settings, even those never envisaged when it was first created.

The models we typically deal with in ML are similar in some ways. They take inputs and give outputs that are often thought of as *predictions*, using rules that are expressible in mathematical notation or code. These predictions could represent the physical world, like "What's the probability that it will rain tomorrow in Seattle?" Or they could represent quantities, like "How many units of yarn will sell next month from our website, *yarnit.ai*?" Or they could even represent abstract human concepts, like "Is this picture aesthetically pleasing to users on average?"

One key difference is that the models we typically use for ML are ones that we cannot write down within a neat little rule like $E = mc^2$, no matter how smart we are. We turn to ML in settings where many pieces of information—often called *features*—need to be taken into account in ways that are hard for humans to specify in advance. Some examples of data that can be processed as features are atmospheric readings from thousands of locations, the color values of thousands of pixels in an image, or the purchase history of all users who have recently visited an online store. When dealing with information with sources that are this complex—much more than one value for mass and one scaling constant—it is typically difficult or impossible for human experts to create and verify reliable models that take advantage of the full range of available information. In these cases, we turn to using vast amounts of previously observed data. In using data to *train a model*, we hope the resulting model will both fit our past data well and generalize to predict well on new, previously unseen data in the future.

A Basic Model Creation Workflow

The basic process that is most widely used for creating ML models right now—formally called *supervised machine learning*—looks like this.

To start, we gather a whole bunch of historical data about our problem area. This might be all of the atmospheric sensor readings from the Pacific Northwest for the last 10 years, or a collection of half a million images, or logs of user browsing history to the online *yarnit.ai* site. We extract from that data a set of *features*, or specific, measurable properties of the data. Features represent key qualities of the data in a way that ML models can easily digest. In the case of numerical data, this might mean scaling down values to fit nicely within certain ranges. For less structured data, we might engineer specific quantities to identify and pull out of raw data. Some examples are input features that might represent things like the atmospheric pressure for each of 1,000 sensor locations, or the specific color and size values for a given yarn ball, or a set of features corresponding to each possible product, with a value of `1` if a given user has viewed that product and `0` if they have not.

For supervised ML, we also require a *label* of some kind, showing the historical outcome that we would like our model to predict, if it were to see a similar situation in the future. This could be the weather result for the given day, like `1` if it rained, and `0` if it did not. Or it could be a score attempting to capture whether a given user found an image to be aesthetically pleasing, such as `1` if they gave it a "thumbs up," and `0` if they did not. Or it could be a value showing the number of units of that given yarn product that happened to be sold in the given month. Or yarn colors and sizes, and so on. We'll record the given label for each entry—and call each one a *labeled example*.

We'll then *train* a model on this historical data, using a chosen model type and a chosen ML platform or service. Currently, many folks choose to use model types based on *deep learning*, also known as *neural networks*, which are especially effective when given very large amounts of data (think millions or billions of labeled examples).[1] Neural networks build connections among layers of nodes based on the examples that are used in training.

In other settings, methods like random forests or gradient boosted decision trees work well when presented with fewer examples.[2] And more complex, larger models are not always preferred, either for predictive power or for maintainability and understandability. Those who are not sure which model type might work best often use methods like automated machine learning (AutoML), which trains many model versions, and try to automatically pick the best one. Even without AutoML, all models have adjustments and settings—called *hyperparameters*—that must be set specifically for each task, through a process called *tuning*, which is really just a fancy name for trial and error. Regardless, at the end, our training process will produce a model, which will take the feature inputs for new examples and produce a prediction as an output. The internals of the model will most likely be treated as opaque and

1 The term *neural networks* is rooted in an analogy between their formulation and the ways that neurons in a brain work. The term *deep learning* became popular in part because so many folks pointed out that the analogy to neurons in the brain is not particularly accurate.

2 Many ML production engineers or SREs do not need to learn the full details of how neural networks, random forests, or gradient boosted decision trees work (although doing so is not as scary or difficult as it seems at first). Reliability engineers working with these systems do, however, need to learn the systems requirements and typical performance of these systems. For this reason, we cover the high-level structure here.

defying direct inspection, often consisting of thousands, millions, or even billions of learned *parameters* that show how to combine the input features (and potentially recombine them in many further intermediate stages) to create the final output prediction.

From the perspective of systems engineering, we might notice something disconcerting. Typically, no human knows what the "right" model is. We can't look at a model produced by training and know whether it is good or bad just from our knowledge. The best we can do is run it through various stress tests and validations and see how it performs,[3] and hope that the stress tests and validations that we come up with are sufficient to capture the range of situations that the model will be asked to handle.

The most basic form of validation is to take some of the training data that we prepared and randomly hold some out to the side, calling it a *held-out test set*. Instead of using this for training, we wait and use it to stress-test our model, reasoning that because the model has never seen this held-out data in training, it can serve well as the kind of previously unseen data for which we hope it will make useful predictions. So we can feed each example in this test set to the model and compare its prediction to the true outcome label to see how well it measures up. It is critical that this validation data is indeed held out from training, because it is all too easy for a model to experience *overfitting*, which happens when the model memorizes its training data perfectly but cannot predict well on new unseen data.

Once we are happy with the validation of our model, it is time to deploy it in the overall production system. This means finding a way to *serve* the model predictions as they are required. One way to do this is by precomputing all possible predictions for all possible inputs and then writing those to a cache that our system can query. Another common method is to create a flow of inputs from the production system and transform them into features so that they can be fed to a copy of our model on demand, and then feed the resulting predictions from the model back into the larger system for use. Both of these options can be understood in context in Chapter 8.

If everything works perfectly from here, we are done. However, the world is full of imperfections, and our models of the world even more so. In the MLOps world, even the smallest of them can potentially create serious issues, and we therefore need to be prepared. Let's take a quick tour of some of these areas that can turn into possible problems.

3 The evaluation of model quality or performance is a complex topic in its own right that is covered in Chapter 5.

Model Architecture Versus Model Definition Versus Trained Model

The term *model* is often used imprecisely to refer to three separate, but related, concepts:

Model architecture
: The general strategy that we are using to learn in this application. This can include both the choice of model family, such as deep neural network (DNN) or random forest, and structural choices such as the number of layers in the DNN or the number of trees in the random forest.

Model definition (or configured model)
: The configuration of the model plus the training environment, and the type and definition of data we will train on. This includes the full set of features that are used, all hyperparameter settings, random seeds used to initialize a model, and any other aspect that would be important for defining or reproducing a model. It is reasonable to think of this as the closure of the entire training environment,[4] although many people don't think through the reliability or systems implications of this (in particular, that a very large set of software and data must be carefully and mutually versioned in order to get reasonable amounts of reproducibility).

Trained model
: A specific snapshot or instantiated representation of the configured model trained on specific data at a point in time, containing a specific set of trained model parameters such as weights and thresholds. Note that some of the software that we use in ML, especially in distributed deployments, has a fair bit of nondeterminism. As a result, the exact same configured model trained twice on the exact same data may or may not produce a significantly different trained model.

Although in this book we try to not confuse these concepts, be aware that they are not carefully differentiated in the industry as a whole and that we also may be occasionally unclear. Most people call all three of these concepts *the model*. Even these terms are ours, as there is no industry-standard terminology to refer to the different uses of *model*.

Where Are the Vulnerabilities?

From a systems reliability standpoint, a system that relies on ML has several areas of vulnerability where things can go wrong. We touch on a range of them briefly in this section, with a much deeper dive coming in later chapters on many of these topics.

4 Another way to say that is "all the metadata and data required to regenerate the training environment."

Note that here we are not referring to security vulnerabilities, but rather to structural or systemic weaknesses that may be the source of failures or model quality problems. This section is an attempt to enumerate some of the most common ways that models fail.

Training Data

Training data is a foundational element that defines much of the behavior of our systems. The qualities of the training data establish the qualities and behavior of our models and systems, and imperfections in training data can be amplified in surprising ways. From this lens, the role of data in ML systems can be seen as analogous to code in traditional systems. But unlike traditional code, which can be described with preconditions, post-conditions, and invariants, and may also be systematically testable, real-world data typically has organic flaws and irregularities. These can lead to a variety of problem types.

Incomplete coverage

The first problem to consider is that our training data may be incomplete in various ways. Imagine that we have atmospheric pressure sensors that stop working at temperatures below freezing, so we have no data on cold days. This creates a blind spot for the model, especially when it is asked to make predictions in this case. One of the difficulties here is that these kinds of problems cannot be detected by using a held-out test set, because by definition the held-out set is a random sample of the data that we already have access to. Detecting these issues requires a combination of careful thought and hard work to gather or synthesize additional data that can help expose these flaws at validation time or correct them at training time.

Spurious correlations

A special form of incomplete coverage happens when correlations in the data do not always hold in the real world. For example, imagine that all images marked "aesthetically pleasing" happen to include white gallery walls in the background, while none of the images marked "not aesthetically pleasing" do. Training on this data would likely result in a model that shows very high accuracy on held-out test data, but that is essentially just a white-wall detector. Deploying this model on real data could have terrible results, even while showing excellent performance on held-out test data.

Again, uncovering such issues requires careful consideration of the training data, and targeted probing of the model with well-chosen examples to stress-test its behavior in a variety of situations. An important class of such problems can occur when societal factors create a lack of inclusion or representation for certain groups of people; these issues are discussed in Chapter 6.

Cold start

Many production modeling systems collect additional data over time in deployment and are retrained or updated to incorporate that data. Such systems can suffer from a *cold-start* problem when there is little initial data. This can happen for systems that were not set up to collect data initially, like if a weather service wants to use a newly created network of sensors that has never been previously deployed. It can also arise in recommender systems, such as those that recommend various yarn products to users and then observe user interaction data for training. These systems have little or no training data at the very start of their lifecycle, and can also encounter cold-start problems for individual items when new products are released over time.

Self-fulfilling prophecies and ML echo chambers

Many models are used within feedback loops, in which they filter data, recommend items, or select actions that then create or influence the model's training data in the future. A model that helps recommend yarn products to users will likely get future feedback only on the few near-the-top rankings that are actually shown to users. A conversational agent likely gets feedback only on the sentences that it chooses to create. This can cause a situation in which data that is not selected never gets positive feedback, and thus never rises to prominence in the model's estimation. Solving this often requires some amount of intentional *exploration* of lower-ranked data, occasionally choosing to show or try data or actions that the model currently thinks are not as good, in order to ensure a reasonable flow of training data across the full spectrum of possibilities.

Changes in the world

It is tempting to think of data as a reflection of reality, but unfortunately it is really just a historical snapshot of a part of the world at a particular time. This distinction can appear a little bit philosophical, but becomes mission critical at times when real-world events cause changes. Indeed, there may be no better way to convince the executive leadership in an organization of the importance of MLOps than to ask the question, "What would happen to our models if tomorrow there was another COVID-style lockdown?"

For example, imagine that our model helps recommend hotels to users, based on its learning on feedback from previous user bookings. It is easy to imagine a scenario in which a COVID-style lockdown creates a sudden sharp drop in hotel bookings, meaning that a model trained on pre-lockdown data is now extremely overly optimistic. As it learned on newer data in which many fewer bookings occurred, it is also easy to imagine that it may do very badly in a world in which lockdowns were then later eased and users wished to book many more hotel rooms again—only to find that the system could not recommend any.

These forms of interactions and feedback loops based on real-world events are not limited to major world disasters. Here are some others that might occur in various settings:

- Election night in a given country causes suddenly different view behavior on videos.
- The introduction of a new product quickly causes a spike in user interest on a certain kind of wool, but our model doesn't have any previous information about it.
- A model for predicting stock price makes an error, overpredicting for a certain stock. The automated hedge fund using this model then incorrectly buys that stock—raising the price in the real market and causing other automated hedge fund models to follow suit.

Labels

In supervised ML, the *training labels* provide the "correct answer," showing the model what it should be trying to predict for a given example. The labels are critical guidance that show the model its objective, often defined as a numerical score. Here are some examples:

- A score of `1` for "spam" and `0` for "not spam" for an email spam filter model
- An amount of daily rainfall, in millimeters, for a given day in Seattle
- A set of labels for each possible word that might complete a given sentence, with `1` if it is the actual word that completes the given sentence and `0` if it is any other word
- A set of labels for each category of object in a given image, with a `1` if that category of object appears prominently in the image and `0` if it does not
- A numerical score showing how strongly a given antibody protein binds to a given virus in a wet lab experiment

Because labels are so important to model training, it is easy to see that problems with labels can be the root cause for many downstream model problems. Let's look at a few.

Label noise

In statistical language, the term *noise* is a synonym for *errors*. If our provided labels are incorrect for some reason, these errors can propagate into the model behavior. Random noise can be tolerable in some settings if the errors balance out over time, although it is still important to measure and assess. More damaging can be errors that occur in certain parts of the data, such as if a human labeler consistently misidentifies frogs as toads for an aquatic image model, or a given set of users is consistently fooled by a certain kind of email spam message, or contamination occurs in an experiment that keeps a given set of antibodies from binding to a given class of viruses.

It is therefore critical to regularly inspect and monitor the quality of the labels and address any issues. In systems in which human experts are used to provide training labels, this can often mean paying excruciating care to documentation of the task specifications and providing detailed training for the humans themselves.

Wrong label objective

ML training methods tend to be extremely effective at learning to predict the labels we provide—sometimes so good that they uncover differences between what we had hoped the labels mean and what they actually represent. For example, if our goal is to make customers at the *yarnit.ai* site happy over time, it can be easy to hope that a "purchase" label correlates with a satisfied user session. This might lead to a model that over-fixates on purchases, perhaps learning over time to promote products that appear to be good deals but in fact are of disappointing quality. As another example, consider the problem of using user clicks as a signal for user satisfaction with news articles—this could lead to models that highlight salacious clickbait headlines or even filter-bubble effects, in which users are not shown news articles that disagree with their preconceptions.

Fraud or malicious feedback

Many systems rely on signals from users or observations of human behavior to provide training labels. For example, some email spam systems allow users to label messages as "spam" or "not spam." It is easy to imagine that a motivated spammer may try to fool such a system by sending many spam emails to accounts under their own control and try to label them as "not spam" in an attempt to poison the overall model. It is also easy to imagine that a model attempting to predict the number of stars a certain product will receive in user reviews may be potentially vulnerable to bad actors who try to overrate their own products—or to underrate those of their competitors. In such settings, careful security measures and monitoring for suspicious trends is a critical part of long-term system health.

In addition to problems with developing a complete and representative dataset, or labeling examples correctly, we can encounter threats to the model during the model-training process. Labels and labeling systems are discussed in much more detail in Chapter 4.

Training Methods

Some models are trained once and then rarely or never updated. But most models will be updated at some point in their lifecycle. This might happen every few months as another batch of wet lab data comes in from antibody testing, or it might happen every week to incorporate a new set of image data and associated object labels, or it might happen every few minutes in a streaming setting to update with new data based on users browsing and purchasing various yarn products. On average, each new update is expected to improve the model overall—but in specific cases, the model might get worse or even completely break. Here are some possible causes of such headaches.

Overfitting

As we discussed in the brief overview of a typical model lifecycle, a good model will generalize well to new, previously unseen data and not just narrowly memorize its training data. Each time a model is retrained or updated, we need to check for overfitting by using held-out validation data. But if we consistently reuse the same validation data, there is a risk that we may end up implicitly overfitting to this reused validation data.

For this reason, it is important to refresh validation data on a regular basis and ensure we are not fooling ourselves. For example, if our validation dataset about purchases on *yarnit.ai* is never updated, while our customers' behavior changes over time to favor brighter wools, our models will fail to track this change in purchase preferences because we will score models that learn this behavior as being of "lower quality" than models that do not. It is important that model quality evaluation include real-world confirmation of a model's performance.

Lack of stability

Each time a model is retrained, there is no guarantee that its predictions are *stable* from one model version to another. One version of a model might perform great recognizing cats and poorly at dogs, while another version might be quite a bit better on dogs and less good on cats—even if both models have similar aggregate accuracy on validation data. In some settings, this can become a significant problem.

For example, imagine a model that is used to detect credit card fraud and to shut down cards that may have been compromised. A model with 99% precision might be very good in general, but if the model is retrained each day and makes mistakes on a different 1% of users each day, then after three months, potentially the entire user base may have been inconvenienced by faulty model predictions. It is important to evaluate model quality on relevant subsections of the predictions that we care about.

Peculiarities of deep learning

Deep learning methods have become so important in recent years because of their ability to achieve extremely strong predictive performance in many areas. However, deep learning methods also come with a specific set of fragilities and potential vulnerabilities. Because they are so widely used, and because they have specific peculiarities and concerns from an MLOps perspective, we go into some detail here on deep learning models in particular.

When deep learning models are trained from scratch—with no prior information—they begin from a randomized initial state and are fed a huge stream of training data in randomized order, most often in small batches. The model makes its best current prediction (which early on is terrible, since the model has not learned very much yet) and then is shown the correct label. Once the math is calculated—computing the *loss gradient*—the internals of the model should be updated. When using this loss gradient, a small update is applied to the model internals that tries to make the model slightly better. This process of small corrections on randomly ordered mini batches is called *stochastic gradient descent* (*SGD*) and is repeated many millions or billions of times. We stop training after we decide that the model shows good performance on held-out validation data.

The key insights about this process are as follows:

Deep learning relies on randomization
: In the initial random state, data is shown in random order. In large-scale parallelized settings, randomness is even inherent in the way that updates are processed because of network and parallel computation effects. Therefore, repeating the process of training the same model with the same settings on the same data can lead to substantially different final models.

It can be difficult to know when training is "done"
: The model performance on held-out validation data generally improves with additional training steps, but sometimes bounces around early on, and later can often get significantly worse if the model starts to overfit the training data by memorizing it too closely. We stop training and choose the best model that we can when performance converges to a good level, often choosing a *checkpoint* version that shows good behavior at an intermediate point. Unfortunately, there is no formal way to know whether the performance we see now is the best we could get if we were to let training continue further. Indeed, a *double-dip* phenomenon discovered relatively recently for a wide range of models shows that our previous conceptions of when to stop may not have been optimal.[5]

Deep learning models sometimes explode in training
: The reason that we take little steps instead of big ones is that it is easy to fall off a mathematical cliff and put the model internals into a state from which it is hard to recover. Indeed, *exploding gradients* is a technical term that connotes

5 OpenAI offers a straightforward description of this phenomenon, along with helpful references to the source papers in "Deep Double Descent" (*https://openai.com/blog/deep-double-descent*) by Preetum Nakkiran.

the appropriate danger. Models that end up in this state often give `NaN` (not a number) values in predictions or intermediate computations. This behavior can also surface as the model's validation performance suddenly worsening.

Deep learning is highly sensitive to hyperparameters
: As has been mentioned, hyperparameters are the various numeric settings that must be tuned for an ML model to achieve best performance on any given task or dataset. The most obvious of these is the *learning rate*, which controls how small each update step should be. The smaller the steps, the less likely the model is to explode or result in strange predictions. But the longer training takes, the more computation is used. Other settings also have significant effects, like how large or complex the internal state is, how large the little batches are, and how strongly to apply various methods that try to combat overfitting. Deep learning models are notoriously sensitive to such settings, which means that significant experimentation and testing is required.

Deep learning is resource intensive
: The training methodology of SGD is effective at producing good models, but relies on a truly enormous number of tiny updates. Indeed, the amount of computation used to train some models can look like thousands of CPU cores running continuously for several weeks.

When deep learning methods are wrong, they might be very wrong
: Deep learning methods extrapolate from their training data, which means that the more unfamiliar a new previously unseen data point is, the more likely we are to have an extreme prediction that might be completely off base or out of the range of typical behavior. These sorts of highly confident errors can be a significant source of system-level misbehavior.

Now that you understand the structure of what can go wrong with model creation, it might be useful to take a broader perspective on the infrastructure required to train models in the first place.

Infrastructure and Pipelines

Models are just one component in larger ML systems, which are typically supported by significant infrastructure to support model training, validation, serving, and monitoring in one or more pipelines. These systems thus inherit all of the complexities and vulnerabilities of traditional (non-ML) pipelines and systems in addition to those of ML models. We will not go into all of those traditional issues here, but will highlight a few areas in which traditional pipeline issues come to the foreground in ML-based systems.

Platforms

Modern ML systems are often built on top of one (or more!) ML frameworks, such as TensorFlow, PyTorch, or scikit-learn, or even an integrated platform such as Azure Machine Learning, Amazon SageMaker, or Google Cloud Vertex AI. From a modeling perspective, these platforms allow developers to create models with a great degree of speed and flexibility, and in many cases to enable the use of hardware accelerators such as GPUs and TPUs or cloud-based computation without a lot of extra work.

From a systems perspective, the use of such platforms introduces a set of dependencies that is typically outside our control. We may be hit with package upgrades, fixes that may not be backward-compatible, or components that may not be forward compatible. Bugs that are found may be difficult to fix, or we may need to wait for the platform owners to prioritize accepting the fixes we propose. On the whole, the benefits of using such frameworks or platforms nearly always outweighs these drawbacks, but costs still must be considered and factored into any long-term maintenance plan or MLOps strategy.

An additional consideration is that because these platforms are typically created as general-purpose tools, we usually need to create a significant number of adapter components, or *glue code*, that can help us transform our raw data or features into the correct formats to be used by the platform, and to interface with the models at serving time. This glue code can quickly become significant in size, scope, and complexity, and is important to support with testing and monitoring at the same level of rigor as the other components of the system.

Feature Generation

Extracting informative features from raw input data is a typical part of many ML systems and may include such tasks as these:

- Tokenizing words in a textual description
- Extracting price information from a product listing
- Binning atmospheric pressure readings into one of five coarse buckets (often referred to as *quantization*)
- Looking up time since last login for a user account
- Converting a system timestamp into a localized time of day

Most of these tasks are straightforward transformations of one data type to another. They may be as simple as dividing one number by another or involve complex logic or requests to other subsystems. In any case, it is important to remember that bugs in feature generation are arguably the *single most common source of errors* in ML systems.

Feature generation is such a hotspot for vulnerabilities for several reasons. The first is that errors in feature generation are often not visible by aggregate model performance metrics, such as accuracy on held-out test data. For example, if we have a bug in the way that our temperature sensor readings are bucketed, it may reduce accuracy by a small amount, but our system may also learn to compensate for this bug by relying more on other features. It can be surprisingly common for bugs to be found in feature-generation code that has been running in production, undetected, for months or years.

A second source of feature generation errors occurs when the same logical feature is computed in different ways at training time and serving time. For example, if our model relies on a localized time of day for an on-device model, that may be computed via a global batch-processing job when training data is computed, but it may be queried directly from the device at serving time. A bug in the serving path could cause errors in prediction that are difficult to detect because of the lack of ground-truth validation labels. We cover a set of monitoring best practices for this important case in Chapter 9.

The third major source of feature-generation errors occurs when our feature generators rely on an upstream dependency that becomes buggy or is hit with an outage. For example, if our model for generating yarn-purchase predictions depends on a lookup query to another service that reports user reviews and satisfaction ratings, our model would have serious problems if that service suddenly went offline or stopped returning sensible responses. In real-world systems, our upstream dependencies often have upstream dependencies of their own, and we are indeed vulnerable to the full stack of them all.

Upgrades and Fixes

One particularly subtle area where upstream dependencies can cause issues in our models arises when the upstream system undergoes an upgrade or bug fix. It may seem strange to say that fixing bugs can cause problems. The principle to remember is *better is not better, better is different—and different may be bad.*

This is because any change to the distribution of feature values that our model expects to see associated with certain data may cause erroneous behavior. For example, imagine that the temperature sensors we use in a weather prediction model have a bug in the code that reports degrees in Fahrenheit when it is supposed to be reporting degrees in Celsius. Our model learns that 32 degrees is freezing, and 90 degrees is a hot summer day near Seattle. If a clever engineer notices this bug and fixes the temperature sensor code to send Celsius values instead, the model will be seeing values of 32 degrees and assuming the world is icy cold when it is hot and sunny.

This form of vulnerability has two key defenses. The first is to arrange for a strong level of agreement with upstream dependencies about being alerted to such changes before they happen. The second is to create monitoring of the feature distributions themselves and alert on change. This is also discussed in more depth in Chapter 9.

A Set of Useful Questions to Ask About Any Model

Researchers and academics tend to focus on the mathematical qualities of an ML model. In MLOps, we can find value in a different set of questions that will help us understand where our models and systems can go wrong, how we can fix issues when they occur, and how we can preventatively engineer for long-term system health:

Where does the training data come from?
: This question is meant conceptually—we need to have a full understanding of the source of the training data and what it is supposed to represent. If we are looking for email spam, do we get access to the routing information, and can it be manipulated by bad actors? If we are modeling user interactions with yarn products, what order are they displayed in, and how does the user move through the page? What important information do we *not* have access to, and what are the reasons? Are there any policy considerations around data access or storage that we need to take into account, especially around privacy, ethical considerations, or legal or regulatory constraints?

Where is the data stored and how is it verified?
: This is the more literal side of the previous question. Is the data stored in one large flat file, or shared across a datacenter? What access patterns are most common or most efficient? Are there any aggregations or sampling strategies applied that might reduce cost but lose information? How are privacy considerations enforced? How long is a given piece of data stored, and what happens if a user wishes to remove their data from the system? And how do we know that the stored data has not been corrupted in some way, that the feeds have not been incomplete, and what sanity checks and verifications we can apply?

What are the features and how are they computed?
: Features, the information we extract from raw data to enable easy digestion for ML models, are often added by model developers with a "more is always better" approach. From an ops perspective, we need to maintain a complete understanding of each individual feature, how it is computed, and how the results are verified. This is important because bugs at the feature-computation level are arguably the most common source of problems at the system level. At the same time, these are often the most difficult to detect by traditional ML

verification strategies—a held-out validation set may be impacted by the same feature computation bug. As suggested previously, most insidious are issues that occur when features are computed by one code path at training time—for example, to optimize for memory efficiency—and by another code path at serving time for real deployment—for example, to optimize for latency. In such cases, the model's predictions can go awry, but we may have no ground-truth validation data to use to detect this.

What kind of examples does the model perform most poorly on?
: Nearly all ML models are imperfect and will make mistakes in their predictions on at least some kinds of examples. It is important to spend time looking at data that our model makes errors on, understanding any commonalities or trends, so that we can identify whether these failure modes will impact the downstream use cases in important ways. This often involves some amount of manual effort, with actual humans looking at actual data to understand it fully, in addition to higher-level summaries.

How is the model updated over time?
: Some models are rarely updated, such as models for automated translation that are trained on huge amounts of data in large batches, and pushed to on-device applications once every few months. Others are updated extremely frequently, such as an email spam filter model that must be kept constantly up-to-date as spammers evolve and develop new tricks to try to avoid detection. However, it is reasonable to assume that all models will eventually need to be updated, and we will need to have structures in place that ensure a full suite of validation checks before any new version of a model is allowed to go live. We will also need to have clear agreements with the model developers in our organization about who makes judgment calls about sufficient model performance, and how problems in predictive accuracy are to be handled.

How does our system fit within the larger environment?
: Our ML systems are important, but as with many complex data processing systems, they are typically only one part of a larger overall system, service, or application. We must have a strong understanding of how our ML system fits into the larger picture in order to prevent issues and diagnose problems if they arise. We need to know the full set of upstream dependencies that provide data to our model, both at training time and serving time, and know how they might change or fail and how we might be alerted if this happens. Similarly, we need to know all of the downstream consumers of our model's predictions, so that we can appropriately alert them if our model should experience issues. We also need to know how the model's predictions impact the end use case, if the model is part of any feedback loops (either direct or indirect), and if there are any cyclic dependencies such as time-of-day, day-of-week, or time-of-year effects. Finally, we need to know how important model qualities like accuracy, freshness, and

prediction latency are within the context of the larger system, so that we can ensure that these system-level requirements are well established and continue to be met by our ML system over time.

What is the worst that could happen?
: Perhaps most importantly, we need to know what happens to the larger system if the ML model fails in any way, or if it gives the worst possible prediction for a given input. This knowledge can help us to define guardrails, fallback strategies, or other safety mechanisms. For example, a stock-price-prediction model could conceivably cause a hedge fund to go bankrupt within a few milliseconds[6] —unless specific guardrails are put in place that limit certain kinds of buying actions or amounts.

An Example ML System

To help ground our introduction to basic models, we will walk through some of the structure of an example production system. We will go through enough detail here so that you can start to see answers to some of the important questions listed previously, but we will also dive deeply into specific areas of this example in later chapters.

Yarn Product Click-Prediction Model

In our imaginary *yarnit.ai* site, ML models are applied to many areas. One of these is in predicting the likelihood that a user will select a given yarn product listing. In this setting, well-calibrated probability estimates are useful for ranking and ordering possible products, including skeins of yarn, various knitting needle types, patterns, and other yarn and knitting accessories.

Features

The model used in this setting is a deep learning model that takes as input the following set of features:

Features extracted from the text of the product description
: These include tokenized words of text, but also specifically identified characteristics such as amount of yarn, size of needles, and product material. Because these characteristics are expressed in a wide variety of ways in product descriptions from different manufacturers, each characteristic is predicted by a separate component model specially trained to identify that characteristic from product

6 That's the time frame for stock-price trading these days. See Michael Lewis's *Flash Boys* (W. W. Norton & Co., 2014) for more information.

description text. These models are owned by a separate team, and provided to our system via a networked service that has occasional outages.

Raw product image data

The raw product image is supplied to the model, after being first normalized to a 32 × 32–pixel square format by squishing the image to fit a square and then averaging pixel values to create the low-resolution approximation. As shown in Figure 3-1, in previous years most manufacturers provided images that were nearly square, and with the product well centered in the image. More recently, some manufacturers have started providing images in a much wider landscape format that must be squished significantly more to become square, and the product itself is often shown in additional settings rather than on a plain, solid-color background.

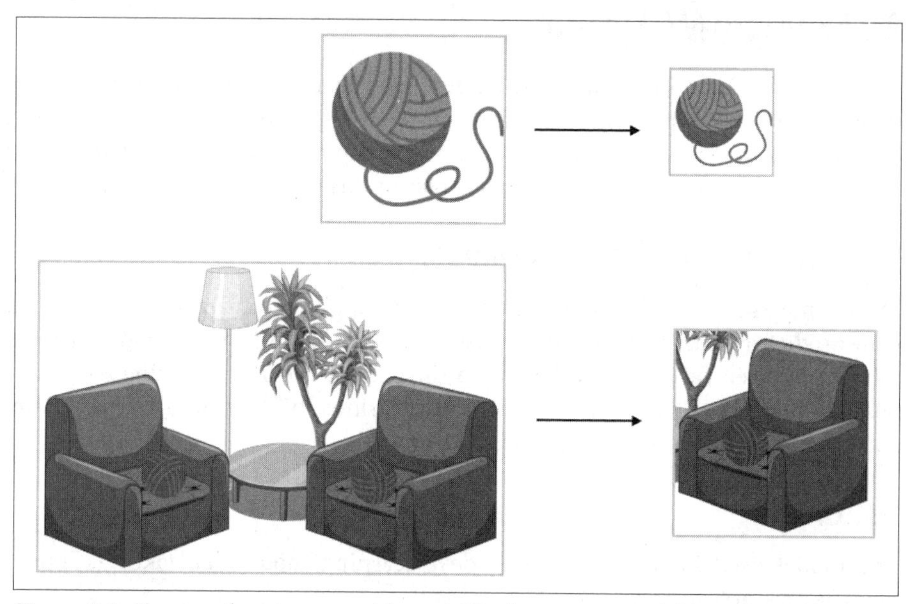

Figure 3-1. Raw product images with variable dimensions and formats (images courtesy of Vecteezy)

Previous user search and click behavior

The logged history of the user, based on user-accepted cookies with appropriate privacy control, is converted into features that show which previous products the user has viewed, clicked, or purchased. Because some users do not accept the use of cookies, not all users have this form of information available at training or prediction time.

Features related to the user's search query or navigation
: The user may enter search queries such as "thick yellow acrylic yarn" or "wooden size 8 needles," or may arrive at a given page by having clicked various topic headings, navigation bars, or suggestions listed on the previous page.

Features related to the product's placement on the page
: Because products that appear higher in the listed results are more likely to be viewed and clicked than products listed farther down, it is important at training time to have features that show where the product was listed at the time the data was collected. Note, however, that this introduces a tricky dependency—we cannot know at serving time the value of these features, because the ranking and ordering of the results on the page depends on our model's output.

Labels for Features

The training labels for our model are a straightforward mapping of `1` if the user clicks the product and `0` if the user does not click the product. However, we need to consider some subtlety in terms of timing—users might click a product several minutes or even a few hours after a result is first returned to them if they happen to get distracted in the middle of a task and return to it later. For this system, we allow for a one-hour window.

Additionally, we have detected that some unscrupulous manufacturers attempt to boost their product listings by clicking their products repeatedly. Other, more nuanced but equally ill-intentioned manufacturers attempt to lower their *competitors'* listings by issuing many queries that place their competitors listings near the top without clicking them.[7] Both of these are attempts at fraud or spam and need to be filtered out before the model is trained on this data. This filtering is done by a large batch-processing job that runs every few hours over recent data to look for trends or anomalies, and to avoid complications, we simply wait until these jobs have completed before incorporating any new data into our training pipeline. Sometimes, however, these filtering jobs fail, and this can introduce significant additional delays into our system.

Model Updating

Our model is often described to executives as "continually updating," in order to adapt to new trends and new products. However, some delays are inherent in the

7 This works because every time a product is shown but not clicked, the model learns that the product was not a good result for that customer in that context, or at least learns that the product was not the best result. Of course, this happens frequently for all products, but if a competitor is able to swamp the system with millions (or more!) of instances of a product being shown but not clicked, the model will learn that customers, in general, just don't like that product.

system. First, we need to wait an hour to see whether a user has really not clicked a given result. Then we need to wait while upstream processes filter out as much spam or fraud as possible. Finally, the feature-extraction jobs that create the training data require batch-processing resources and introduce several more hours of delay. In practice, then, our model is updated about 12 hours after a given user has viewed or clicked a given product. This is done by incorporating batches of data that update the model from its most recent checkpoint.

Fully retraining the model from scratch requires going back to historical data and revisiting that data in order, with the goal of mimicking the same sequence of new data coming in over time. This is done from time to time by model developers when new model types or additional features are added to the system. For more details, see Chapter 10.

Model Serving

Fortunately, our online store is quite popular—we get a few hundred queries per second, which is not shabby for the worldwide knitting products market. For every query, our system needs to quickly score candidate products so that a set of product listings can be returned in the two- to three-tenths of a second before a user begins to perceive waiting time. Our ML system, then, needs to have low serving latency. To optimize for this, we create a large number of serving replicas in a cloud-based computation platform, so that requests are not queued up unnecessarily. Each replica reloads to get the newest version of the model every few hours. (We talk more about the ins and outs of model serving in Chapter 8.)

Because the model is refreshed on the go and serves live, our system has several areas that need to be monitored to ensure that performance in real time continues to be good. These areas include the following:

Model freshness
: Is the model actually getting updated on a regular basis? Are the new checkpoints being successfully picked up by the serving replicas?

Prediction stability
: Over time, do the aggregate statistics look roughly in line with recent history? We can look at basic verifiables, such as the number of predictions made per minute, and the average prediction values. We can also monitor whether the number of clicks that we predict over time matches the number of clicks we then later see.

Feature distributions
: Our input features are the lifeblood of our system. We monitor basic aggregate statistics on each input feature to ensure that these remain stable over time. Sudden changes in the values for a given feature may indicate an upstream outage or other issue that needs to be managed.

Of course, this is just a starter set—we go into significantly more details on monitoring options in Chapter 9.

Common Failures

Thinking through worst-case scenarios helps us be prepared. Each of these tends to be related to the overall product needs and requirements, because in the end it is the impact to our overall product ecosystem that really matters. These worst-case scenarios are as follows:

Infinite latency
: One bad thing that could happen is that our model never returns values. Maybe the serving replicas are overloaded or are all down for some reason. If the system hangs indefinitely, users will never get any results. A basic time-out and a reasonable fallback strategy will help a lot here. One fallback strategy can be to use a much cheaper, simpler model in case of time-outs. Another might be to precompute some predictions for the most popular items and use these as fallback predictions.

All predictions are zero
: If our model were to score `0` for all products, none would be shown to users. This would, of course, indicate an issue with the model, but we need to monitor average predictions and fall back to another system if things go awry.

All predictions are bad
: Imagine that our model is corrupted, or an input feature signal becomes unstable. In this case, we might get arbitrary or random predictions, which would confusingly show random products to users, creating a poor user experience. To counter this, one approach might be to monitor both the aggregate variance of predictions along with the averages. This can happen for just a subset as well, so we may need to monitor predictions for relevant subsets of the predictions.

Model favors just a few
: Imagine that a few products happen to get very high predictions, and everything else gets low predictions. Naively, this might create a setting in which those other products, or new products over time, never get a chance to be shown to users and thus never get click information that would help them rise in the rankings. A small amount of randomization can be helpful in these situations to ensure that the model gets some amount of exploration data, which will allow the model to learn about products that had not previously been shown. This form of exploration data is also useful for monitoring, as it allows us to reality-check the model's predictions and ensure that when it says a product is unlikely to be clicked, this holds true in reality as well.

Clearly, many other things might go wrong—and fortunately, MLOps folks tend to have strong and active imaginations in this regard. Thinking through scenarios in this way allows us to robustly prepare our systems and ensure that any failures are not catastrophic and can be quickly addressed.

Conclusion

In this chapter, we have gone through some of the basics of ML models and how they fit into overall systems that rely on ML. Of course, we have only begun to scratch the surface. In later chapters, we dive into more depth on several of the topic areas that we touched on here.

Here are a few things that we hope you take away:

- Model behavior is determined by data, rather than by a formal (program) specification.
- If our data changes, our model behavior will change.
- Features and training labels are critical inputs to our models. We should take the time to understand and verify every input.
- ML models may vary in their degree of understandability, but we can still monitor and assess their behavior.
- Disasters do happen, but their impact can be minimized with careful forethought and planning.
- MLOps folks need to have strong working relationships and agreements with model developers in their organization.

CHAPTER 4

Feature and Training Data

By Robbie Sedgewick and Todd Underwood

It should be clear by this point that models come from data. This chapter is about the data: how it is created, processed, annotated, stored, and ultimately used to create the model. You will see that managing and handling the data creates specific challenges for repeatability, manageability, and reliability, and we will make some concrete recommendations about how to approach those challenges. For background, make sure to see (if you haven't already) Chapters 2 and 3.

This chapter covers the infrastructure that accepts data from a source and readies it for use by the training system. We will discuss three fundamental functional subsystems involved in this task: a feature system, a system for human annotations, and a metadata system. We discussed features a little in the previous chapter; another way of thinking about them is that they are characteristics of the input data, especially characteristics that we have determined are predictive of something we care about. Labels are specific cases of the output that we want from the model that we ultimately train. They are used as examples to train that model. Another way to think about labels is that they are the target or "correct" values for a specific data instance that the model will learn. Labels can be extracted from logs by correlating the data with another independent event, or they can be generated by humans. We'll discuss the systems needed for generation of human labels at scale, often called *annotations*. And finally, we'll briefly cover metadata systems, which keep track of the details about how the other systems work and are critical to making them repeatable and reliable.

Several aspects of these systems are usually shared between the feature and data system—most notably the metadata system. Since *metadata* (data about the data that we are collecting and annotating) is best understood after we know what we're doing with the data, we will discuss those systems after we have explored the requirements and characteristics of the feature and labeling systems.

Features

The data is just the data. Features are what make it useful for ML.[1] A *feature* is any aspect of the data that we determine is useful in accomplishing the goals of the model. Here, "we" includes the humans building the model or many times can now include automatic feature-engineering systems. In other words, a feature is a specific, measurable aspect of the data, or a function of it.

Features are used to build models. They are the structure that connects the model back to the data used to assemble the model. Previously, we have said that a model is a set of rules that takes data and uses it to generate predictions about the world. This is true of a model architecture and a configured model. But a trained model is very much a formula for combining a collection of features (essentially, feature definitions used to extract feature values from real data—we cover this definition more completely later in this chapter). Features frequently are more than just pieces of the raw data, but rather involve some kind of preprocessing.

These concrete examples of features should provide some intuition about what they are:

- From a web log, information about the customer (say, browser type).
- Individual words or combinations of words from text a human enters into an application.
- The set of all the pixels in an image, or a structured subset thereof.
- Current weather at the customer's location when they loaded the page.
- Any combination or transformation of features can itself be a feature.

Typically, features contain smaller portions of structured data extracted from the underlying training data. As modeling techniques continue to develop, it seems likely that more features will start to resemble raw data rather than these extracted elements. For example, a model training on text currently might train on words or combinations of words, but we expect to see more training on paragraphs or even whole documents in the future. This has significant implications for modeling, of course, but people building production ML systems should also be aware of the impact on systems of having the training data grow larger and less structured.

1 We are aware that one of the promises of deep learning is that it is sometimes possible to use raw data to train a model without specifically identifying features. This promise is partly true for some use cases but applies mostly, right now, to perceptual data like images, video, and audio. In those cases, we may not need to extract specific features from the underlying data, although we may still get good value from metadata features. For other cases, featurization is still required right now. So even deep learning modelers will benefit from an understanding of features.

At YarnIt we have several models, and one of them is to make recommendations for different or additional products while a customer is shopping on the site. This recommendations model is called by the web application during a shopping session from both the main product page where a customer is viewing an individual product and from the shopping cart confirmation page, when a customer has just added a product to their shopping cart. The web server calls the model to ask which additional products it should show to this customer under these circumstances, in hopes of showing the customer something else they might need or want and increasing our sales to that customer.

In this context, we might consider some of the following useful features that we would want available to the model as it is queried for additional products:

- Product page or shopping cart confirmation page. Is the user browsing or buying?
- The current product the user is looking at, including information from the name of the current product, possibly information from the picture of the current product, the product category, manufacturer, and price.
- Customer average purchase size or total purchases per year, if we know it. This might be an indication of just how spendy a customer is.
- Knitter or crocheter? Some customers never crochet, and others only crochet. Knowing this might plausibly help us recommend the right yarns, needles, and patterns. We might have customers self-identify this characteristic or might infer from previous purchases or browsing behavior.
- Customer country. Some products may be popular in only some places. Also some countries are colder than others, and that might be a signal.

These are just a few ideas to give a flavor of what a feature might be. It is important to note that, absent any actual data, we have no idea whether any of these features is useful. They certainly seem like they might be helpful, but many features that seem like they might work simply do not. Most commonly, such features are only slightly predictive, and the things they predict might be predicted better by another feature we already have. In those cases, the new feature adds cost and complexity but no value at all. Remember that features add maintenance costs in many parts of the system, and those costs exist until the feature is removed. We should be cautious in adding new features, especially those that depend on new data sources or feeds that, themselves, will now have to be monitored and maintained.

Before we go too much further, we need to clarify two completely distinct uses of the term *feature* and how we differentiate them:

Feature definition
: This is code (or an algorithm or other written description) that describes the information we are extracting from the underlying data. But it is not any specific instance of that data being extracted. For example, "Source country of the current customer obtained by taking the customer's source IP address and looking it up in the geolocation data service" is a feature definition. Any particular country—for example, the "Dominican Republic" or "Russia"—would not be a feature definition.

Feature value
: This is a specific output of a feature definition applied to incoming data. In the preceding example, the "Dominican Republic" is a feature value that might be obtained by determining that the current customer's source IP address is believed to be most commonly used in that country.

This is not (yet) standard industry terminology, but we adopt it in this section to distinguish between figuring out how we're trying to extract information from the data and figuring out what we have extracted so far.

Feature Selection and Engineering

Feature selection is the process by which we identify the features in the data that we will use in our model. *Feature engineering* is the process by which we extract those features and transform them into a usable state. In other words, through these activities, we figure out what matters (for the ML task we are trying to accomplish) and what doesn't. Building and managing features has changed over the years and will continue to evolve, as it moves from an entirely manual process to a mostly automated one, and in some cases a completely automated one. In this chapter, we refer to the processes of selecting and transforming features jointly as *feature engineering*.

In human-driven feature engineering, the process normally starts with human intuition based on an understanding or the domain of the problem, or at least detailed consultation with experts. Imagine trying to build a predictive model of what customers would buy from *yarnit.ai* without knowing what yarn, or needles, or knitting are, or worse, the very idea of how online retail works at all. Understanding the underlying problem area is key, and more specificity is better. After that, an ML engineer spends time with a dataset and a problem and uses a set of statistical tools to evaluate the likelihood that a particular feature, or several features in combination with one another, will be useful in the task. Next, ML engineers typically brainstorm to generate a list of possible features. The range is pretty large. Time of day could be a feature predicting which yarns customers will buy. Local humidity could be a feature.

Price could be a feature. Of these three features, one of them is much more likely to be useful than the others. It is up to humans to generate and evaluate those ideas by using the ML platform and model metrics.

For algorithmic feature engineering, sometimes included as part of AutoML, the process is considerably more automatic and data bound. AutoML, outlined in Chapter 3, is capable of not only selecting from identified features but also being able to programmatically apply common transforms (like log scaling or thresholding) to existing features. There are other ways that we can algorithmically learn something (an embedding) about the data without explicitly specifying it. Still, algorithms are generally able to identify only features that exist in the data, whereas humans can imagine new data that could be collected that might be relevant. Subtly, but perhaps even more importantly, humans understand the process of data collection, including any likely systemic bias. This might have a material impact on the value of the data. Nonetheless, especially when constrained to particular types of problems, algorithmic feature engineering and evaluation can be as effective or more effective than human feature engineering. This is an evolving set of technologies, and we should expect the balance between humans and computers here to continue to develop.

Lifecycle of a Feature

The distinction between feature definitions and feature values becomes especially important when considering the lifecycle of a feature. Feature definitions are created to fill a need, evaluated against that need, and eventually discarded as either the model is discarded or better features are found to accomplish the same goal. Here is a simplified version of the lifecycle of a feature (both the definition and the representative values):

1. *Data collection/creation*
   Without data, we have no features. We need to collect or create data in order to create features.

2. *Data cleaning/normalization/preprocessing*
   Although even the process of feature creation could be considered some kind of data normalization, here we're referring to coarser preprocessing: eliminating obviously malformed examples, scaling input samples to a common set of values, and possibly even deleting specific data that we should not train on for policy reasons. This might seem outside the feature engineering process, but no features can exist until the data exists and is in a usable form. The topic of data normalization and preprocessing is huge and beyond the scope of this book, but building the infrastructure to consistently perform that preprocessing and monitor it is an important area of responsibility.

3. *Candidate feature definition creation*
   Using either subject-matter expertise plus human imagination, or automated tools, develop a hypothesis for which elements or combinations of the data are likely to accomplish our model's goals.

4. *Feature value extraction*
   We need to write code that reads the input data and extracts the features that we need from the data. In some simple situations, we might want to do this inline as part of the training process. But if we expect to train on the same data more than a few times, it's probably sensible to extract the feature from the raw data and store it for later efficient and consistent reading. It is important to remember that if our application involves online serving, we need a version of this code to extract the same features from the values that we have available at serving in order to use them to perform inference in our model. Under ideal circumstances, the same code can extract features for training and for serving, but we may have additional constraints in serving that are not present in training.

5. *Storage of feature values in a feature store*
   And this is where we save the features. A feature store is just a place to write extracted feature values so that they can be quickly and consistently read during training a model. This is covered in detail in "Feature store" on page 72.

6. *Feature definition evaluation*
   Once we have extracted a few features, we will most likely build a model using them or add new features to an existing model in order to evaluate how well they work. In particular, we will be looking for evidence that the features provide the value that we were hoping they would. Note that this evaluation of feature definitions comes in two distinct phases that are connected. First, we need to determine whether the feature is useful at all. This is a coarse-grained evaluation; we are simply trying to decide whether to continue working on integrating the feature into our model. The next phase occurs if we decide to keep that feature. At that point, we need a process to continuously evaluate the quality and value (compared to cost) of the feature. This will be necessary so that we can determine that it is still working the way we expect and providing the value we expect even several years from now.

7. *Model training and serving using feature values*
   Perhaps this is obvious, but the entire point of having features is to use them to train models, and use the resulting models for a particular purpose.

8. *(Usually) Update of feature definitions*
   We frequently have to update the definition of a feature, either to fix a bug or simply to improve it in some way. If we add and keep track of versions for feature definitions, this will be much easier. We can update the version and then,

optionally, reprocess older data to create a new set of feature values for the new version.

9. *(Sometimes) Deletion of feature values*
   Sometimes we need to be able to delete feature values from our feature store. This can be for policy/governance reasons; for example, we may no longer be allowed to store these feature values because a person or government has revoked that permission. It can also be for quality/efficiency reasons: we may decide those values are bad in some way (corrupted, sampled in a biased way, for example) or just too old to be useful.

10. *(Eventually) Discontinuation of a feature definition*
    Everything comes to an end, including useful features. At some point in the lifecycle of the model, we will either find a better way to provide the value that this feature definition provides or find that the world has changed enough that this feature no longer provides any value. Eventually, we will decide to retire the feature definition (and values) entirely. We will need to remove any serving code that refers to the feature, remove the feature values in the feature store, cancel the code that extracts the feature from the data, and proceed to delete the feature code.

Feature Systems

To successfully manage the flow of data through our systems, and to turn data into features that are usable by our training system and manageable by our modelers, we need to decompose the work into several subsystems.

As was mentioned in the introduction to this chapter, one of these systems will be a metadata system that tracks information about the data, datasets, feature generation, and labels. Since in most cases this system will be shared with any labeling systems, we will discuss it at the end of this chapter. For now, let's walk through the feature systems starting with raw data and ending up with features stored in a format ready to be read by a training system.

Data ingestion system

We will have to write software that reads raw data, applies our feature extraction code to that data, and stores the resulting feature values in the feature store. In the case of one-time extraction, even for a very large amount of data, this may be a relatively ad hoc process with code designed to be run once. But in many cases, the process of extracting data is a separate production system all its own.

When we have many users or repeated use of the data ingestion system, it should be structured as an on-demand, repeatable, monitored data processing pipeline. As is the case for most pipelines, the biggest variable is user code. We will need a system whereby feature authors write code that identifies features and extracts them to store in the feature store. We can either let feature authors run their code themselves, which imposes a substantial operational burden on them, or we can accept the challenge and provide them a development engineering environment. This helps them write reliable feature-extraction code so that we can run that code reliably in our data ingestion system.

We need to build a few systems to facilitate feature authors writing reliable and correct features. To begin with, we should note that features should be versioned. We'll likely want to substantially change a feature over time, perhaps because of changes in data that it merges with or other factors related to the data we are collecting. In these cases, a feature version helps keep the transition clear and avoids unintended consequences of the change.

Next, we'll need a test system that checks feature-extraction code for basic correctness. Then, we'll need a staging environment for running proposed feature extraction on a certain number of examples and provide those to the feature authors along with basic analysis tools to ensure that the feature is extracting what it is expected to extract. At this point, we may want to allow the feature to be run or may want additional human review for reliability concerns (dependence on external data, for example). The more work we do here, the more productive feature authors will be.

Finally, it is important to note that some labels can be effectively calculated or generated from the very data we have at data ingestion time. A good example of this at YarnIt is suggested products and sales. We suggest products in order to sell more of them. The features are the characteristics of the product or characteristics about the customer, and the label is whether the customer bought it or not. As long as we can join the suggestion logs against the orders, we will have this label when we construct the features. In cases like this, the data ingestion system will also generate labels for those features as well as the features themselves, and both can be stored together in a common datastore. We will talk much more about labeling systems in "Labels" on page 50.

Feature store

A *feature store* is a storage system designed to store extracted feature (and label) values so that they can be quickly and consistently read during training a model and even during inference. Most ML training and serving systems have some kind of a feature store even if it is not called that.[2] They are most useful, however, in larger, centrally managed services, especially when the features (definitions and values both) are shared among multiple models. Recognizing the importance of putting

---

2 For example, ML training that occurs on mobile devices will still need to train on some data, but will not have a structured, managed feature store, since there's no need to mediate that data for other on-device users.

our feature and label data in a single, well-managed place has significantly improved the production readiness of ML in the industry. But feature stores do not solve every problem, and many people come to expect more of them than they can deliver. Let's review the problems that a feature store solves and the benefits it does provide for your training system.

The most important characteristic of a feature store is its API. Different commercial and open source feature stores have different APIs, but all of them should provide a few basic capabilities:

Store feature definitions
: Usually these are stored as code that extracts the feature in a raw data format and outputs the feature data in the desired format.

Store feature values themselves
: Ultimately, we need to write features in a format that is easy to write, easy to read, and easy to use. This will be largely determined by our proposed use cases and most commonly is divided into ordered and unordered data, but we cover nuances in "Lifecycle Access Patterns" on page 75.

Serve feature data
: Provide access to feature data quickly and efficiently at a performance level suitable to the task. We absolutely do not want expensive CPUs or accelerators stalled, waiting on the I/O of reading from our feature store. This is a pointless way to waste an expensive resource.

Coordinate metadata writes with the metadata system
: To get the most out of our feature store, we should keep information about the data we store in it in the metadata system. This helps model builders. Note that metadata about *features* is somewhat different from metadata about *runs of the pipeline*, although both are useful in troubleshooting and reproducing problems. Feature metadata is most useful for model developers, while pipeline metadata is most useful for ML reliability or production engineers.

Many feature stores also provide basic normalization of data in ingestion as well as more sophisticated transformations on data in the store. The most common transformations are standardized in-store bucketing and built-in transforming features.[3]

---

3 Bucketing is just the process of placing continuous data into discrete categories. *Transforming features* are those features that are the result of a computed combination of one or more other features. Simple examples include ideas like returning the day of the week when provided a date feature, or returning the country name from a feature that stores a latitude and longitude of a point on Earth. More complex examples might include fixing the color balance of a picture or choosing a particular projection for 3D data. Perhaps one of the most common examples is to convert 32-bit numbers (either integers or, worse, floating-point numbers) into 8-bit numbers in order to significantly reduce the space and computational resources required to process them.

The feature store API needs to be carefully calibrated for the use case. We should consider asking the following questions as we think about what we need in a feature store:

- Are we reading data in a particular order (log lines that are timestamped) or in no order that matters (a large collection of images in a cloud storage system)?
- Will we read the features frequently or only when training a new model? In particular, what is the ratio of bytes/records read to bytes/records written?
- Is the feature data ingested once, never appended to, and seldom updated? Or is the data frequently appended to while older data is continuously deleted?
- Can feature values be updated, or is the store append-only?
- Are there special privacy or security requirements for the data we are storing? In most cases, the extracted features of a dataset with privacy and use restrictions will also have privacy and use restrictions.

After thinking about these questions, we should be able to determine our needs for a feature storage system. If we're lucky, we will be able to use one of the existing commercial or open source feature stores on the market. If not, we'll have to implement this functionality ourselves, whether we do it in an uncoordinated fashion or as part of a more coherent system.

Once we are clear on the requirements for an API and have a clearer understanding of our data access needs, in general we will find that our feature store falls into one of two buckets:[4]

Columns
: The data is structured and decomposable into columns, not all of which will be used in all models. Typically, the data is also ordered in some way, often by time. In this case, column-oriented storage is the most flexible and efficient.

Blobs
: This is an acronym for *binary large objects*, although the common English word is also descriptive. In this case, the data is fundamentally unordered, mostly unstructured, and is best stored in a manner that's more efficient at storing a bunch of bytes.

4 Some examples will contain features from both buckets. For example, the pictures in an image are blobs that are best accessed as unstructured data, but the metadata about the image (camera that took it, date, exposure information, location information) is structured and might even be ordered if it includes fields like `date`.

Lifecycle Access Patterns

As we make choices about our feature store, it is critical to keep in mind how, and how often, we will use the data that we are storing. Consider two cases: a small amount of data that we retrain on constantly versus a large amount of data that we use a few times and then delete.

In the former case, we will absolutely want to spend time extracting columnar data, preprocessing those columns, and making access as efficient as possible, even when it slows data ingestion. The payoff will be in the faster and cheaper reads that we will get later. In the latter case, significant preprocessing is a waste of time and processing. We should ingest the data as cheaply and automatically as possible, and worry about processing when we read the data.

In all cases, thinking about when we plan to delete the data is important. As covered in Chapter 2, reliable and effective deletion can be accomplished, but only if we design for it. For example, we need to determine whether we will delete by time period (e.g., all of last month's data) or by end user (e.g., all of the data for customer number 8723423). Once we know that, we can design the storage layout for effective deletion. One of the best techniques for this involves encrypting the data based on the scope we plan to use for deletion (per day, per data source, per customer, or other). We then store only the keys in a storage system with guaranteed access protection and the ability to securely delete data (usually with multiple overwrites on all persistent storage media). When we get a verified request to delete data by scope, we simply delete the relevant keys. Thus, although the data is still very much available, and may even be replicated to multiple targets or backed up, it is inaccessible and effectively deleted with little effort or latency.

Many feature stores will need to be replicated, partially or completely, in order to store data near the training and serving stacks. Since ML computation requirements are generally considerable for training, we often prefer to replicate data to the place or places where we can get the best value for our training dollar. Having the feature store implemented as a service facilitates the management of this replication.

Transforming Features

Previously, we mentioned transforming features as one desirable piece of functionality in a feature store. For those of us who have worked on relational database systems, *transforming features* are the stored procedures of feature stores—they are fixed transformations of the data, written in code, but stored in the storage system.

Transforming features implement functionality as simple as bucketing a feature that arrives as a continuous one (e.g., bucketing age into 0–4, 5–17, 18–35, 36–50, 51–65, 66+, or whatever buckets make sense for our use case). But transforming features may also combine multiple feature columns and consistently return a different value that is a function of the multiple columns. They can even be computed across multiple rows if features are dependent on the distribution of data, although this can be complicated and expensive. In all cases, the key is to remember that transforming features offer a consistent and deterministic value that is the programmatic result of the code we have written being applied to the data in the feature store.

Historically, transforming features were often implemented in other locations in an ML training pipeline, commonly during the training phase. But then recall that features also need to be looked up during serving in most ML applications. So transforming features were implemented in training and in serving. And, as inevitably happens, code drift occurs. Some transforming features might be implemented slightly differently in some models compared to others, or in training compared to serving. Moving the transforming feature into the feature store eliminates this class of failure.

Putting transforming features into the feature store carries one other potential advantage: performance. If a transforming feature is computationally intensive and we intend to read that column frequently, a feature store might choose to materialize that feature, by processing new data as it arrives and writing out the result of the transforming feature into a new column. This has two obvious risks. First, we may change the definition of the feature, which would require recomputing the column over all of the data. Second, the materialized column could become out of sync with the data if we have any bugs in the system that processes new data. Nevertheless, if materializing a feature saves us considerable computation and I/O time, it might be worth it.

Feature quality evaluation system

As we develop new features, we need to evaluate what, if anything, those features add to the quality of the overall model, in combination with existing features. This topic is covered extensively in Chapter 5. The general idea to know at this point is that we can combine the approaches of using slightly different models, A/B testing, and a model quality evaluation system in order to effectively evaluate the benefit of each new feature under development. We can do this quickly and at relatively low cost.

One common approach is to take an existing model and retrain it by using a single additional feature. In the case of a web application like at YarnIt, we can then direct a fraction of our user requests to the new model and evaluate its performance on a task. For example, if we add a feature for `Country the User Is In` to a model suggesting new products for a user to try, we can direct 1% of the user requests (either by request or by user) to the new model. We can evaluate whether the new model has a higher likelihood of recommending products that users actually buy, and doing so can inform the choice of whether to keep the new feature or to eliminate it and try other ideas.

Keep in mind that even a feature that adds value may not be worth the cost to collect, process, store, and maintain it. For all but the most trivially obvious features, it is a good habit to calculate a rough return on investment (ROI) for every new feature added to the system. This will help us avoid useful features that are still more expensive than the value that they add.

Labels

Although features seem like the most important aspects of the data, one thing is more important: labels. By this point, you should have a solid understanding of what features are for and what systems considerations should be taken into account when managing large numbers of features. But supervised learning models, in particular, require labels.

Labels are the other main category of data used in training an ML model. While features serve as the input to the model, labels are examples of the correct model output. They are used in the training process to set the (many!) internal parameters of the model, to tune the model so that it will produce the desired outputs on the features it gets at inference time. Held-out examples and labels (labels not used in training) are also used in model evaluation to understand model quality.

As we discussed earlier, for some classes of problems, like our recommendation system, the labels can be generated algorithmically from the system's log data. These features are almost always more consistent and more accurate than human-labeled features. Since this data is generated from the system's log data often with the feature data, these labels are most commonly stored in the feature system for model training and evaluation. In fact, all labels can be stored in the feature store, although labels generated by humans need additional systems to generate, validate, and correct these labels before storing them for model training and evaluation. We discuss these systems in the next section.

Human-Generated Labels

So let's turn our attention to the large classes of problems requiring human annotations to provide the model training data. For example, building a system to analyze and interpret human voice needs human annotation to ensure that the transcription is accurate and to understand what the speaker meant. Image analysis and understanding often needs example annotated images for image classification or detection problems. Getting these human annotations at the scale needed to train an ML model is challenging from both implementation and cost perspectives. Effort must be dedicated to designing efficient systems to produce these annotations. We will now focus on the primary components of these human annotation systems.

For a concrete example involving our fictional yarn shop, YarnIt, consider the case of an advanced new feature that, given an image of crocheted fabric, can predict the crochet stitch that was used to produce that fabric. Such a model requires someone to design the set of crochet stitches for the model to predict, providing the set of classes that model can output. Then large quantities of images of crocheted fabric, covering all of these stitches, must be labeled by crochet experts as to what stitch produced this fabric. Using these expert labels, a model can be trained to classify new images of crocheted fabric and determine the stitch used.

This is not cheap. Humans need to be trained on the task, and each image may need to be labeled multiple times to ensure that we have trustworthy results. Because of the large cost associated with acquiring human-generated labels, training systems should be designed to get as much mileage from them as possible. One technique commonly used is data augmentation: the feature data is "fuzzed" in a way that changes the features but doesn't change the correctness of the label.[5] For example, consider our stitch classification problem. Common image operations like scaling, cropping, adding image noise, and changing the color balance don't change the classification of the stitch in the image but can greatly increase the number of images available for the model to train on. Similar techniques can be used in other classes of problems. Care must be taken, however, not to train and test on two images fuzzed from the same source image, and for this reason any data augmentation of this sort should be done by the training system and not in the labeling system (or not, and be careful if you have a good reason to do otherwise, like a very expensive fuzzing algorithm).

One important note: while some kinds of tremendously complex data can best be labeled by humans, other types of data definitely cannot be labeled by humans at all. Typically, this is abstract data with high dimensionality that makes it difficult for a human to determine the correct answer quickly. Sometimes humans can be provided with augmentation software to assist in these tasks, but other times they are just the wrong option for performing the labeling.

Annotation Workforces

The first question that often comes up with human annotation problems is who will do the labeling. This is a question of scale and equity.[6] For simpler models, for which a small amount of data is sufficient to train the model, typically the engineer building the model will do their own labeling, often with hacky, homebuilt tools (and a significant chance of biased labels). For more complex models, dedicated annotation teams are used to provide human annotations at a scale not otherwise possible.

---

5 This is also a technique used to expand the training dataset algorithmically.

6 Organizations need to ensure that human labelers are treated fairly and paid reasonably for their work.

These dedicated annotation teams can be colocated with the model builder or remotely provided by third-party annotation providers. They range in size from a single person to hundreds of people, all generating annotated data for a single model. The Amazon Mechanical Turk service was the original platform used for this, but since then a proliferation of crowdsourcing platforms and services have developed. Some of these services use paid volunteers, and others use teams of employees to label the data.

Cost, quality, and consistency trade-offs arise in the choice of labeling. Crowdsourced labeling often requires additional effort to verify quality and consistency, but paid labeling staff can be expensive. The costs for these annotation teams can easily exceed the computational costs of training the models. We discuss some organizational challenges of managing a large annotation team in Chapter 13.

Measuring Human Annotation Quality

As the quality of any model is only as good as the data used to train the model, quality must be designed into the system from the start. This becomes increasingly true as the size of the annotation team grows. Quality can be achieved in multiple ways depending on the task, but the most frequent techniques used include the following:

Multiple labeling (also called consensus labeling)
: The same data is given to multiple labelers to check for agreement among them.

Golden set test questions
: Trusted labelers (or the model builder) produce a set of test questions that are randomly included in the unlabeled data to evaluate the quality of the produced labels.

A separate QA step
: A fraction of the labeler data is reviewed by a more trusted QA team. (Who QAs the QA team? Perhaps the model builder, but depending on context, this could be a separate QA team, policy expert, or someone else with domain expertise.)

Once they're measured, quality metrics can be improved. Quality issues are best addressed by managing the annotation team with humility and by understanding that they will produce higher-quality results when they have the following:

- More training and documentation
- Recognition for quality and not just throughput
- A variety of tasks over the workday
- Easy-to-use tools

- Tasks with a balanced set of answers (no needle-in-a-haystack tasks)
- An opportunity to provide feedback on the tools and instructions

Annotation teams managed in this way can provide high-quality results. However, even the best, most conscientious labeler will miss things occasionally, so processes should be designed to detect or accept these occasional errors.

An Annotation Platform

A labeling platform organizes the flow of data to be annotated and the results of the annotations while providing quality and throughput metrics of the overall process. At their heart, these systems are primarily work-queuing systems to divide the annotation work among the annotators. The actual labeling tool that allows the labelers to view the data and provide their annotations should be flexible to support any arbitrary annotation task.

With a team or organization that is working on multiple models simultaneously, the same annotation team may be shared among multiple annotation projects. Furthermore, each annotator might have different sets of skills (e.g., language skills or knowledge of crochet stitches), and the queuing systems can be relatively complex and require careful design to avoid problems such as scalability issues or queue starvation. Pipelines enabling the output of one annotation task to serve as the input to another can be useful for complex workflows. Quality measurement using the techniques discussed previously should be designed into the system from the start, so project owners can understand the labeling throughput and quality of all their annotation tasks.

Although historically many companies have implemented their own labeling platforms with their own set of these features, many options exist for prebuilt labeling platforms. The major cloud providers and many smaller startups offer labeling platform services that can be used with arbitrary annotation workforces, and many annotation workforce providers have their own platform options that can be used. This is a rapidly changing area with new features being added to existing platforms all the time. Publicly available tools are moving beyond simple queuing systems and are starting to provide dedicated tools for common tasks, including advanced features like AI-assisted labeling (see the following section). When deciding on any new technology platform, data security and integration costs must be considered along with the platform capabilities.

As mentioned in "Feature store" on page 72, in many cases the most sensible place to store completed labels is in the feature store. By treating human annotations as their own columns, we can take advantage of all the other functionality provided by the feature store.

Active Learning and AI-Assisted Labeling

Active learning techniques can focus the annotation effort on the cases in which the model and the human annotators disagree or the model is most uncertain, and thereby improve overall label quality. For example, consider an image detection problem where the labeler must annotate all the occurrences of a particular object in an image. An active-learning labeling tool might use an existing model to pre-label the image with proposed detections of the object in question. The labeler would then approve correct proposed detections, reject bad ones, and add any missing ones. While this can greatly increase labeler throughput, it must be done with care to not introduce bias to the models. Such active learning techniques can actually increase overall label quality since the model and humans will often have their best performance on different kinds of input data.

A semi-supervised system allows the modeler to bootstrap the system with *weak heuristic functions* that imperfectly predict the labels of some data, and then use humans to train a model that takes these imperfect heuristics to high-quality training data. Systems like this can be particularly valuable for problems with complex, frequently changing category definitions requiring models to be retrained quickly and frequently.

Efficient annotation techniques for particularly complex labeling tasks is an ongoing field of research. Particularly if you are doing a common annotation problem, a quick review of available tools from cloud and annotation providers is well worth your time, as they are often adding new capabilities for AI-assisted annotation.

Documentation and Training for Labelers

Documentation and labeler training systems are some of the most commonly overlooked parts of an annotation platform. While labeling instructions often start simply, they inevitably get more complex as data is labeled and various corner cases are discovered. To continue our preceding YarnIt example, perhaps some crochet stitches are not mentioned in the labeling instructions, or the fabric is made from multiple different stitches. Even conceptually simple annotation tasks such as "marking all the people in an image" can end up with copious instructions on proper handling of various corner cases (reflections, pictures of people, people behind car windows, etc.).

Labeling definitions and directions should be updated as new corner cases are discovered, and the annotation and modeling teams should be notified about the changes. If the changes are significant, previously labeled data might have to be re-annotated to correct data labeled with old instructions. Annotation teams often have significant turnover, so investing in training for using annotation tools and for understanding labeling instructions will almost always give big wins in label quality and throughput.

Metadata

Feature systems and labeling systems both benefit from efficient tracking of metadata. Now that you have a relatively complete understanding of the kinds of data that will be provided by a feature or labeling system, you can start to think about what metadata is produced during those operations.

Metadata Systems Overview

A *metadata system* is designed to keep track of what we're doing. In the case of features and labels, it should minimally keep track of the feature definitions we have and the versions used in each model's definitions and trained models. But it is worth pausing for a minute and trying to see into the future: what are we eventually going to expect out of a metadata system, and is there any way to anticipate that?

Most organizations start building their data sciences and ML infrastructure without a solid metadata system, only to regret it later. The next most common approach is to build several metadata systems, each targeted at solving a particular problem. This is what we're about to do here: make one for tracking feature definitions and mappings to feature stores. Even within this very chapter, you're about to see that we're going to need to store metadata about labels, including their specification and when particular labels were applied to particular feature values. Later, we're going to need a system for mapping model definitions to trained models, along with data about the engineers or teams responsible for those models. Our model serving system is also going to need to keep track of trained model versions, when they were put into production. Any model quality or fairness evaluation systems will need to be read from all of these systems in order to identify and track the likely contributing causes of changes in model quality or violations of our proposed fairness metrics.

Our choices for a metadata system are relatively simple:

One system
: Build one system to track metadata from all of these sources. This makes it simple to make correlations across multiple subsystems, simplifying analysis and reporting. Such a large system is difficult to get right from a data schema perspective. We will be constantly adding columns when we discover data we would like to keep track of (and backfilling those columns on existing data). It will also be difficult to stabilize such a system and make it reliable. From a systems design perspective, we should ensure that our metadata systems are never in the live path of either model training or model serving. But it's difficult to imagine how feature engineering or labeling can take place without the metadata system being functional, so it can still cause production problems for our humans working on those tasks.

Multiple systems (that work together)

We can build separate metadata systems for each task we identify. Perhaps we would have one for features and labels, one for training, one for serving, and one for quality monitoring. Decoupling the systems provides the standard advantages and costs that decoupling always does. It allows us to develop each system separately without concern for the others, making them nimbler and simpler to modify and extend. Additionally, an outage of one metadata system has limited production impact on others. The cost, though, is the added difficulty of analysis and reporting across those systems. We will have processes that need to join data across the features, labeling, training, and serving systems. Multiple systems should always be designed in a way that allows their data to be joined, which either means creating and sharing unique identifiers or establishing a meta-metadata system that tracks the relationships of data fields across the metadata systems.

If the needs are simple and well understood, prefer a single system.[7] If the area is rapidly developing and our teams expect to continue extending what they track and how they work, multiple systems will simplify the development over time.

Dataset Metadata

For metadata about features and labels, here are a few specific elements that we should ensure are included:

Dataset provenance

Where did the data come from? Depending on the source of our data, we might have a lookup table of logs from various systems, a key for an external data provider with data about when we downloaded the data, or even a reference to the code that generated the data.

Dataset location

For some datasets, we will store raw, unprocessed data. In this case, we should store a reference to where we keep that dataset, as well as perhaps information about where we got it from. Some data we create for ourselves on an ongoing basis, such as logs from our systems, and so in those cases we should store the log or datastore reference where that data is stored, or where we are permitted to read from it.

7 The industry is littered with organizations that have multiple metadata systems, each of which believes itself to be the one single system. If the needs are simple and well understood, prefer one system, but take action to ensure that it's the only system unless you outgrow the needs of it.

Dataset responsible person or team
: We should track which person or team is responsible for the dataset. In general, this is the team that chose to download or create the dataset, or the team that owns the system that produces the data.

Dataset creation date or version date
: It is often useful to know the first date a particular dataset was used.

Dataset use restrictions
: Many datasets have restrictions on their use, either because of licensing or governance constraints. We should document that in the metadata system for easy analysis and compliance later.

Feature Metadata

Keeping track of metadata about our feature definitions is part of what will enable us to reliably use and maintain those features. This metadata includes the following:

Feature version definition
: The feature definition is a reference to code or another durable description to what data the feature reads and how it processes the data to create the feature. This should be updated for every updated version of the feature definition. As was previously described, versioning these definitions (and restricting the versions in use) will make the resulting codebase more predictable and maintainable.

Feature definition responsible person or team
: There are two good use cases for storing this information: figuring out what a feature is for and finding someone who can help resolve an incident when the feature might be at fault. In both cases, it is useful to store authorship or maintainer information about that feature.

Feature definition creation date or current version date
: This may be fairly obvious, but it's useful to get a change history of when a feature was most recently updated and when it was originally created.

Feature use restrictions
: This is important but trickier to store. Features may be restricted from use in some contexts. For example, it may be illegal to use a particular feature in some jurisdictions. Age and gender may be a reasonable predictor of automobile insurance risk models, but insurance is highly regulated, and we may not be permitted to take those fields into account. Banning particular fields only for specific uses is difficult to track and implement, but the restrictions might be even more subtle. For example, age may be able to be taken into account, but only with certain, specific bucketing (like `under 25`, `25-64`, `65-79`, and `over`

80). In that specific case, it's easier to just define a transforming feature built on top of the age column that meets these bucketing requirements and prohibit the general age feature from being used for insurance purposes while allowing the insurance_bucketed_age feature to be used. But the general case of storing and applying feature restrictions based on governance requirements is extremely difficult, and no great designs or solutions exist at the time of writing.

Label Metadata

We should also track metadata about labels. This is intended to help with the maintenance and development of the labeling system itself, but might also be used by the training system as it uses the labels:

Label definition version
: Switching to metadata specific to labels and analogously to features, we must store the version of any label definitions to understand which labeling instructions the labels were made with.

Label set version
: In addition to label definition changes, changes to the labels may occur because of incorrect labels getting corrected or new labels being added. If the dataset is being used for comparison with an older model, using an older version of the labels may be desirable, to make the comparison more apples-to-apples.

Label source
: Although not typically needed for training, it is sometimes necessary to know the source of each label in a dataset. This may be the source that particular label was licensed from, the human who produced the label (along with any QA that was applied to the label), or the algorithm that produced the label if an automated labeling approach was used.

Label confidence
: Depending on how the labels are produced, we might have different estimates of the confidence of correctness for different labels. For example, we might have lower confidence in labels that are produced by an automated approach, or labels produced by a newer labeler. Users of these labels might choose different thresholds to decide which labels to use in training their models.

Pipeline Metadata

This section covers a final type of metadata that we won't spend as much time on: metadata about the pipeline processes themselves. This is data about the intermediate artifacts we have, which pipeline runs they came from, and which binaries produced them. This type of metadata is produced automatically by some ML training systems; for example, ML Metadata (MLMD) is automatically included in TensorFlow

Extended (TFX), which uses it to store artifacts about training runs. These systems are either integrated into those systems or are somewhat difficult to implement later. As a result, we don't cover them much here.

More generally, metadata systems are often overlooked or deprioritized. They should not be. They are one of the most effective and direct contributors to productive use of the data in an ML system. Metadata unlocks value and should be prioritized.

Data Privacy and Fairness

Feature and labeling systems give rise to profound privacy and ethical considerations. While many of these topics are covered much more completely in Chapter 6, calling out a few specific topics here explicitly is worthwhile.

Privacy

Both the data that we receive and human annotation of that data has the significant possibility of containing PII. While the simplest approach to dealing with private information is to simply prohibit private or sensitive information from entering into our feature storage system, this is often not practical.

PII data and features

If we plan to have PII data in features, we will want to plan to do at least three things in advance:

- Minimize the PII data we process and store
- Restrict and log access to the feature store containing the private features
- Plan to delete private features as soon as possible

It is best to plan for correct handling of PII data in advance. Before even considering collecting PII data, a clear process should be in place for obtaining consent from users: they should know what data they are providing and how it will be used. Many organizations find it valuable to write a plan documenting exactly the data that will be collected, how it will be processed, where it will be stored, and the circumstances under which it can be accessed and will be deleted. This allows for an internal (and possibly, eventually, external) review of the procedures to ensure compliance with relevant laws and regulations. Most organizations will want to document their procedures about planning for PII data and train staff on these procedures regularly.

Remember that from an organizational point of view, private data is much more of a liability than an asset. Therefore, we should be completely convinced that the features containing private data are required—that they produce sufficient value to outweigh the risks of processing and storing private data. As is always the case, different pieces of nonprivate data may be combined to create a private data element.

Private data and labeling

Human annotation of PII data introduces a host of legal or reputational hazards if not carefully and properly handled. The details about how to properly manage this kind of data used in human annotation systems is extremely context specific and is beyond the scope of this book. Often the best way to handle PII data is to split the data so that the human doing the annotation has access to only the non-PII parts of the data. This is specific to the problem at hand. Any labeling of PII data should be done with the utmost care and awareness from project leadership of the risks involved.

Use of human annotators also introduces an additional risk of the model unintentionally learning the cultural biases of the annotation instructions or the team itself. This potential bias is best combated by thoughtful documentation and consideration of potential areas for confusion, strong lines of communication with the annotation team, and the hiring of a diverse annotation team. On the positive side, a well-trained annotation team can be one of the most effective ways to filter sourced data with potential biases to understand and remove bias.

Fairness

Fairness is a significant topic that is covered much more broadly and thoroughly in Chapter 6. Suffice it to say here that considering fairness is important while thinking of features and their labels. It is not easy to select features and sets of features that ensure that the resulting ML system can be used in only a fair fashion. While it is true that we need to avoid selecting features and datasets that are unrepresentative and biased, this alone will not be sufficient to ensure fairness overall. This would be a good time for those with a particular interest in fairness to read (or reread) Chapter 6.

Conclusion

Production ML systems require mechanisms to efficiently and consistently manage training data. Training data almost always consists of features, so having a structured feature store significantly facilitates writing, storing, reading, and ultimately deleting features. Many ML systems also have a component of human annotation of data. Humans annotating data require their own systems to facilitate rapid, accurate, and verified annotations, which ultimately need to be integrated into the feature store.

We hope we have given a clearer understanding of these components and the considerations that should go into selecting or, in the worst case, building, them.

CHAPTER 5

Evaluating Model Validity and Quality

OK, so our model developers have created a model that they say is ready to go into production. Or we have an updated version of a model that needs to be swapped in to replace a currently running version of that model in production. Before we flip the switch and start using this new model in a critical setting, we need to answer two broad questions. The first establishes model *validity*: will the new model break our system? The second addresses model *quality*: is the new model any good?

These are simple questions to ask but may require deep investigation to answer, often necessitating collaboration among folks with various areas of expertise. From an organizational perspective, it is important for us to develop and follow robust processes to ensure that these investigations are carried out carefully and thoroughly. Channeling our inner Thomas Edison, it is reasonable to say that model development is 1% inspiration and 99% verification.

This chapter dives into questions of both validity and quality, and provides enough background to allow MLOps folks to engage with both of these issues. We will also spend time talking about how to build processes, automation, and a strong culture around ensuring that these issues are treated with the appropriate attention, care, and rigor that practical deployment demands.

Figure 5-1 outlines the basic steps of model development and the role that quality plays in it. While this chapter focuses on evaluation and verification methods, it is important to note that these processes will likely be repeated over time in an iterative fashion. See Chapter 10 for more on these topics.

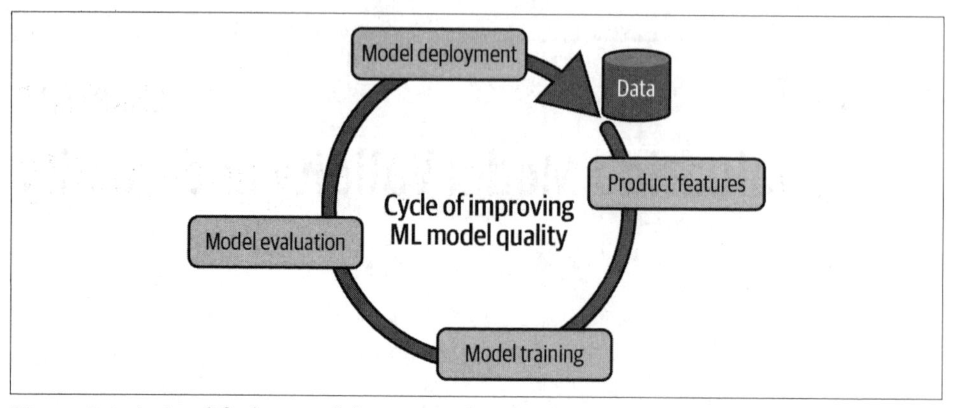

Figure 5-1. A simplified view of the repeated cycle of model development

Evaluating Model Validity

It has been said that all humans crave validation of one form or another, and we MLOps folks are no different. Indeed, *validity* is a core concept for MLOps, and in this context we cover the concept of whether a model will create system-level failures or crashes if put into production.

The kinds of things that we consider for validity checks are distinct from model quality issues. For example, a model could be horribly inaccurate, incorrectly guessing that every image shown should be given the label `chocolate pudding` without causing system-level crashes. Similarly, a model might be shown to have wonderful predictive performance in offline tests, but rely on a particular feature version that is not currently available in the production stack, or use an incompatible version of some ML package, or rarely give values of `NaN` that cause downstream consumers to crash. Testing for validity is a first step that allows us to be sure that a model will not cause catastrophic harm to our system.

Here are some things to test for when verifying that a model will not bring our system to its knees:

Is it the right model?
: Surprisingly easy to overlook, it is important to have a foolproof method for ensuring that the version of the model we are intending to serve is the version we are actually using. Including timestamp information and other metadata within the model file is a useful backstop. This issue highlights the importance of automation and shows the difficulties that can arise with ad hoc manual processes.

Will the model load in a production environment?
: To verify this, we create a copy of the production environment and simply try to load the model. This sounds pretty basic, but it is a good place to start because it

is surprisingly easy for errors to occur at this stage. As you'll learn in Chapter 7, we are likely to take a trained version of a model and copy it to another location where it will be used either for offline scoring by a large batch process, or for online serving of live traffic on demand. In either case, the model is likely to be stored as a file or set of files in a particular format that are then moved, copied, or replicated for serving. This is necessary because models tend to be large, and we also want to have versioned checkpoints around that can be used as artifacts both for serving and for future analysis or as recovery options in case of an unforeseen crisis or error. The problem is that file formats tend to change slightly over time as new options are added, and there is always at least some chance that the format a model is saved in is not a format compatible with our current serving system.

Another issue that can create loading errors is that the model file may be too large to be loaded into available memory. This is especially possible in memory-constrained serving environments, such as on-device settings. However, it can also occur when model developers zealously increase model size to pursue additional accuracy, which is an increasingly common theme in contemporary ML. It is important to note that the size of the model file and the size of the model instantiated in memory are correlated, but often only loosely so, and are absolutely not identical. We cannot just look at the size of the model file to ensure that the resulting model is sufficiently small to be successfully loaded in our environment.

Can the model serve a result without crashing the infrastructure?

Again, this seems like a straightforward requirement: if we feed the model one minimal request, does it give us a result back of any kind, or does the system fail? Note that we say "one minimal request" as opposed to many requests intentionally, because these sorts of tests are often best done with single examples and single requests to start. This both minimizes risk and makes debugging easier if failures do arise.

Serving the result for one request might cause a failure, for several reasons:

Platform version incompatibility

Especially when using third-party or open source platforms, the serving stack could easily be using a different version of the platform than the training stack used.

Feature version incompatibility

The code for generating features is often different in the training stack from the serving stack, especially when each stack has different memory, compute cost, or latency requirements. In such cases, it is easy for the code that generates a given feature to get out of sync in these different systems, causing failures—one form of a general class of problems often referred to as *training-serving skew*. For example, if a dictionary is used to map word tokens to integers, the serving stack might be using a stale version of that dictionary even after a newer one was created for the training stack.

Corrupted model
: Errors happen, jobs crash, and in the end our machines are physical devices. It is possible for model files to become corrupted in one way or another, either through error at write time or by having `NaN` values written to disk if there were not sufficient sanity checks in training.

Missing plumbing
: Imagine that a model developer creates a new version of a feature in training, but neglects to implement or hook up a pipeline that allows that feature to be used in the serving stack. In these cases, loading in a version of the model that relies on that feature will result in crashes or undesirable behavior.

Results out of range
: Our downstream infrastructure might require the model predictions to be within a given range. For example, consider what might happen if a model is supposed to return a probability value between 0 and 1, not inclusive, but instead returns a score of exactly 0.0, or even –0.2. Or consider a model that is supposed to return 1 of 100 classes to signify the most appropriate image label, but instead returns 101.

Is the computational performance of the model within allowable bounds?
: In systems that use online serving, models must return results on the fly, which typically means tight latency constraints must be met. For example, a model intended to do real-time language translation might have a budget of only a few hundred milliseconds to respond. A model used within the context of high-frequency stock trading might have significantly less.

Because such constraints are often in tension with each other, for changes made by model developers in the search for increased accuracy, it is critical to measure latency before deployment. In doing so, we need to keep in mind that the production settings are likely to have peculiarities around hardware or networking that create bottlenecks that might differ from a development environment, so latency testing must be done as close to the production setting as possible. Similarly, even in offline serving situations, overall compute cost can be a significant limiting factor, and it is important to assess any cost changes before kicking off huge batch-processing jobs. Finally, as discussed previously, storage costs, such as size of the model in RAM, are another critical limitation and must be assessed prior to deployment. Checks like this can be automated, but it may also be useful to verify manually to consider trade-offs.

For online serving, does the model pass through a series of gradual canary levels in production?

Even after we have some confidence in a new model version based on the validation checks, we will not want to just flip a switch and have the new model suddenly take on the full production load. Instead, our collective stress level will be reduced if we first ask the model to serve just a tiny trickle of data, and then gradually increase the amount after we have assurance that the model is performing as expected in serving. This form of canary ramp-up is a place where model validation and model monitoring, as discussed in Chapter 9, overlap to a degree: our final validation step is a controlled ramp-up in production with careful monitoring.

Evaluating Model Quality

Ensuring that a model passes validation tests is important, but by itself does not answer whether the model is good enough to do its job well. Answering these kinds of questions takes us into the realm of evaluating model quality. Understanding these issues is important for model developers, of course, but is also critical for both organizational decision makers and for MLOps folks in charge of keeping systems running smoothly.

Offline Evaluations

As discussed in the whirlwind tour of the model development lifecycle in Chapter 2, model developers typically rely on offline evaluations, such as looking at accuracy on a held-out test set as a way to judge how good a model is. Clearly, this kind of evaluation has limitations, as we will talk about later in this chapter—after all, accuracy on held-out data is not the same thing as happiness or profit. Despite their limitations, these offline evaluations are the bread and butter of the development lifecycle because they offer a reasonable proxy while existing in a sweet spot of low cost and high efficiency that allows developers to test many changes in rapid succession.

Now, what is an evaluation? An *evaluation* consists of two key components: a performance metric and a data distribution. *Performance metrics* are things like accuracy, precision, recall, area under the ROC curve, and so on—we will talk about a few of these later on if they are not already familiar. *Data distributions* are things like "a held-out test set that was randomly sampled from the same source of data as the training data" that we have talked about before, but held-out test data is not the only distribution that might be important to look at. Others might include "images specifically from roads in snowy conditions" or "yarn store queries from users in Norway" or "protein sequences that have not previously been identified by biologists."

An evaluation is always composed of both a metric and a distribution together—the evaluation shows the model's performance on the data in that distribution, as computed by the chosen metric. This is important to know because folks in the ML world sometimes use shorthand and say things like "this model has better accuracy" without clarifying what the distribution is. This sort of shorthand can be dangerous for our systems because it neglects to mention which distribution is used in the evaluation, and can lead to a culture in which important cases are not fully assessed. Indeed, many of the issues around fairness and bias that emerged in the late 2010s can likely be tracked down to not giving sufficient consideration to the specifics of the data distribution used at test time during model evaluations. Thus, when we hear a statement like "accuracy is better," we can always add value by asking *on what distribution?*

Evaluation Distributions

Perhaps no question is more important in the understanding of an ML system than deciding how to create the evaluation data. Here are some of the most common distributions used, along with some factors to consider as their strengths and weaknesses.

Held-out test data

The most common evaluation distribution used is *held-out test data*, which we covered in Chapter 3 when reviewing the typical model lifecycle. On the surface, this seems like an easy thing to think about—we randomly select some of our training data to be set aside and used only for evaluation. When each example in the training data has an equal and independent chance of being put into the held-out test data, we call this an *IID test set*. The *IID* term is statistics-speak that means *independently and identically distributed*. We can think of the IID test set process as basically flipping a (maybe biased) coin or rolling a die for each example in the training data, and holding each one out for the IID test set based on the result.

The use of IID test data is widely established, not because it is necessarily the most informative way to create test data, but because it is the way that respects the assumptions that underpin some of the theoretical guarantees for supervised ML. In practice, a purely IID test set might be inappropriate, though, because it might give an unrealistically rosy view of our model's expected performance in deployment.

As an example, imagine we have a large set of stock-price data, and we want to train a model to predict these prices. If we create a purely IID test set, we might have training data from 12:01 and 12:03 from a given day in the training data, while data from 12:02 ends up in the test data. This would create a situation in which the model can make a better guess about 12:02 because it has seen the "future" of what 12:03 looks like. In reality, a model that is guessing about 12:02 would be unable to have access to this kind of information, so we would need to be careful to create our evaluations in a way that does not allow the model to train on "future" data. Similar examples might exist in weather prediction, or yarn product purchase prediction.

The point here is not that IID test distributions are always bad, but rather that the details here really do matter. We need to apply careful reasoning and common sense to the creation of our evaluation data, rather than relying on fixed recipes.

Progressive validation

In systems with a time-based ordering to data—like our preceding stock-price prediction example—it can be quite helpful to use a progressive validation strategy, sometimes also called *backtesting*. The basic idea is to simulate how training and prediction would work in the real world, but playing the data through to the model in the same order that it originally appeared.

For each simulated time step, the model is shown the next example and asked to make its best prediction. That prediction is then recorded and incorporated to the aggregate evaluation metric, and only then is the example shown to the model for training. In this way, each example is first used for evaluation and then used for training. This helps us see the effects of temporal ordering, and ask questions like, "What would this model have done last year on election day?"

The drawback is that this setup is somewhat awkward to adapt if our models require many passes over the data to train well. A second drawback is that we must be careful to make comparisons between models based on evaluation data from exactly the same time range. Finally, not all systems will operate in a setting in which the data can meaningfully be ordered in a canonical way, like time.

Golden sets

In models that continually retrain and evaluate using some form of progressive validation, it can be difficult to know whether the model performance is changing or whether the data is getting easier or harder to predict on. One way to control this is to create a *golden set* of data that models are not ever allowed to train on, but that is from a specific point in time. For example, we might decide that the data from October 1 of last year might be set aside as golden set data, never to be used for training under any circumstance, but held aside.

When we set aside the golden set of data, we also keep with it either the results of running that set of data through our model or, in some cases, the result of having humans evaluate the golden set. We might sometimes treat these results as "correct" even if they are really just the predictions for those examples from a specific process and at a particular point in time.

Performance on golden set data like this can reveal any sudden changes in model quality, which can aid debugging greatly. Note that golden set evaluations are not particularly useful for judging absolute model quality, because their relevance to current performance diminishes as their time period recedes into the past. Another problem can arise if we are not able to keep golden set data around for very long (for example, to respect certain data privacy laws or to respond to requests for deletion or expiration of access). The primary benefit of golden sets is to identify changes or bugs, because, typically, model performance on golden set data changes only gradually as new training data is incorporated into the model.

Stress-test distributions

When deploying models in the real world, one worry is that the data they may encounter in reality may differ substantially from the data they were shown in training. (These issues are sometimes described by different names in the literature, including *covariate shift*, *nonstationarity*, or *training-serving skew*). For example, we might have a model trained largely on image data from North America and Western Europe, but that is then later applied in countries across the globe. This creates two possible problems. First, the model may not perform well on the new kinds of data. Second, and even more important, we may not know that the model would not perform well because the data was not represented in the source that supplied the (supposedly) IID test data.

Such issues are especially critical from a fairness and inclusion standpoint. Imagine we are building a model to predict the yarn color preferred by a user, given a provided portrait image. If our training data does not include portrait images with a wide range of skin tones, an IID test set might not have sufficient representation to uncover problems if the model does not do well for images of people with especially dark skin tones. (This example harkens back to seminal work by Buolamwini and Gebru.)[1] In cases like this, it's important to create specific stress-test distributions in which carefully constructed test sets each probe for model performance on different skin tones. Similar logic applies to testing any other area of model performance that might be critical in practice, from snowy streets for navigation systems developed in temperate climates, to a broad range of accents and languages for speech-recognition systems developed in a majority English-speaking workplace.

---

1 See "Gender Shades: Intersectional Accuracy Disparities in Commercial Gender Classification" (*https://oreil.ly/g37lG*) by Joy Buolamwini and Timnit Gebru.

Sliced analysis

One useful trick to consider is that any test set—even an IID test set—can be sliced to effectively create a variety of more targeted stress-test distributions. By *slicing*, we mean filtering the data based on the value of a certain feature. For example, we could slice images to look at performance on images with only snowy backgrounds, or stocks from companies that were only in their first week of trading on the market, or yarns that were only a shade of red. So long as we have at least some data that conforms to these conditions, we can evaluate performance on each of these cases through slicing. Of course, we need to take care not to slice too finely, in which case the amount of data we would be looking at would be too small to say anything meaningful in a statistical sense.

Counterfactual testing

One way to understand a model's performance at a deeper level involves learning what the model would have predicted if the data had been different. This is sometimes called *counterfactual testing* because the data that we end up feeding to the model runs counter to the actual data in some way. For example, we might ask what the model would predict if the dog in a given image were not on a grassy background, but instead shown against a background of snow or of clouds.[2] We might ask if the model would have recommended a higher credit score if the applicant had lived in a different city, or if the model would have predicted a different review score if the pronoun for the lead actor in a movie had been switched from *he* to *she*.

The trick here is that we might not have any examples that match any of these scenarios, in which case we would take the step of creating synthetic counterfactual examples by manipulating or altering examples that we do have access to. This tends to be most effective when we want to test that a given alteration does *not* substantially change model prediction. Each of these tests might reveal something interesting about the model and the kinds of information sources it relies on, allowing us to use judgment about whether it is appropriate model behavior and whether we need to address any issues prior to launch.

A Few Useful Metrics

In some corners of the ML world, there is a tendency to look at a single metric as the standard way to view a model's performance on a given task. For example, accuracy was, for years, the one way that models were evaluated on ImageNet held-out test

2 See, for example, "Noise or Signal: The Role of Image Backgrounds in Object Recognition" (*https://arxiv.org/abs/2006.09994*) by Kai Xiao. The What-If Tool (*https://oreil.ly/M07ff*) is also an excellent example of tooling that allows for counterfactual probing.

data. And indeed, this mindset is most often seen in benchmarking or competition settings, in which the use of a single metric simplifies comparisons between different approaches. However, in real-world ML, it is often ill-advised to myopically consider only a single metric. It is better to think of each metric as a particular perspective or vantage point. Just as there is no one best place to watch the sun rise, there is no one best metric to evaluate a given model, and the most effective approach is often to consider a diverse range of metrics and evaluations, each of which has its own strengths, weaknesses, blind spots, and peculiarities.

Here, we will try to build up our intuition around some of the more common metrics. We divide them into three broad categories:

Canary metrics
: Are great at indicating that something is wrong with a model, but are not so effective at distinguishing a good model from a better one.

Classification metrics
: Help us understand the impact of a given model on downstream tasks and decisions, but require fiddly tuning that can make comparisons between models more difficult.

Regression and ranking metrics
: Avoid this tuning and make comparisons easier to reason about, but may miss specific trade-offs that might be available when some errors are less costly than others.

Canary metrics

As we've noted, this set of metrics offers a useful way to tell when something is horribly wrong with our model. Like the fabled canary in the coal mine, if any of these metrics is not singing as expected, then we definitely have a problem to deal with. On the flip side, if these metrics look good, that does not necessarily mean that all is well or that our model is perfect. These metrics are just a first line of detection for potential issues.

Bias. Here we use *bias* in the statistical sense rather than the ethical sense. Statistical bias is an easy concept—if we add up all the things we expect to see based on the model's predictions, and then add up all the things we actually see in the data, do we get the same amount? In an ideal world, we would, and typically a well-functioning model will show very low bias, meaning a very low difference between the total expected and observed values for a given class of predictions.

One of the nice qualities of bias as a metric is that unlike most other metrics, there is a "correct" value of 0.00 that we do expect most models to achieve. Differences here of even a few percent in one direction or another are often a sign that something is wrong. Bias often couples well with sliced analysis as an evaluation strategy to uncover problems. We can use sliced analysis to identify particular parts of the data that the model is performing badly on as a way to begin debugging and improving the overall model performance.

The drawback of bias as a metric is that it is trivial to create a model with perfectly zero bias, but that is a terrible model. As a thought experiment, this could be done by having a model that just returns the average observed value for every example—zero bias in aggregate, but totally uninformative. A pathologically bad model like this can be detected by looking at bias on more fine-grained slices, but the larger point remains. Bias as a metric is a great canary, but just having zero bias is not by itself indicative of a high-quality model.

Calibration. When we have a model that predicts a probability value or a regression estimate, like a probability that a user will click a given product or a numerical prediction of tomorrow's temperature, creating a *calibration plot* can provide significant insight into the overall quality of the model. This is done essentially in two steps, which can be thought of roughly as first bucketing our evaluation data in a set of buckets and then computing the model's bias in each of these buckets. Often, the bucketing is done by model score—for example, the examples that are in the lowest tenth of the model's predictions go in one bucket, the next lowest tenth in the next bucket, and so on—in a way that brings to mind the idea of sliced analysis discussed previously. In this way, calibration can be seen as an extension of the approach of combining bias and sliced analysis, in a systematic way.

Calibration plots can show systemic effects such as overprediction or underprediction in different areas, and can be an especially useful visualization to help understand how a model performs near the limits of its output range. In general, calibration plots can help show areas where a model may systematically overpredict or underpredict, by plotting the observed rates of occurrence versus their predicted probabilities. This can be useful to help detect situations where the model's scores are either more or less reliable. For example, the plot in Figure 5-2 shows a model that gives good calibration in the middle ranges, but does not do as well at the extremes, overpredicting on actual low probability examples and underpredicting on actual high-probability examples.

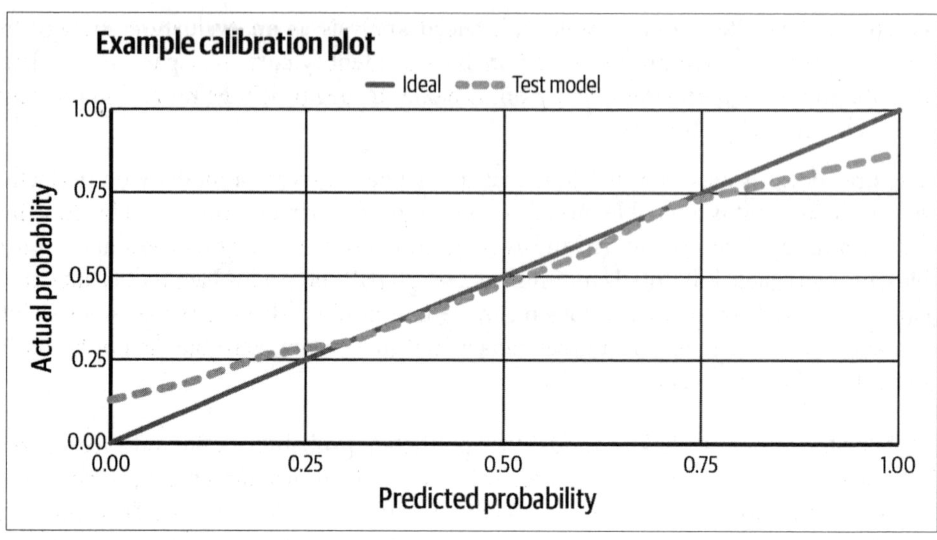

Figure 5-2. An example calibration plot

Classification metrics

When we think of model evaluation metrics, classification metrics like accuracy are often the first ones that come to mind. Broadly speaking, a *classification metric* helps measure whether we've correctly identified that a given example belongs to a specific category (or *class*). Class labels are typically discrete—things like `click` or `no_click`, or `lambs_wool`, `cashmere`, `acrylic`, `merino_wool`—and we tend to judge a prediction as a binary correct or incorrect on getting a given class label right.

Because models typically report a score for a given label, such as `lambs_wool: 0.6`, `cashmere: 0.2`, `acrylic: 0.1`, `merino_wool: 0.1`, we need to invoke a decision rule of some kind to decide when we are going to predict a given label. This might involve a threshold, like "predict acrylic whenever the score for `acrylic` is above 0.41 for a given image," or it might ask which class label gets the highest score out of all available options. Decision rules like these are a choice on the part of the model developer, and are often set by taking into account the potential costs of different kinds of mistakes. For example, it may be significantly more costly to miss identifying stop signs than to miss identifying merino wool products.

With that background, let's look at a couple of classic metrics.

Accuracy. In conversation, many folks use the term *accuracy* to mean a general sense of goodness, but accuracy also has a formal definition that shows the fraction of predictions for which the model was correct. This satisfies an intuitive desire—we want to know how often our model was right. However, this intuition can sometimes be misleading without appropriate context.

To place an accuracy metric into context, we need to have some understanding of how good a naive model that always predicts the most prevalent class would be, and also to understand relative costs of different types of errors. For example, 99% accuracy sounds pretty good, but may be completely terrible if the goal is figuring out when it is safe to cross a busy street—we would be almost sure to be in an accident quite soon. Similarly, 0.1% accuracy sounds horrible but would be an amazingly good performance if the goal was to predict winning lottery number combinations. So, when we hear an accuracy value quoted, the first question should always be, "What is the base rate of each class?" It is also worth noting that seeing 100% accuracy—or perfect performance on any metric—is most often cause for concern rather than celebration, as this may indicate overfitting, label leakage, or other problems.

Precision and recall. These two metrics are often paired together, and are related in an important way. Both metrics have a notion of a *positive*, which we can think of as "the thing we are trying hard to find." This can be finding spam for a spam classifier, or yarn products that match the user's interest for a yarn store model. These metrics answer the following related questions:

Precision
: When we said an example was a positive, how often was that indeed the case?

Recall
: Out of all of the positive examples in our dataset, how many of them were identified by our model?

These questions are especially useful to ask and answer when positives and negatives are not evenly split in the data. Unlike accuracy, the intuitions of what a precision of 90% or a recall of 95% might mean scale reasonably well even if positives are just a small fraction of the overall data.

That said, it is important to notice that the metrics are in tension with each other in an interesting way. If our model does not have sufficient precision, we may be able to increase its precision by increasing the threshold it uses to make a decision. This would cause the model to say "positive" only when it is even more sure, and for reasonable models would result in higher precision. However, this would also mean that the model refrains from saying "positive" more often, meaning that it identifies fewer of the total possible number of positives and results in lower recall because of the increased precision. We could also trade off in the other direction, lowering thresholds to increase recall at the cost of less precision. This means that it is critical to consider these metrics together, rather than in isolation.

AUC ROC. This is sometimes just referred to as *area under the curve* (*AUC*). *ROC* is an abbreviation for *receiver operating characteristics*, a metric that was first developed to help measure and assess radar technology in the Second World War, but the acronym has become universally used.

Despite the confusing name, it has the lovely property of being a threshold-independent measure of model quality. Remember that accuracy, precision, and recall all rely on classification thresholds, which must be tuned. The choice of threshold can impact the value of the metric substantially, making comparisons between models tricky. AUC ROC takes this threshold tuning step out of the metric computation.

Conceptually, AUC ROC is computed by creating a plot showing the true-positive rate and the false-positive rate for a given model at every possible classification threshold, and then finding the area under that plotted curved line; see Figure 5-3 for an example. (This sounds expensive, but efficient algorithms can be used for this computation that don't involve actually running a lot of evaluations with different thresholds.) When the area under this curve is scaled to a range from 0 to 1, this value also ends up giving the answer to the following question: "If we randomly choose one positive example and one negative from our data, what is the probability that our model gives a higher prediction score to the positive example rather than the negative?"

No metric is perfect, though, and AUC ROC does have a weakness. It is vulnerable to being fooled by model improvements that change the relative ordering of examples far away from any reasonable decision threshold, such as pushing an already low-ranked negative example even lower.

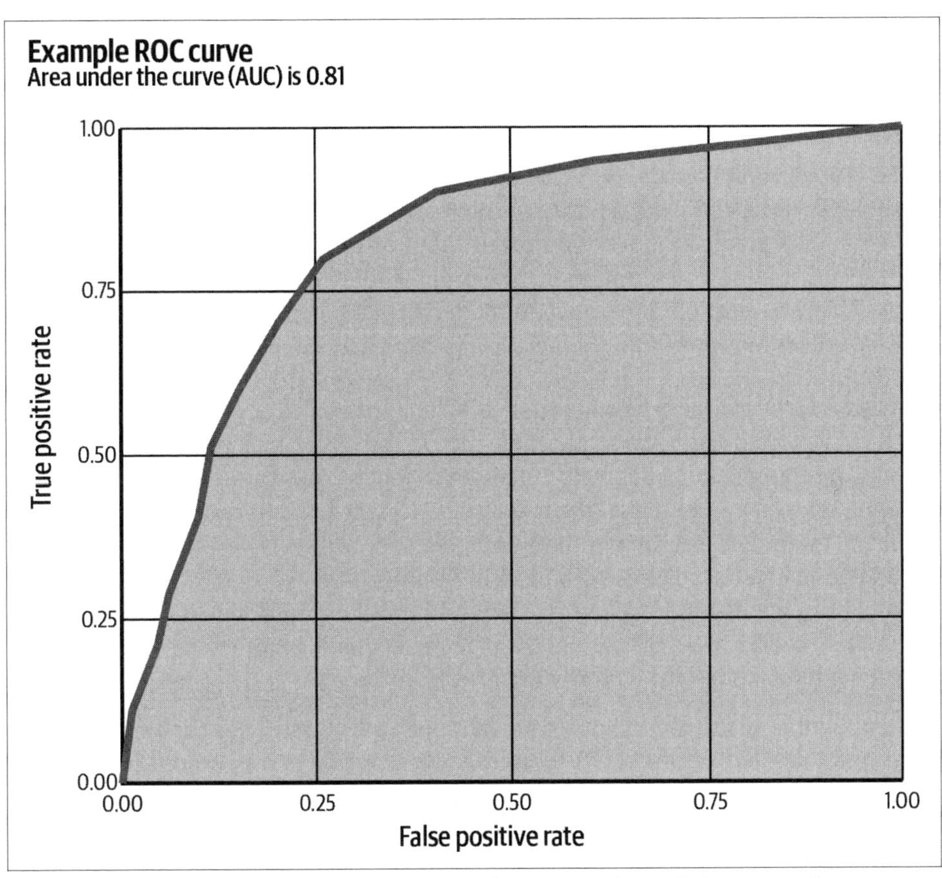

Figure 5-3. An ROC curve demonstrating performance of a model on a set of classifications

Precision/recall curves. Just as an ROC curve maps out the space of trade-offs between true-positive rate and false-positive rate at different decision thresholds, many folks plot precision/recall curves that map out the space of trade-offs between precision and recall at different decision thresholds. This can be useful to get an overall sense of comparison between two models across a range of possible trade-offs.

Unlike the AUC ROC, the computed area under the precision/recall curve does not have a theoretically grounded statistical meaning, but is often used in practice as a quick way to summarize the information nonetheless. In cases of strong levels of class imbalance, there is a case to be made that the area under the precision/recall curve is a more informative metric.[3]

Regression metrics

Unlike classification metrics, regression metrics do not rely on the idea of a decision threshold. Instead, they look at the raw numerical output that represents a model's prediction, like predicted price for a given skein of yarn, number of seconds a user might spend reading a description, or the probability that a given picture contains a puppy. They are most often used when the target value itself is continuously valued, but have utility in discrete valued label settings like click-through prediction as well.

Mean squared error and mean absolute error. When comparing predictions from a model to a ground-truth value, the first metric we might look at is the difference between our prediction and reality. For example, in one case, our model might predict 4.3 stars for an example that had 4 stars in reality, and in another case it might predict 4.7 stars for an example that had 5 stars in reality. If we were to aggregate those values without thinking about it, so we could look at averages over many values, we would run into the mild annoyance that in the first example the difference was 0.3 and in the second it was –0.3, so our average error would appear to be 0, which feels misleading for a model that is clearly imperfect.

One fix for this is to take the absolute value of each difference—creating a metric called *mean absolute error* (*MAE*) to average these values across examples. Another fix is to square the errors—raising them to the power of two—to create a metric called *mean squared error* (*MSE*). Both metrics have the useful quality that a value of 0.0 shows a perfect model. MSE penalizes larger errors much more than smaller errors, which can be useful in domains where you do not want to make big mistakes. MSE can be less useful if the data contains noise or outlier examples that are better ignored, in which case MAE is likely a better metric. It can be especially useful to compute both metrics and see if they yield qualitatively different results for a comparison between two models, which can provide clues into a deeper level of understanding their differences.

Log loss. Some people think of *log loss* as an abbreviation for *logistic loss*, because it is derived from the *logit* function, which is equivalent to the log of the odds ratio between two possible outcomes. A more convenient way to think of it might be as *the loss we want to use when we think about our model outputs as actual probabilities*. Probabilities are not just numbers restricted to the range from 0.0 to 1.0, although

3 See, for example: "Precision-Recall Curve Is More Informative Than ROC in Imbalanced Data: Napkin Math & More," (*https://oreil.ly/wZA17*) by Tam D. Tran-The.

this is an important detail. Probabilities also meaningfully describe the chance that a given thing will happen to be true.

Log loss is great for probabilities because it will highlight the difference between a prediction of 0.99, 0.999, and 0.9999, and will penalize each more confident prediction significantly more if it turns out to be incorrect—and the same thing happens at the other end of the range for predictions like 0.01, 0.001, and 0.0001. If we do indeed care about using the model outputs as probabilities, this is quite helpful. For example, if we are creating a risk-prediction model predicting the chance of an accident, there is an enormous difference between an operation being 99% reliable and 99.99% reliable—and we could end up making very bad pricing decisions if our metrics did not highlight these differences. In other settings, we might just loosely care how likely a picture is to contain a kitten, and 0.01 and 0.001 probabilities might both be best interpreted as "basically unlikely," in which case log loss would be a poor choice of final metric. Lastly, it is important to note that log loss can give infinite values (which show up as `NaN` values and destroy averages) if our models were to predict values of exactly 1.0 or 0.0 and be in error.

Operationalizing Verification and Evaluation

We have just taken a whirlwind tour through the world of evaluating model validity and model quality. How do we turn this knowledge into something actionable?

Assessing model validity is something that anyone who cares about production should know how to do. This is possible, even if you don't do model evaluation daily, with a combination of training, checklists/processes, and automated support code for the simpler cases (which itself saves human expertise and judgment for more demanding cases).

For questions of model quality evaluations, things are perhaps a little more ambiguous. Obviously, it is highly useful for MLOps folks to have a working knowledge of the various distributions and metrics that are most critical for assessing model quality for our system. An organization may go through a few phases.

In the earliest days of model development for an organization, the biggest questions are often much more around getting something working rather than about how to evaluate it. This can lead to relatively coarse strategies for evaluation. For example, the main problems in developing the first version of a yarn store product recommendation model are much more likely to be around creating a data pipeline and a serving stack, and model developers might not have bandwidth to choose carefully between varying classification or ranking metrics. So our first standard evaluation might just be AUC ROC for predicting user clicks within a held-out test set.

As the organization develops, a greater understanding develops of the drawbacks or blind spots that a given evaluation might have. Typically, this results in additional metrics or distributions being developed that help shed light on important areas of model performance. For example, we might notice a cold-start problem in which new products are not represented in the held-out set, or we might decide to look at calibration and bias metrics across a range of slices by country or product listing type to understand more about our model's performance.

At a later stage, the organization may start to go back and question basic assumptions, such as whether the chosen metrics reflect the business goals with sufficient veracity. For example, in our imaginary yarn store, we may come to realize that optimizing for clicks is not actually equivalent to optimizing for long-term user satisfaction. This may require a full reworking of the evaluation stack and careful reconsideration of all associated metrics.

Are these questions within the realm of model developers or MLOps folks? Opinions here may vary, but we believe that a healthy organization will encourage multiple points of view and rich discussions on these questions.

Conclusion

This chapter has focused on establishing an initial viewpoint of model validity and model quality, both of which are critical to assess before moving a new version of a model into production.

Validity tests help establish that a new version of the model will not break our system. These include establishing compatibility with code and formats, and making sure that the resource requirements of computation, memory, and latency are all within acceptable limits.

Quality tests help give assurance that a new version of the model will improve predictive performance. This most often involves assessment of the model's performance on some form of held-out or test data, with the appropriate choice of evaluation metric suited for the application task.

Together, these two forms of testing establish a decent level of trust in a model and will be a reasonable starting point for many statically trained ML systems. However, systems that deploy ML in a continuous loop will require additional verification, as detailed in Chapter 10.

CHAPTER 6

Fairness, Privacy, and Ethical ML Systems

By Aileen Nielsen

This chapter is devoted to topics related to ethical considerations and legal obligations when creating or deploying ML systems. We cannot offer an exhaustive source of guidance on these topics, but this resource can point you in the right direction. At the end of this chapter, you should have a good sense of fundamental ethical considerations for ML deployment as well as concrete language and conceptual categories that will get you started in educating yourself more thoroughly in the domains most immediately applicable to your own work.

Editor's note: When we put together the list of topics for what MLOps folks truly need to know, issues of fairness, privacy, and ethical concerns in AI and ML systems were right at the top of the list. However, we also knew that it was difficult for a group of authors with strong industry affiliations to provide truly unbiased views on these complex issues. Therefore, we invited Aileen Nielsen, author of *Practical Fairness* (*https://oreil.ly/tsjGP*) (O'Reilly, 2020), to contribute this chapter independently. While we gave feedback on drafts for clarity, the views here are entirely hers, and she had full editorial control on this chapter. You're getting it straight from a world-class expert!

We should also note from the outset that fairness and ethics in AI remain highly contested topics. Indeed, one reasonable position right now is that a quite viable approach to promoting fairness in computing systems is *not* to use AI/ML. However, for those who find themselves compelled to do so, or who believe that this skepticism of algorithmic solutions can be overcome in their specific use cases, this chapter provides a starting point for understanding the challenges of how to do it right.

What's more, it should be recognized that in some cases existing or otherwise traditional solutions to problems in the pre-algorithmic age (e.g., having designated human decision makers, or no clear decision makers) haven't always been great.[1] For example, a great deal of empirical research suggests that race influences judges' sentencing decisions and even informal decisions by store clerks as to whether to allow customers to return items. Therefore, when we think about fairness and ethics in AI and ML, we also have to recognize that they may very well provide improvements in some cases relative to human decision makers. So, despite the gloom and doom you will find in this chapter, we also recognize from the outset that some uses of algorithms have been tremendously successful in terms of increasing overall fairness, even if there isn't yet a clear solution to making AI and ML fair and just in a guaranteed or global sense.

In this chapter, you will find sections that focus on specific hot topics—notably, fairness, privacy, and Responsible AI. We recognize the primacy of these topics both in terms of public awareness and concern as well as in terms of attention devoted from industry and scholarly groups alike. A lot of development is occurring in these areas, and we want to empower you to dive into these topics with a good initial background that you can pick up from this chapter. Within this chapter, we also provide notes on how you might consider refactoring work at your organization, in practical ways, so as to enhance fairness and ethics in your own AI/ML work.

Fairness (a.k.a. Fighting Bias)

Algorithmic fairness, and other variations of this term, are a hot topic in ML and have been for many years. When you read about this subject, you'll most often see fairness used as a concept directly related to bias—that is, as fairness being the absence of bias, and bias being the condition of unfairness.[2]

For quite some time, ML researchers, legal scholars, and activists alike have manifested growing concern about the possibility that ML could perpetuate existing social biases, or even create new forms of bias. In framing such discussions, many have emphasized that this could happen because data used to train ML systems could be taken from biased systems or collected in a biased way. In short, people have made arguments akin to the notion of garbage in, garbage out.

1 Examples of designated human decision makers include judges as the formal decision makers in legal adjudication, or university professors as the formal decision makers in grading decisions. Examples of no clear decision makers include "flat" organizations, in which it can sometimes be unclear who holds final decision-making authority.

2 As a general matter, we take issue with this undue narrowing of the concept of fairness, and for the sake of clear communication, we address a host of other issues that we view as related to *fairness* in "Responsible AI" on page 127. In this section, we focus on the mainstream use of *fairness* as exclusively related to bias.

But you should be careful to realize that the *garbage in* does not necessarily refer to bad data. The garbage is the *outcome* that results from unrepresentative training data or algorithmic bias. This is a key point: it's important to realize that the *fundamental* source of bias can come from many steps in the ML pipeline, including bad data, but also including bad modeling choices. Here's a nonexhaustive list of some commonly cited sources of bias:

Sampling bias

In sampling bias, the process of collecting data itself has bias built into a system. A commonly given example is that marijuana use among white and Black Americans is thought to be roughly equal, but the rate of arrests for marijuana possession is far higher for Black Americans than for white Americans. This is almost certainly due to (among other reasons) sampling bias. Because of racism manifesting at both the level of individual decisions and at a systemic level, Black Americans are far more likely to find themselves searched by police for marijuana possession than are white Americans (some may be familiar with a related manifestation of racism, driving while Black (*https://oreil.ly/4w8oQ*)).

ML-relevant example:

Crime-prediction algorithms, such as those built to influence police patrolling allocations, likely suffer from sampling bias.[3] Across many nations and continents, it remains a consistent pattern that policing is directed at communities of low socioeconomic status. Data coming from such a process oversamples crime in certain areas or among certain demographics and so likely misrepresents the underlying base rates of crime in different communities due to this sampling bias. Yet, this data, sampled in a biased fashion, then creates new biased inputs for more ML modeling to allocate future police patrols. The algorithm likely keeps sending police back to the same areas too often because other areas are not oversampled in the same way.

Disparate treatment

Bias can result from individuals explicitly being treated differently. Disparate treatment is banned both by the government and by some regulated areas of the

3 "Machine Bias" (*https://oreil.ly/38he8*) by Julia Angwin et al. is the seminal study, published by ProPublica, that identified this bias. This work, discussing the use of algorithmic risk scores in the US criminal justice system, is foundational in the understanding of the kind of harms that AI can create and in launching the Responsible AI movement. The article discusses the use of algorithmic risk scores in the criminal justice system in the US. The most common use of these biased risk scores is to determine whether defendants should be allowed free while awaiting trial in an overly burdened criminal justice system. The factor being predicted in this case should be whether the accused will show up for trial and whether they will commit crimes in the meantime, not generally at any point in the future. It is not clear that a single prediction about "will be accused of a crime in the future" can be useful in decisions about bail, sentencing, and parole. This article discusses the COMPAS algorithm, which is discussed in some detail later in this chapter.

private sector when there is no justified reason for it. In almost all examples of disparate treatment on the basis of race (such as segregated schools or policies of hiring only whites), courts have not found a justified reason for such discrimination (after the Civil Rights movement). For disparate treatment on the basis of gender, courts have sometimes found that reasons offered to treat genders differently (such as different performance thresholds for physical fitness tests) could be justified by compelling interests.

ML-relevant example:

ML algorithms are often celebrated for their ability to spot patterns that elude humans. However, sometimes these patterns are plain old sexism. Amazon famously developed but did not deploy (*https://oreil.ly/zyKHl*) an in-house hiring algorithm that applied a strong negative parameter for attending a woman's college.[4] Thus we see a case of an algorithm using irrelevant factors to make a decision in a way that directly discriminates against a group (in this case, women going to women's colleges).[5] If the algorithm had been used, this would have looked like a case of disparate treatment.

Systemic bias

This source of bias can be challenging to both identify and mitigate as compared to the preceding examples in this list. Broadly, we can think of systemic bias as a host of factors that likely cannot be identified in individual cases as explanatory features but that, on an aggregate level, clearly influence differences in outcomes for individuals due to the structural limitations (limitations that are baked into our social, educational, and employment systems, among others). AI systems cannot be separated from the societal contexts in which they are built, so AI built without an understanding of context is likely to worsen systemic bias. Indeed, even the particular problems to which AI is applied are often highlighted as themselves manifestations of systemic bias. For example, some have asked why predictive policing has been actively used for so long to predict crimes among socioeconomically vulnerable populations, while algorithmic systems to identify or predict *police* misbehavior are relatively uncommon.

---

4 This was one of several factors that led some governments, notably that of New York City, to propose regulation of AI when used in hiring. See "New York City Proposes Regulating Algorithms Used in Hiring" (*https://oreil.ly/xjN4P*) by Tom Simonite.

5 A larger debate arises about whether discrimination is "taste based" or "statistical." Some argue that the decision to attend a women's college might in some way be indicative of a personality type that is relevant to a hiring decision. However, such a hypothesis seems unlikely to apply. Those interested in learning more can look to research literature in both law and economics about the mechanisms of and motivations for discrimination, including disparate treatment.

ML-relevant example:

In training an ML system, sometimes features are selected because they seem to provide "common sense" examples of "merit" as some employers or educational institutions might conceptualize that notion. For example, middle class high school students are often told to participate in extracurricular activities to show that they are highly motivated leaders. However, using data about participation in extracurricular activities (in an algorithm or in human decision making) without additional context would contribute to a system that rewards middle class students who have access to such possibilities. This system would likewise disadvantage low-income students whose schools may not offer such activities or who are not able to participate in such activities because they need to work or contribute to family obligations. Thus what "common sense" dictates to people unfamiliar with different social contexts and unversed in the consequences of systemic bias could look quite biased once this additional information is incorporated into the analysis.

Tyranny of the majority

This source of bias relates to the form of training process that is used. In many commonly used modeling systems, the numerical majority category typically affects training loss the most through total numbers. Therefore, if the modeling process does not account for this, many kinds of models will de facto favor the majority and minimize the error for majority groups more directly than for minority groups. This same concern sometimes motivates the design of political systems so as to protect minority interests specifically rather than trusting to majority rule.

ML-relevant example:

It is well known that training ML systems with data that does not involve relatively equal distributions of characteristics (imbalanced datasets) can be quite challenging if we seek to achieve good performance for all classes. Such a situation is likely to arise when working with datasets that include different kinds of people, as most contain some kind of imbalance in data, be it gender imbalance, racial imbalance, or other forms of human diversity, such as geography or language. Therefore, modeling human behavior should require paying particular attention to make sure that the model is giving everyone a fair shake rather than modeling everyone according to a prototype that in fact is reasonably accurate only for someone in the majority class.

Tragically, it remains indisputable that these forms of bias (and many others) continue to pop up in a wide range of ML applications, even with increasing media attention about algorithmic fairness. There are many reasons for this. Social biases are widespread and are not always apparent at the individual level. Someone might think they are making an unbiased decision both because they do not have all the details about the system in which they are making the decision, and because of their own unconscious biases. This is not a problem that will be solved by a book chapter, but one that a book chapter can make readers aware of and motivated to do something about.

The ML/AI research community has been developing methods and techniques to systematically identify and remediate some of these biases. Practitioners who are trying to develop responsible and fair AI systems should be aware of these emerging tools that might help. What's more, ML/AI may very well provide a way forward to make systems fairer than they have been with the use of unaccountable humans, so long as ML development is done carefully and with appropriate safeguards, when aimed at appropriate problems.

A Note on Solutionism

This chapter strongly emphasizes how complicated and difficult it can be to implement AI in a fair and responsible fashion. One neat trick can help you avoid the entire problem: don't use ML/AI (if you can get away with it). Although many systems undoubtedly work better with some degree of automation, that's not always the case. Sometimes the human touch is needed. Sometimes the human touch is even more efficient, as well as fairer, than a machine touch.

And, if you do conclude that some degree of automation and algorithm use is justified, make sure you ask yourself exactly *how* using that automation makes sense. Most situations allow many degrees of use between full algorithm and full human, so think carefully about *what* an algorithm should be doing even after you decide *whether* an algorithm should be doing anything at all.

One of the bigger problems with ML is that it often adds complexity, risk, and cost, with little value and no net advantage. We call these uses *solutionism*: merely because we have a readily available (and trendy) solution (ML), we are trying to figure out what problem it solves.

ML systems are necessarily complicated. They can accomplish amazing things when applied well to problems that fit, but they can add enormous risk (and significant bias or other forms of unfairness) to other kinds of applications. The smartest thing an ML team can do is to consistently and skeptically consider whether ML is necessary or useful at all. By doing so, we can confine our efforts to ML applications that truly add significant value, and therefore are worth the effort required to ensure privacy and fairness as we implement the system.

Definitions of Fairness

In the ML community, we have not cohered around a single definition of fairness, but common categories exist that have intuitive and appealing descriptions. For example, some definitions of *fairness* emphasize individual fairness. These notions of fairness argue that, for two individuals who are "the same" other than irrelevant factors (race, gender, etc.), these individuals should be treated the same.[6] Other definitions of fairness emphasize that ML should look to fairness at the level of groups. That is, error rates should be the same and of the same quality across groups, and overall the level of performance should likewise be equally high among groups. Still other definitions of fairness might get more complicated and look to establish causal mechanisms to understand what might drive individuals to success or failure, before looking to categorize them algorithmically.

Two common categories of fairness that are intuitive and appealing are group parity and calibration-based assessments. These are far from being the only notions of fairness, but we select these two to discuss for two reasons. First, each of these definitions is straightforward and has a strong intuitive appeal. Second, each highlights two distinct notions of fairness.

As its name suggests, *group parity* emphasizes fairness as matter to be compared on the basis of comparing groups. It puts emphasis on recognizing that outcomes for individuals should be looked at, at least in part, on the basis of sensitive attributes of their identity, and most particularly on legally protected attributes, such as gender and race, but also on attributes that may not be legally protected but that some consider morally important, such as economic status. There are good reasons for this, including that when we look at the world, such features often turn out to be horribly influential in actual outcomes.

On the other hand, *calibration* emphasizes individual fairness, which is another value most of us find quite intuitive and appealing. We likewise feel strongly that individuals should be treated on the basis of who they are and what they do as individuals, and not based on where they come from.

It would seem that these two definitions of fairness need not be at odds philosophically. In a perfect world, where groups and individuals are treated fairly, we would get the same outcomes. Unfortunately, this turns out not to be the case in our imperfect world, for a variety of reasons, including fundamental mathematical limitations.[7] For this reason—at least for now—practitioners must choose their definition of fairness. As we will describe, different use cases will have different reasonable definitions of fairness.

6 We use scare quotes in this sentence because, of course, defining what or who is "the same" or "of equal merit" is itself an exercise of judgment and not an objective truth.

7 For an accessible demonstration of the conflict between calibration and group parity, we strongly recommend "Fair Prediction with Disparate Impact: A Study of Bias in Recidivism Prediction Instruments" (*https://arxiv.org/abs/1610.07524*) by Alexandra Chouldechova.

While applying different notions of fairness in different situations may seem strange, mature readers will realize that we likewise apply the same principles in real life. Consider the US, a strongly market-driven economy. For the most part—and even in our current era of populism—there tends to be a pervasive individualized notion of fairness when it comes to the labor market. Americans seem to think people should get what they can get on the open market, albeit with some concerns that wages not go too low (thus, there is a legal minimum wage) and likewise with some concerns that the highest rates of pay have gone too high (demonstrated by increasing concerns about rising inequality and the rise of the billionaire class). On the other hand, when it comes to healthcare, most Americans think everyone should get good medical care regardless of differences in individual health or inborn genetics.

It seems (at least to us armchair anthropologists) that at some level Americans are comfortable with an allocation of economic benefits based, likely in part, on inborn ability but are not comfortable with this notion when it comes to an allocation of medical benefits. Thus, ordinary people in fact seem to commonly apply different notions of fairness to different domains of life, and we propose that the same could plausibly be true in different contexts or use cases of ML.

It might seem that the ML ethics community should be attempting to converge on a single definition of *fairness*. This might seem especially true when we can see that different definitions of fairness are divergent or even contradictory in practice. At present, given the state of maturity of the industry and this field, it is almost certainly not a goal to converge on a single definition of fairness. This is for two reasons.

First, we don't need to. The state of the art of most models deployed in the real world is still fundamentally unfair by any metric (or at least of unexamined fairness). We have many useful definitions of fairness to select from, and picking one or several to work toward provides us with the opportunity for rapid progress right away without needing to await further theoretical or legal developments.

Second, and more structurally, the field of AI and ML theory is currently largely controlled by people from the dominant groups of society. These are the very people most likely to miss the ways that AI and ML systems can negatively impact disempowered or underrepresented groups. Surely, these same people shouldn't be the arbiters of what will be set down as the proper definition of fairness to the exclusion of all others in the future.

Group parity fairness definitions require that relevant rates of algorithmic performance be the same across various groups. A group parity requirement could, for example, require that the hiring rate across all ethnic groups be the same, but might also require that the accuracy for diagnosing low oxygen (*https://oreil.ly/wU0mJ*) be the same regardless of skin color. This idea is compelling and intuitive because it paints a picture of the society many of us would like to see—one in which bad luck or opportunity are distributed equally among communities.

In contrast to group parity, *calibration* fairness definitions require that a ML model work equally well for all individuals. While this cannot be directly measured, what can be measured is that a model score means the same thing for an individual regardless of their group membership. So a calibration-oriented definition of fairness would require that for any group, the ML score means the same thing.

This sounds intimidating and highly technical, but it can be easier to digest with a specific example. A commonly given example of calibration is that of the Correctional Offender Management Profiling for Alternative Sanctions, or COMPAS (*https://oreil.ly/KEVqN*), recidivism scoring algorithm, which is used by criminal courts in many US states. Because this algorithm has been shown to be calibrated correctly across racial categories, the same COMPAS score (say, 0.5) means the same thing for both Black and white data subjects: that is, a white offender and a Black offender with the same risk score have the same probability of reoffending.[8] When understood in plain English, this seems intuitive as well—that a model score should mean the same thing for everyone, no matter what group they belong to.

You may wonder whether having many definitions of fairness is a problem. After all, there are many correct and intuitive ways of describing the world in other contexts, so why not in the case of fairness? The difficulty comes from the fact that—at least in the world we currently live in—these definitions of fairness come into conflict. The COMPAS algorithm is a compelling example. That algorithm was not necessarily implemented in a careless fashion, but was, in fact, sometimes adopted as part of a movement to reduce bias in the criminal justice system. The algorithm passed calibration checks that were considered to be the gold standard in ensuring fairness and nondiscrimination.

However, reporters at ProPublica, a nonprofit newsroom, later demonstrated that—despite this calibration—the algorithm had higher rates of false positives in labeling Black defendants high risk, and higher rates of false negatives in labeling white defendants low risk. In other words, Black defendants were more likely than whites to be labeled as having a high probability of a violent re-offense and then not go on to commit such a re-offense, while whites were more likely to be labeled as having a low probability of re-offending but then going on to in fact re-offend.

This led to a public outcry, prompting many mathematicians and computer scientists to work on the problem. However, scholars quickly realized that their goal—of

8 See the Chouldechova paper (*https://arxiv.org/abs/1610.07524*) referenced in the preceding footnote.

making a risk score that demonstrated both calibration and statistical parity–was impossible with most real-world data. Statistical parity and calibration cannot be mathematically satisfied at the same time, unless base rates of events are the same in each group. However, equality of base rates is a condition rarely met in the real world.[9]

Mathematically, not all fairness definitions can be satisfied at the same time, given real world conditions. We have to decide which fairness goals to pursue in an ML context, and likewise decide whether focusing on a specific fairness metric might even tend to reduce the tendency toward a holistic viewpoint that might otherwise be more helpful in enhancing fairness and other ethical values.

For now, we consider that for a specific ML tool, it may make sense to choose and emphasize a particular fairness metric, while recognizing that it will not be possible to satisfy all intuitive and normatively desirable notions of fairness at the same time.

We do not want to allow perfection to become the enemy of the good. Rather, practitioners have found that having many definitions of fairness is more actionable than having none, and that working toward a metric of fairness is broadly likely to enhance many other forms of fairness as well. What's more, some researchers have recognized that different situations may call for different fairness metrics, given the relative harms or policy goals of a particular use case of an algorithm, and so interested readers can find useful guidance as to how to choose a fairness metric for a specific task given specific concerns about the likely consequences of various kinds of mistakes.

What Are Normative Values?

Normative values (or, in short, *norms*) are deliberate choices that individuals or societies make as to what is morally or socially desirable, and also as to what is not acceptable. Normative values are distinct from empirical realities. A normative statement is about what *should* happen in the world, while an empirical statement is about what *does* happen in the world.

9 If we look for and interpret questions about algorithmic fairness broadly, other impossibility theorems are related to fairness concerns. For example, Arrow's impossibility theorem demonstrates that three intuitive fairness criteria for a specific form of voting cannot all be met at the same time. These three criteria, stated quite roughly, are that (1) universally shared individual preferences between two options will necessarily translate into an election outcome reflecting that preference, (2) stable individual preferences will translate into stable election outcomes, and (3) no single voter will possess the power to determine the election outcome. Thus it is a common problem across disciplines and technological or organizational mechanisms that we cannot always have our fairness cake and eat it too. Many thanks to Niall Murphy for this example.

An empirical statement relating to algorithmic bias could be that algorithmic bias causes economic inefficiency because, for example, deserving job candidates are weeded out and thus lost as hiring opportunities. Such an empirical statement does not make a value judgment but simply observes the world as it is. In contrast, an example normative statement could be that algorithmic bias is wrong because all equally qualified people should have an equal chance at a job as a matter of fundamental fairness. This example normative statement doesn't look to describing the results of a social phenomenon but rather purports to judge that social phenomenon or offer proscriptions about what should be changed about that social phenomenon.

Of course, normative values can be difficult to articulate and defend when it comes to implementing them in practice. For example, the statement that all equally qualified people should have an equal chance at a job obscures the tricky question of what "equally qualified" means and whether jobs should always go to the most equally qualified applicant. There might be some other notion of a more deserving applicant in some scenarios. Consider that in a family business, many would consider it fairer to hire a reasonably well qualified family member rather than a slightly more qualified stranger. On the other hand, others would argue against this policy as entrenching nepotism and inequality. And so, we can see how different sets of similar but not identical normative values can lead to conflicts that will continuously surface in real-world debates and decisions about the best way to enhance fairness in society.

Reaching Fairness

Concretely, a classic ML setup has three modes of working toward fairer (less biased) outcomes. Here we provide a flavor of how these work, so you can approach the literature with an overview and understand the benefits and costs of different approaches:

Preprocessing

These methods intervene in data rather than in models. Preprocessing methods take multiple approaches to reducing unfairness in the inputs into ML training. Most bluntly, data points can be relabeled. For example, some approaches, such as that of Kamiran and Calders (2011) (*https://oreil.ly/I2fDD*), offer ways of identifying data that should be relabeled because the data suggests a biased outcome.[10]

Another, less drastic approach, is to seek a representation of the data that reduces information about the group to which an individual belongs; an approach pioneered by Zemel et al. (*https://oreil.ly/vW9fa*) (2013), for example, proposes describing the data such that an individual's sensitive attributes can no longer be guessed accurately. Because these methods look to the data as the point of intervention, they are model agnostic. A general rule of thumb is that *intervening*

10 For example, individuals from a favored group who receive an improbably favorable outcome given their merit.

at the earliest possible stage in the ML pipeline is most likely to yield best results (and leave options open for subsequent additional interventions).

But, there are, of course, valid concerns about preprocessing approaches. Changing the data labels aggressively challenges a data-driven discipline, such as ML, in fundamentally removing information. Likewise, methods that look to "transform" the data rather than directly manipulate the labels also challenge the fundamental tenets of a data-driven approach when data is deliberately removed or changed. Also it can be difficult to identify exactly what about the model changes as a result of changing the data, since in most cases it will not be possible to know all that clearly affects the relabeling a few data points.

In-processing

These methods intervene during model training. This can manifest in any way in which the actual step-by-step training of a model is affected by fairness considerations. In many cases, this has been addressed through adjusting the loss function used during model training. Various "penalties" can be added to reflect the fairness costs imposed by biased outcomes, as is reflected in the work of Kamishima et al. (*https://oreil.ly/rEc4P*) (2011). Such methods are similar to regularization techniques.

Another method is similar to that described in preprocessing (learned fair representations) in terms of motivation to remove identifying information from the model's knowledge about the sensitive attribute. An adversarial model is trained simultaneously with the model of interest, such that the adversarial model's target is to guess which sensitive category an output is associated with. The model of interest is trained to a task but also optimized to reduce the information that its outputs transmit to the adversarial model. Some of these methods, such as the work of Zhang et al. (*https://oreil.ly/SVmNt*) (2018), can be model agnostic.

Post-processing

These methods intervene on the model's labels rather than on the model directly. In this way, they correct outcomes from the model based on meeting certain targets. An examples of post-processing is introducing randomization, as in Hardt et al. (2016) (*https://oreil.ly/b9EhM*), in the case where—without such randomization—false negatives might be different among different groups (an example of applying group parity).

Another example of post-processing is to set different thresholds—say, for a credit score (*https://oreil.ly/DzYYL*) or college admissions scores—for different groups such that the predictions or decisions made with the score can be equally accurate for different groups.[11] Because these methods intervene after a model has run, they are model agnostic.

Which method of intervention to choose depends on a variety of factors. In some cases, an organization may need to apply a specific stage of intervention because that is the intervention over which they have control. For example, an organization may have received a pretrained neural network that it will fine-tune. Since the organization doesn't have access to the original training data or method, post-processing might prove a more viable option.

On the other hand, another organization might find that the options offered by post-processing are normatively problematic because they violate certain fundamental values held by some people. For example, people may be uncomfortable with the idea of having explicitly different score cutoffs for different groups, and they also might be uncomfortable with the notion of eliminating, and replacing with randomized numbers, those outputs from an algorithm that are, in fact, likely to be correct.

To date, there are no well-established guidelines from regulators or prominent ethical leaders regarding the best way to intervene,[12] and we suspect that this will have to be a highly context-sensitive analysis. But the need for highly contextualized case-by-case decision making is no different from other elements of running an organization or making decisions that affect other people's lives. Algorithms will help increase efficiency and uniformity, but there will never be a single go-to algorithmic fairness solution for all scenarios.

11 This can sometimes be understood as a way of correcting where calibration is lacking. For example, it is well documented that credit scores used in the US do not seem to mean the same thing across racial groups with respect to propensity to repay. So banks that rely exclusively on the score and do not adjust the thresholds could inadvertently label credit seekers with false-negative rates that are different by racial group. However, different thresholds for different groups necessarily raises other legal and ethical concerns. And so, the reader can see that fairness in the real world can be quite challenging indeed.

12 At the time this book is going to press (August 2022), Europe has put forth a wide range of groundbreaking legislation on a host of issues that relate to fairness in digital environments, including AI. Most notably, the EU published a proposed AI regulatory framework in April 2021 (*https://oreil.ly/b5DZn*) that would entail a risk-based set of regulations, with different legal requirements for different levels of risk from an AI use case. The proposed law would also ban certain high-risk use cases, including social credit scoring. Likewise, China brought sweeping new AI regulations (*https://oreil.ly/U5xVq*) into force as of March 2022 to regulate a variety of common practices that disadvantage consumers, including price discrimination on the basis of personal information and content-aggregation algorithms.

Algorithmic Fallbacks for Fairness?

As you will read about in other chapters of this book, sometimes an algorithm's performance will be obviously or catastrophically bad. Other times an algorithm may be unavailable. In such cases, you'll need a fallback. A technical fallback can be a helpful idea for fairness-related concerns too.

Here are a few scenarios where you might want to suddenly cut access to an algorithm and have an algorithmic fallback in place:

- A human user reports an algorithmic output so offensive to basic notions of fairness that the system should be completely unavailable until an audit determines the source of the problem and corrects that source.
- Continuous monitoring systems you have wisely put in place indicate a drift of results beyond acceptable fairness thresholds (for whichever fairness metric you have chosen).
- The algorithm recommends such a drastic (or possibly harsh) output that you send the output for human review prior to making the output available to the decision subject.

In these and other such cases, you could have a variety of algorithmic fallbacks in place, including the following:

- Temporarily using a nonalgorithmic solution. For example, consider giving everyone the same output—perhaps the average output from the algorithm—while you sort out the algorithmic issue.
- Shunting the algorithmic outputs to a human reviewer and having a prompt to "please wait until we can contact you."
- Radical transparency in which an appropriately tailored error message is provided: "Sorry our algorithm is having some trouble, and we don't want to give you irresponsible outputs. Check back later."

Fairness as a Process Rather than an Endpoint

This is a difficult set of topics, and it can be strongly discouraging to know that AI can and does cause so many harms. Don't let this bad news dissuade you from all of the advantages that can be achieved with a thoughtfully engineered ML/AI system. The news about fairness doesn't have to be bad.

Let's consider two problems:

- Fairness wise, you can't please everyone. There is no "perfectly" fair solution to any algorithmic fairness challenge (at least none has been identified so far).
- Likewise, there is no perfect way to enforce fairness even if you do settle on a particular definition of fairness (that is, defining a metric doesn't guarantee you can make a model perfectly conform to that metric or indicate how you should attempt to do so).

Let us rebut both of these points in terms of their relevance to your choices in your ML modeling:

- You have many good fairness definitions to choose from.
- Any process to work toward fairness is far better than a world in which we ignore these issues—because ignoring hasn't led to good outcomes.

It can be helpful to think of fairness as a process rather than a specific endpoint. While this can be disappointing to those who like to think they can build an ML algorithm to solve a business problem and move on, the reality is more complicated. Products take upkeep for many reasons. The world changes, and so deployment conditions change, and so models must evolve. Fairness is no different.

A Quick Legal Note

It is far beyond the scope of this work to provide legal guidance; that is what your company's legal department is for! Nonetheless, it is worth mentioning that despite the seeming newness of algorithmic fairness as a topic, the concept of discrimination has had an important and explicit place in law for centuries in one form or another. The topics most in the forefront today, such as racial and gender equality, have likewise occupied a prominent role in law for decades. This has especially been the case in areas of civic life that are understood as core to human life, such as employment, education, healthcare, housing, and access to financial credit services. If you are doing ML work that touches on these core areas, you likely need to be deeply concerned about fairness—and, specifically, about fairness as it is implemented in your nation's antidiscrimination laws.

Privacy

Privacy is a notion that has proven notoriously difficult to define in scholarship. Privacy measures have therefore proven difficult to create, particularly with respect to future-proofing. The rise of big data has proven a dramatic example of this.

Experts used to think that de-identified data was appropriate and safe to release. Such data would have removed what was thought to be identifying information, such as name or address, that could easily be matched to a person. However, information about other factors associated with, but seemingly not directly related to, a specific person's identity was still released, such as birthday, race, or zip code.

With the advent of big data, many datasets were compiled, often about overlapping groups of people, and it then became possible to use different datasets together to identify people from de-identified information. For example, for most Americans, it turned out to be possible to identify them in a de-identified dataset merely from knowing their birthday, zip code, and gender.[13] And, the world is increasingly full of new datasets, sometimes resulting from data breaches but other times resulting from people voluntarily sharing information about themselves—information that becomes very easy to access by any casual creepy stalker (*https://oreil.ly/AOVNs*). We might like to think the world is an anonymous place, but it simply isn't the case today.

Two key ideas about privacy are described here, both in terms of how they relate to a notion of privacy, but also in how they relate to taking an existing dataset with personal information and turning it into a dataset that can be used or released without compromising individual identity:

k-anonymity

The idea behind k-anonymity is that in a given dataset, for any given combination of categories of interest, there should be at least *k* individuals (externally specified) who fall into any given bucket. So, for example, we could apply k-anonymity to a dataset listing individuals in a town by requiring that data be bucketed such that for any zip code / birth information / gender category, there were at least 10 individuals. There might be many ways to accomplish this. Two potential opportunities would be to report birthdays at the month or year level and to report only the first four digits of a zip code rather than all five.

This notion of privacy and preventive measure essentially conceptualizes privacy as not being in too small a group within a dataset. Recommended size (*https://oreil.ly/zl6MO*) is at least 5 in the medical domain, although *k* sizes far larger (even greater than 50) have been reported. The appropriate size for *k* will depend on the particular domain, which could have different implications regarding sensitivity of the data, possibility to build useful datasets with large *k*, and possibility of reidentification given other known potentially linkable datasets.

13 For an early example of reidentification, see "'Anonymized' Data Really Isn't—And Here's Why Not" (*https://oreil.ly/7nFhr*) by Nate Anderson, describing Latanya Sweeney's work on the topic.

Differential privacy

This mathematical method adds noise to data such that probabilistic guarantees can be made regarding the possibility (or more importantly, lack of possibility) to make inferences about a specific individual when given access to that noisified data. The idea behind differential privacy is that it would be a privacy problem for one individual's inclusion, or non-inclusion, in a dataset to influence operations, such as the calculation of averages or other statistics, so that the dataset's information could be inferred based on aggregate reporting. So for example, if the mean age of a class is reported and the size of the group, with and without an individual, it is possible to know that individual's age. However, if differential privacy is applied, it would, in probability, not be possible to infer that individual's age at a pre-specified level of precision and given a pre-specified query budget.[14] Differential privacy can apply quite broadly, not just to the computation of aggregate statistics but also to the training of ML models, with methods to ensure that a model's outputs or training is not contingent on inclusion of a particular data point.

These notions of privacy may seem quite technical, but are relevant to real privacy problems, such as managing the responsible and privacy-preserving release of open datasets or ensuring that legal rights have a meaningful technical translation. For example, the EU's GDPR gives an individual a right to data deletion. However, if an ML model has already been trained, that right to deletion may not be entirely meaningful. For example, does an individual's right to delete their data mean that they can also force a company to prove that its ML models have also "forgotten" their data?[15]

Some have explored how training models with differential privacy might address this concern. However, technical challenges remain because producing ML models with differential privacy guarantees is quite a technical challenge. It is challenging to ensure both differential privacy and a high level of performance at the same time.[16]

14 Querying a differentially private dataset many times can lead to violations of the differential privacy guarantees. This is because such guarantees are valid with only a given privacy budget, which limits the number of queries that can be made while ensuring differential privacy guarantees.

15 In theory, we could retrain a model entirely every time a bit of training data was marked for deletion. In practice, that would be wildly wasteful and impractical—and also possibly unnecessary. More work needs to be done in this area to understand how technical needs, sustainability concerns, and individual rights to privacy can successfully coexist.

16 One reason that differentially private modes do not achieve the highest levels of accuracy relates to the methods used to create differentially private models. One key technique is noise injection, which by definition reduces accuracy.

Methods to Preserve Privacy

The definitions and scenarios described previously with respect to k-anonymity and differential privacy refer to very specific notions of privacy and computational measures. And, indeed, privacy is itself a highly technical and specialized field, perhaps best left to experts in terms of implementation.[17] However, a variety of accessible privacy-enhancing measures are transparent, straightforward to implement, and meaningful. You should include them in your own workflow and will likely need to customize them.

These are likely to already be familiar concepts to systems administrators and those who deal with compliance issues. However, these sometimes are painfully unfamiliar to data scientists and ML engineers. It is my hope that they will become part of basic project specification and daily practice and consideration in the future.

Technical measures

The following are some technical measures to take:

Access controls
: A key way to preserve privacy and reduce threats is to implement robust access controls. Any data about people in a database should be treated as a "need to know" resource, with ML engineers requesting access for specific purposes rather than being able to freely access or browse data. Likewise, access should be revisited periodically to make sure that staff members do not retain access to data for which they no longer have a valid active reason for access.

Access logging
: Keep track of who is accessing specific forms of data and when. This makes it possible to understand data use patterns, see when someone might be inappropriately accessing data, and preserve evidence in case allegations of inappropriate use are made later. An analysis of such logs may also indicate ways in which data storage schema could be refactored to reduce the extent of data that different use patterns can access. For example, if an ML model calls for access to a sensitive table of data merely to access one column, consider splitting off that column of data rather than granting access to a full table of additional but unnecessary pieces of information.

Data minimization
: The collection and use of data should be minimized. Data should not be collected merely because it "could be useful" in the future. Data should be logged only

[17] One recommended starting point is TensorFlow Privacy (*https://github.com/tensorflow/privacy*), which includes training algorithms for differentially private models.

when there is an immediate use case for this data (preferably, with some benefit to those about whom the data is collected).

Data separation

The data needed for legitimate business uses should be separated from sensitive data (such as names and addresses) that is unlikely to be relevant to creating an ML model. For example, to predict users' clicks, there doesn't seem to be a justification for knowing a user's name or address.[18] Therefore, there is no reason for that information to be stored with information that might be useful for that particular prediction task, such as past browsing history or demographic information.[19] As noted previously, studying your data access logs can help you identify ways in which data storage can be refactored to minimize exposure of data to ML applications.

Institutional measures

Here are some institutional measures available as well:

Ethics training

Everyone needs ethics training when they enter a new domain, and the same goes for designing ML products. ML engineers should be given—but usually are not—a thorough review not only of general ethics training (such as the possibility of bias in data) but also domain-specific training when building algorithms for a specific use case. Too often organizations do not have any formal discussions or training about privacy or ethics more generally, and even basic training, no matter how "corny," serves to bring the issues to the fore.

Data access guidelines

In addition to technical measures already described, it makes sense to have explicit rules that are readily available regarding what constitutes appropriate access to data and use of that data and what use cases are expressly prohibited. A lack of clear and explicit ethics rules can lead to an institutional culture without

---

18 A separate question is when click prediction is a useful activity and when it raises troubling AI concerns. No doubt, predicting someone's preferences or interests has beneficial uses, as is the case for many examples of click prediction. On the other hand, increasing research both on the mutability of human preferences generally and on the particular scenario of manipulation of behavior and preferences in online environments also points to the danger of confusing click prediction that represents preference *accommodation* (giving people what they want) with preference *manipulation* (making people want what you have). The latter obviously has lots of fairness concerns and is related to a growing literature on *dark patterns*, which are digital design patterns that tend to lead users of digital products to do things that are against their interests (but in the interests of the designers of those products).

19 Of course, the inclusion of demographic information may prove problematic, as some datasets may have a strong prediction relationship between seemingly anonymous information and PII, and this should be factored into the way data separation is conducted.

accountability. Organizations should have clear data access and appropriate data use guidelines in place in a location that is readily apparent and accessible.

Privacy by design
: Privacy by design is a set of design principles that can apply to any digital product, including ML pipelines and ML-driven products. The notion of privacy by design is that privacy is something that shouldn't be tacked on to the end of a process that already exists. Rather, privacy should be intrinsic to design considerations from the start and should be a question and concern addressed at all working stages. Privacy by design can provide a flexible but holistic way to ensure privacy in all elements of ML development.

While these may all seem like basic and obvious notions, few organizations, large or small, do these basic things, or even pursue anything akin to a privacy or fairness agenda. Whether you are at a startup, an academic institution, or a large corporation, there is almost surely something you can contribute to enhance privacy in your organizer's ML pipeline.

> **Data Access Guidelines**
>
> If you're in an organization that lacks formal data access guidelines, here's a short list to get you started:
>
> - No personal use of data (i.e., no looking up whether your cousin has ever ended up in the database)
> - No examining data of someone you know (same point about your cousin)
> - No analyses you wouldn't want to see the light of day or that could cause a PR problem (think of Uber's misjudged blog post (*https://oreil.ly/578qJ*) on "Rides of Glory")

In short, there are a variety of ways to protect privacy by taking concrete steps in your organization. Unfortunately, these steps are rarely taken, but they can be simple and effective. You have nothing to lose (and lots to gain) by initiating such simple efforts in your own workflow.

A Quick Legal Note

It is far beyond the scope of this work to give an extensive review of privacy laws that usually affect digital products. Here we highlight a few key categories of laws that are related to privacy and ML:

Data breach notification laws
: Data breach notification laws require those who hold data to notify people whose data they hold if that data was compromised in a data breach. To ensure compliance, such laws usually apply strict penalties if a holder of data becomes aware of a data breach and does not make the appropriate notifications. Such laws can sometimes also apply to a data holder who should have been aware of a data breach, to ensure that firms cannot simply choose to remain ignorant. Empirical research to date suggests that such laws have not prevented data breaches from becoming increasingly, even exponentially, more common over time. Nonetheless, such laws are useful in providing notice to consumers as to when they may be most at risk because of exposure of their personal data.

Data protection and personal privacy laws
: The most prominent example of data protection law is the EU's GDPR. Many countries around the world have data protection laws that give people basic rights, such as the right to know what data is collected about them by online venues, what the venues do with that data, and whether the data is even correct. Some laws even go further, giving consumers the power to opt out of data collection, or even giving consumers the right to have data deleted or blocked from data sales. Unfortunately, empirical research has shown that such laws appear to be widely disregarded in consumer-facing applications. Nevertheless, such laws give consumers active ways that they can take steps to protect their privacy.

Laws against unfair and deceptive practices
: This category of more general consumer protection law is important in the US context, since the US otherwise lacks a comprehensive national personal data privacy regime. The US Federal Trade Commission, an important source of consumer protection enforcement, has sometimes brought actions finding that companies infringed basic expectations of fairness or honesty in their business practices as a means of protecting consumer privacy. A key example of this is companies that do not even respect the terms of the privacy policies they themselves author and leave on their websites. Thus it is—at the least—essential to ensure that, as your organization builds out data collection and ML modeling capabilities, you ensure that such practices are consistent with public-facing privacy policies and terms of service.

Responsible AI

Responsible AI has come to be used as a catchall for ethical concerns we should contemplate when training or deploying an ML system. This is a growing area, and it's safe to assume that neither industry nor the academic community yet has a firm grasp on the scope of harms that should be, or can be, addressed by Responsible AI. Here, under this rubric, we address some additional questions that have received a good deal of attention in recent years. However, we emphasize that the issues here are highlights. We do not purport to offer an exhaustive list of Responsible AI values.

Explanation

ML model *explanation* is the process of analyzing and presenting information about an ML system to describe how that system works. This process and the goal of making the model amenable to human understanding is often discussed in shorthand as *explainability*, and it's a key area of interest with respect to ML. Many people have a desire to understand why an ML system is working the way that it does as part of understanding why it has reached a particular outcome that affects them.[20]

For both technical and ethical reasons, it is desirable to have methods that can "explain" how an ML model works or why it reached a particular outcome in a particular case. The technical motivation for explainable AI is related to controlling model quality and possibly learning about the data through the model. Technologists who develop explanations of their models can gain insights into why their model is working well or poorly, and may even learn something about the underlying domain of the data. Ethicists and advocates who seek explainable ML have similar concerns but for different reasons. They find it empowering to know how a model came to a conclusion, so that the conclusion can make sense to those affected by it, or even be challenged where the ML conclusion doesn't make sense.

However, explanation is not a simple thing. Explanations can serve any of a number of purposes and therefore quite a wide variety of information can constitute an explanation, depending on the purpose for which it is sought and the audience for whom it is prepared. In this way, model explanation can feel a lot like thinking through the problems that highly trained professions, such as doctors or lawyers, face in trying to give advice or a diagnosis of a real-world situation to a particular audience. For example, when doctors explain a recommended treatment and its associated risks, they will tune the explanation to their audience. They'd probably give one explanation to a fellow physician, another less technical but still rigorous explanation to a patient known to have a bioengineering degree, and still another to a person who was suffering from dementia but who might still be competent to make their own medical decisions. Thus, we can see that an explanation depends on the *audience* or *consumer* of that explanation.

The explanation that is provided also depends on its *purpose*. If the purpose of an ML explanation is to inspect model quality, such as to make sure the model is making the right decisions for the right reasons, then a global explanation might be preferred. A

20 A good starting point for understanding the centrality of explainability to Responsible AI, but also the complexity of getting it right, is "Explaining Explanations in AI" (*https://oreil.ly/OOeoT*) by Brent Mittelstadt et al.

global explanation indicates how the model works generally and why certain general decision rules will be followed. On the other hand, if the purpose of the explanation is to enable a specific person who was refused credit (or had another undesired outcome) to know why in a way that could help her improve her chances in the future, that person would want a local explanation. A *local* explanation explains why that person in particular was refused and the most actionable adjustments to possibly make in the future.

We won't get into more detail on explanations for the moment. For our purposes here, we want you to walk away understanding these key points:

- There is no one single correct explanation of a model.
- Explanations need to be tuned to audience and purpose.
- Explanations need to be useful and sometimes actionable.

In terms of what you should do concretely, the following could be the minimum steps to take at the start of your journey to get better acquainted with ML explanation techniques:

- At the least, it is helpful for an end user to know the inputs used for an algorithm (even this minimal information is often not available). It is even better if you can provide a list of relative feature importances. This would be most meaningful if placed in a prominent location and in accessible language so that ordinary people can see and understand the model.
- Another easy way to get some intuitive and concrete information that can be explanatory is to generate test inputs, perhaps counterfactual pairings, and see what these look like. For example, some counterfactual pairings shouldn't matter for some use cases (e.g., gender counterfactuals should not change credit decisions), while others most certainly should (e.g., body weight counterfactuals probably will often change medical intervention decisions). These can constitute a basic smell test as well as a way of providing example explanations to those who are the decision subjects.
- Contemplate whether you can offer global (explaining the model overall) or local (explaining a particular ML model decision/classification) explanations and explore a few techniques for doing so. Is a certain technique particularly appropriate for your audience and the kind of decisions your ML product is making?

As various explanation techniques have emerged, some researchers have offered useful guidance as to what systems might serve different purposes and be appropriate to use.[21] The most important consideration is to determine the level of sophistication of the end user of the explanation and the purpose of providing an explanation. From there, you will often be able to identify at least one, and usually more, technical options, some open source and already implemented by experts.[22]

Effectiveness

ML *effectiveness*—that is, an ML product actually achieving its desired target, and for the right reasons—is key to responsibly deploying ML. Yet, as highlighted quite famously by Cathy O'Neil's *Weapons of Math Destruction* (Crown, 2016), a particularly worrying element of ML is that in many deployment scenarios, ML can be a self-fulfilling prophecy: ML might appear to work, but this could be for entirely wrong reasons.

Consider, as O'Neil did, ML products for hiring purposes, in which candidates can be flagged to be rejected automatically. Perhaps the ML algorithm is correct that someone would be a bad hire, but, also quite likely, we may never know whether someone was a good hire because we didn't hire them. But if a variety of ML algorithms are all using the same logic to not hire a particular candidate, that candidate never gets the chance at a job, and we never actually know whether that candidate could have done a good job (systematically missing data).

If someone is labeled by a ML algorithm in a particular way, sometimes that label is trusted to settle the issue, even if humans are supposed to be exercising some level of supervision.[23] In the employment scenario, someone could be labeled a poor candidate and not be hired by the human who is using the algorithmic assistance. That job seeker is perhaps not hired only because the algorithm labeled them as a poor option rather than because they actually were a poor option. And perhaps the use of the same or similar algorithms across many potential employers makes the situation even worse. Perhaps that job seeker will face the same algorithm deployed by many potential employers, thus facing an extended or even indefinite period of joblessness. At some point, their extended period of unemployment will itself become another factor that an ML algorithm is likely to use as a flag against hiring someone.

---

21 "Explainable Machine Learning for Public Policy: Use Cases, Gaps, and Research Directions" (*https://arxiv.org/pdf/2010.14374.pdf*) by Kasun Amarasinghe et al. provides a great and accessible example of considerations that go into what kind of explanation is likely to be useful.

22 IBM's AI Explainability 360 library (*https://aix360.mybluemix.net*) is an easy-to-use open source library that includes a wide variety of cutting-edge research methods for model explanation. The toolkit provides an API as well as tutorials that provide a wide range of example use cases for applying methods from the library via the explainability API.

23 Consider as an extreme example Poland's rollout of a classification algorithm to categorize unemployed job seekers into three categories. An initial algorithmic assessment was made, from which staff could, in theory, deviate. In practice, they deviated from the algorithmic label in only 0.58% of cases. See "Profiling the Unemployed in Poland" (*https://oreil.ly/eBvcv*) by Jędrzej Niklas et al.

This concern isn't just limited to scenarios that O'Neil identifies in potential anecdotes. Lawmakers too are concerned. For example, in the recently proposed Algorithmic Justice and Online Platform Transparency Act of 2021 (*https://oreil.ly/iVwyj*), US Senator Ed Markey and US Representative Doris Matsui have proposed that only safe and effective algorithms be legal. The effectiveness of an algorithm would be established by showing that the ML algorithm "has the ability to produce its desired or intended result."[24]

Given this law, or similar requirements of efficacy, we can think through how O'Neil's example of self-fulfilling hiring algorithms could play out. The designer of a hiring ML algorithm should show that the algorithm actually predicted who would be a good employee, rather than simply who was unlikely to be hired. One way of thinking about this is *external validity*—that is, the notion that the findings or conclusions that power an algorithm translate to a general and real-world concept that matters. A requirement of showing efficacy is akin to a requirement that the algorithm's logic is externally valid, that it generalizes to some reasonable degree.

Self-fulfilling algorithms are, of course, not the only concern related to algorithmic effectiveness. Some other terms related to this concept are briefly described here:

Robustness
: Algorithms can be effective only where they are resistant to foreseeable attacks or misuse and are designed to limit or prevent such foreseeable abuse.

Validated performance
: Models must work in their deployed use cases. Models should be checked for good performance anytime they are deployed in a new situation, such as on a new population, or even simply being applied over a lengthy period of time.

Logic
: While some celebrate ML on the basis that such models can "find patterns that humans can't see," sometimes the patterns identified don't make any sense, or make sense for the wrong reason. If the logic of a particular input's relevance doesn't pass a basic smell test, it's time to ask questions. Requiring some degree of logic is also consistent with Responsible AI goals related to efficacy.

Social and Cultural Appropriateness

Another element of consideration for the responsible use of ML relates to the acceptable use of technology more generally in social situations or with social ramifications. Are all roles, as observer or decider, appropriate for a machine? For example, would

24 The full quotation from the bill is, "An algorithmic process is effective if the online platform employing or otherwise utilizing the algorithmic process has taken reasonable steps to ensure that the algorithmic process has the ability to produce its desired or intended result."

you want to bc told that you were going to die of a terrible illness…by an algorithm? Likewise, do you want your child "watched" by an ML product or by a human babysitter? Perhaps the answer is that you don't care, and perhaps the answer is that you do. Sometimes, when people refuse algorithmic products it is because of ideas related to human dignity or social respect, rather than safety concerns about an algorithmic product. That is, just because something can be automated well doesn't guarantee that people will feel respected when they are interacting with an algorithm.

These cultural concerns about human dignity can apply to the users of an ML product, rather than merely the subjects. Consider, for example, a recent Twitter Responsible AI study in which the team ultimately concluded that the best way to make a feature fair (in this case, a photo-cropping feature) was to remove (*https://oreil.ly/YsWXs*) the feature rather than refine it, in part to encourage autonomy and agency by those who were posting photos. Sometimes the best solution is the remove a technical "solution."

Responsible AI Along the ML Pipeline

The various specific concerns related to Responsible AI that we've discussed previously will necessarily overlap in the course of a real-world ML pipeline. In this section, we include specific points to consider relating to pragmatic questions you should pose to yourself and your team, depending on where along the ML pipeline you find yourself.

Use Case Brainstorming

If you are brainstorming potential use cases, perhaps because you see a new business opportunity or see the possibility of obtaining new data, you should be thinking about the following fundamental questions related to your potential project:

- Is this a use case that will undermine privacy, and if so, how will you take precautions from the start to build in privacy protections?
- Is this a use case that touches on fundamental concerns about human dignity or social expectations, which could add additional limitations to the scope of appropriate use of algorithmic approaches?
- Is the decision or classification to be made by ML an important decision, where fairness should be particularly guarded, and if so, are there indicators that this can be accomplished? What constitutes *important* will vary, but sometimes these are understood to be decisions with legal effect or legal-like effects (such as hiring or education). Another way of defining *important* would be to first identify those models in your own organization as areas where a mistaken decision would have the most significant impact for a decision subject.

Data Collection and Cleaning

At this point in the pipeline, you have decided on a use case and are looking for data and preparing data for modeling. Now, the following concerns should be addressed:

- Will data be acquired in a way that respects informed consent?[25] Have you disclosed the purposes for which you will use the data to the subjects in a reasonably informative and transparent way?
- Have you stored data in a manner that promotes privacy and minimizes the likelihood of unintentional disclosures?
- Have you done exploratory analyses of the data to look for potential bias in that data?

Model Creation and Training

Now you are in receipt of data, and it's time to do some modeling:

- Have you come up with an affirmative plan to monitor and address bias? How will you choose from various forms of fairness interventions based on the potential harms of bias and based on the particular normative values or legal restrictions of your use case?
- Are you training in a manner that will reduce data leaks from the trained model and enhance robustness against malicious attacks (*https://arxiv.org/abs/1412.6572*)?
- Have you considered in your loss function the relative degree of harm from different mistakes, rather than lazily setting all mistakes to the same loss value?
- Have you chosen your model architecture (say, opaque neural network versus interpretable linear function) by understanding the relative importance of accuracy and explainability for your particular application?
- If your domain has something like scientific laws with strong predictive value, have you included this domain knowledge in your model architecture and training choices?

Model Validation and Quality Assessment

At this stage of the pipeline, you may receive a model that you are told is as good as it is going to be in terms of accuracy. Your job is to kick the tires in a reasoned way and decide whether to give this model approval to go forward, or send it back to training:

---

25 It's worth noting that defining *consent* can be complicated. After all, no one appreciates all the pop-up consent notices that have proliferated after GDPR. *Consent* here should be interpreted in a broad rather than narrow sense of the word.

- Have you asked whether the model was trained on a proxy, and if so, what data is available to justify the use of that proxy?
- Have you tested the model robustly, with a fair selection of held-out data in realistically challenging situations? Basic accuracy remains an ethical obligation as well as a business target and technical measure.
- Are you able to identify and understand the logic driving the model globally and to generate individual explanations in case an individual might ask for them? (See Chapter 9 for more information.)

Model Deployment

Now it's time to make the model available for its planned use:

- Have you put in place monitoring programs to continually assess the performance of the system in actual use? Such performance could be assessed with respect to traditional performance metrics (such as accuracy) as well as with respect to fairness metrics (as discussed earlier).
- Have you established ex ante criteria to assess whether the model is working as expected?[26]
- Have you tried running the model in an online mode so that you can watch how it performs, counterfactually, in advance of an actual product launch? Or in the alternative, you can maintain a shadow model, which makes predictions on productive data even when users cannot see the results. This can be an intermediate step to understanding likely deployment performance without running the risks associated with actual deployment.

Products for the Market

Whether models are intended for internal or external use, they need to meet the same standards for fairness and privacy protections. But models that will be directly accessed by external users generally have an additional set of requirements that must be met.

If the end goal for your model is to be directly accessed by human users, consider the following:

- Will you make available a recourse or method to challenge a decision, and if so how?

26 It's good to specify these in advance so you can't "cheat" or otherwise rationalize poor performance after the fact. This doesn't have to involve rigid adherence to those criteria, but it will keep you honest about what constitutes a good job even if you do reassess those criteria over time.

- Will you make available explanations about how the ML system works, possibly even specific guidance for individuals about how their outcomes were formed?[27]
- How will you detect events or problems you haven't anticipated? How will you learn about what you currently do not know that you do not know?

Conclusion

We have reviewed many fairness, privacy, and other ethical considerations that affect the design, training, and deployment of real-world ML systems. These topics are all highly complex. This chapter should give you and your organization a sampling of key issues you should factor into designing and deploying ML systems. To progress further, many excellent learning materials and academic research papers are out there on all these topics for you to look for more information.[28]

In the meantime, this complexity shouldn't discourage you! We recommend the following steps:

- Create basic institutional rules and safeguards in your workplace.
- Human-readable changes can be easier to understand and more transparent than computational techniques.
- Institutional changes encourage people to bring up ethical concerns at meetings. It's better for fairness not to be off in a dark corner somewhere. Try raising your hand and mentioning a fairness concern you have. Start with a small one (data access, a recourse interface, statistical parity), and you can work your way up to more sophisticated targets over time. Over time, organizations can move even further—for example, by creating a *red team* that intentionally and proactively tries to manipulate products or identify harmful outcomes.
- Ethical guidelines can also serve as a high-level but practical checklist for initiating new projects or approving the launch of completed products.
- Address fairness, privacy, and ethics concerns routinely at product inception.
- Most fairness, privacy, and other ethical problems in ML originate at the level of product conception and creation. Start asking relevant questions when contemplating whether a project is even a good idea at the outset.
- Are you using fair data that you obtained in a legal and ethical way? Guidance from Gebru et al. (2021) (*https://oreil.ly/6Vo2R*) can provide a great framework for documenting your data and its appropriate use.

---

27 Model Cards, as previously mentioned, are one system to accomplish just this.

28 As a shameless self-promotion, consider Aileen Nielsen's book-length introduction to hands-on ML ethics: *Practical Fairness* (*https://oreil.ly/tsjGP*), another O'Reilly title.

- Are appropriate working conditions in place to maintain the security and privacy of data?
- If you succeed at the goal you propose, will that be a good (or at least neutral) thing for the world at large?
- Create a virtuous cycle of Responsible AI practices.
- The more you learn and the longer you keep Responsible AI concerns in mind, the more your ML pipeline and ML offerings will reflect good fairness practices.
- If you can make a commitment to learning a little about Responsible AI each week—and implementing it slowly in your own work—at the end of a year you'll note a remarkable degree of progress.

We wish you good luck as you begin practical steps to make the world a better place, starting with your ML products (including, sometimes, not using them).

CHAPTER 7
Training Systems

ML training is the process by which we transform input data into models. We take a set of input data, almost always preprocessed and stored in an efficient way, and process it through a set of ML algorithms. The output is a representation of that data, called a *model*, that we can integrate into other applications. For more details on what a model is, see Chapter 3.

A *training algorithm* describes the specific steps by which software reads data and updates a model to try to represent that data. A *training system*, on the other hand, describes the entire set of software surrounding that algorithm. The simplest implementation of an ML training system is on a single computer running in a single process that reads data, performs some cleaning and imposes some consistency on that data, applies an ML algorithm to it, and creates a representation of the data in a model with new values as a result of what it learns from the data. Training on a single computer is by far the simplest way to build a model, and the large cloud providers do rent powerful configurations of individual machines. Note, though, that many interesting uses of ML in production process a significant amount of data and as a result might benefit from significantly more than one computer. Distributing processing brings scale but also complexity.

In part, because of our broad conception of what an ML training system is, ML training systems may have less in common with one another across different organizations and model builders than any other part of the end-to-end ML system. In Chapter 8, you will see that even across different use cases, many of the basic requirements of a serving system are broadly similar: they take a representation of the model, load it into RAM, and answer queries about the contents of that model sent from an application. In serving systems, sometimes that serving is for very small models (on phones, for example). Sometimes it is for huge models that don't even all fit on a single computer. But the structure of the problem is similar.

In contrast, training systems do not even necessarily live in the same part of our ML lifecycle (see Figure 1-1). Some training systems are closest to the input data, performing their function almost completely offline from the serving system. Other training systems are embedded in the serving platform and are tightly integrated with the serving function. Additional differences appear when we look at the way that training systems maintain and represent the state of the model. Because of this significant variety of differences across legitimate and well-structured ML training systems, it is not reasonable to cover all of the ways that organizations train models.

Instead, this chapter covers a somewhat idealized version of a simple, distributed ML training system. We'll describe a system that lives in a distinct part of the ML loop, next to the data and producing artifacts bound for the model quality evaluation system and serving system. Although most ML training systems that you will encounter in the real world will have significant differences from this architecture, separating it out will allow us to focus on the particularities of training itself. We will describe the required elements for a functional and maintainable training system and will also describe how to evaluate the costs and benefits of additional desirable characteristics.

Requirements

A training system requires the following elements, although they might appear in a different order or combined with one another:

Data to train on
: This includes human labels and annotations if we have them. This data should be preprocessed and standardized by the time we use it. It will usually be stored in a format that is optimized for efficient access during training. Note that "efficient access during training" might mean different things depending on our model. The data should also be stored in an access-protected and policy-enforcing environment.

Model configuration system
: Many training systems have a means of representing the configuration of an individual model separate from the configuration of the training system as a whole.[1] These should store model configurations in a versioned system with some metadata about the teams creating the models and the data used by the models. This will come in extremely handy later.

1 In many modern frameworks (notably, TensorFlow, PyTorch, and JAX), the configuration language used is most commonly actual code, usually Python. This is a significant source of headaches for newcomers to the ML training system world, but does offer advantages of flexibility and familiarity (for some).

Model-training framework

Most model creators will not be writing model-training frameworks by hand. It seems likely that most ML engineers and modelers will eventually be exclusively using a training systems framework and customizing it as necessary. These frameworks generally come with the following:

Orchestration

Different parts of the system need to run at different times and need to be informed about one another. We call this *orchestration*. Some of the systems that do this include the following two elements as well, but these functions can be assembled separately, so they are broken out here.

Job/work scheduling

Sometimes part of orchestration, job scheduling refers to actually starting the binaries on the computers and keeping track of them.

Training or model development software

This software handles the ordinary boilerplate tasks usually associated with building an ML model. Common examples right now include TensorFlow, PyTorch, and many others. Disagreements rivaling religious wars start over which of these is best, but all of them accomplish the same job of helping model developers build models more quickly and more consistently.

Model quality evaluation system

Some engineers don't think of this as part of the training system, but it has to be. The process of model building is iterative and exploratory. Model builders try out ideas and discard most of them. The model quality evaluation system provides rapid and consistent feedback on the performance of models and allows model builders to make decisions quickly.

This is the most commonly skipped portion of a training system but really is mandatory.

If we do not have a model quality evaluation system, each of our model developers will build a more ad hoc and less reliable one for themselves and will do so at a higher cost to the organization. This topic is covered much more extensively in Chapter 5.

Syncing models to serving

The last thing we do with a model is send it to the next stage, usually the serving system but possibly another kind of analysis system.

If we have a system that provides for these basic requirements, we will be able to offer a minimally productive technical environment to model developers. In addition to these basic elements, though, we will want to add infrastructure specifically geared toward reliability and manageability. Among these elements, we should include some careful thoughts about monitoring this multistage pipeline, metadata about team ownership of features and models, and a full-fledged feature storage system.

Basic Training System Implementation

Figure 7-1 depicts a proposed architecture for a simple but relatively complete and manageable ML training system.[2]

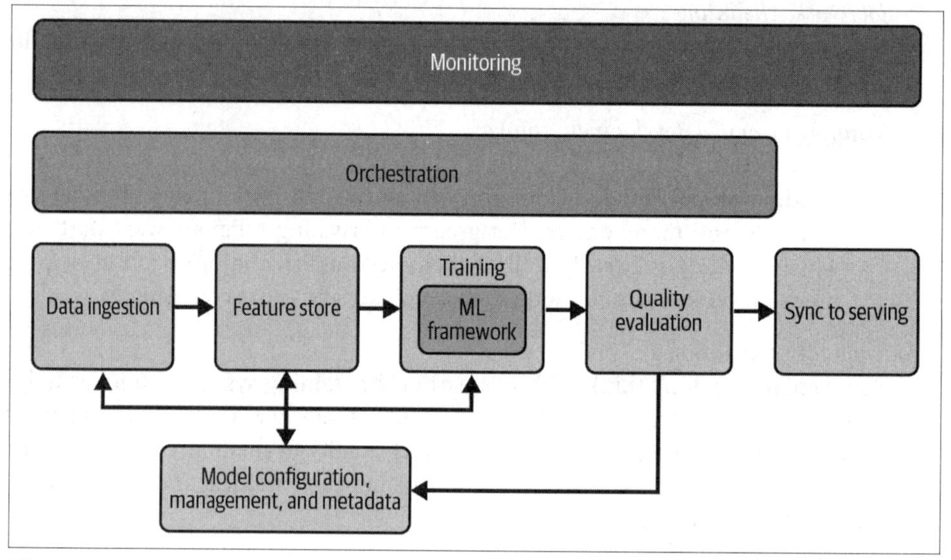

Figure 7-1. Basic ML training system architecture

In this simplified training system, the data flows in from the left, and models emerge on the right. In between we clean up, transform, and read the data. We use an ML framework that applies a training algorithm to turn the data into a model. We evaluate the model that we just produced. Is it well formed? Is it useful? Finally, we copy a servable version of that model into our serving system so we can integrate it into our application. All the while, we keep track of our models and data in a

2 Several types of ML systems (notably, reinforcement learning systems) are quite different from this architecture. They often have additional components, like an agent and simulation, and also put prediction in what we call "training" here. We're not ignorant of these differences, but chose the most common components to simplify this discussion. Your system may have these components in a different order or might have additional ones.

metadata system, we make sure the pipeline continues to work, and we monitor the whole thing. Next, we'll go into detail about the roles of each of these components.

Features

Training data is data about events or facts in the world that we think will be relevant to our model. A *feature* is a specific, measurable aspect of that data. Specifically, features are those aspects of the data that we believe are most likely to be useful in modeling, categorizing, and predicting future events given similar circumstances. To be useful, features need a consistent definition, consistent normalization, and consistent semantics across the whole ML system, including the feature store, training, quality evaluation, and serving. For significantly more detail, see Chapter 4.

If we think of a feature for YarnIt purchase data like "purchase price," we can easily understand how this can go badly if we're not careful. First of all, we probably need to standardize purchases in currency, or at least not mix currencies in our model. So let's say we convert everything to US dollars. We need to guarantee that we do that conversion based on the exchange rate at a particular point in time—say, the closing rate at the end of the trading day in London for the day that we are viewing data. We then have to store the conversion values used in case we ever need to reconvert the raw data. We probably should normalize the data, or put it into larger buckets or categories. We might have everything under $1 in the 0th bucket, and $1–$5 in the next bucket, and so on in $5 increments. This makes certain kinds of training more efficient. It also means that we need to ensure we have standard normalization between training and serving and that if we ever change normalization, we update it everywhere carefully in sync.

Features and feature development are a critical part of how we experiment when we are making a model. We are trying to figure out which aspects of the data matter to our task and which are not relevant. In other words, we want to know which features make our model better. As we develop the model, we need easy ways to add new features, produce new features on old data when we get a new idea for what to look for in existing logs, and remove features that turned out to not be important. Features can be complicated.

Feature Store

We need to store the features and, not surprisingly, the most common name for the system where we do that is the *feature store*. The characteristics of a feature store will exist, even if our model training system reads raw data and extracts features each time. Most people will find it convenient, and an especially important reliability bonus, to store extracted features in a dedicated system of some kind. This topic is covered extensively in Chapter 4.

One common data architecture for this is a bunch of files in a directory (or a bunch of buckets in an object storage system). This is obviously not the most sophisticated data storage environment but has a huge advantage, giving us a fast way to start training, and it appears to facilitate experimentation with new features. Longer term, though, this unstructured approach has two huge disadvantages.

First, it is extremely difficult to ensure that the system as a whole is consistently functioning correctly. Systems like this with unstructured feature-engineering environments frequently suffer from training-serving feature skew (whereby features are defined differently in the training and serving environments) as well as problems related to inconsistent feature semantics over time, even in the training system alone.

The second problem is that unstructured feature-engineering environments can actually hinder collaboration and innovation. They make it more difficult to understand the provenance of features in a model and more difficult to understand who added them, when, and why.[3] In a collaborative environment, most new model engineers will benefit enormously from understanding the work of their predecessors. This is made easier by being able to trace backward from a model that works well, to the model definition, and ultimately to the features that are used. A feature store gives a consistent place to understand the definition and authorship of features and can significantly improve innovation in model development.

Model Management System

A *model management system* can provide at least three sets of functionality:

Metadata about models
: Configuration, hyperparameters, and developer authorship

Snapshots of trained models
: Useful for bootstrapping new variants of the same model more efficiently by using transfer learning, and tremendously useful for disaster recovery when we accidentally delete a model

Metadata about features
: Authorship and usage of each feature by specific models

While these functions are theoretically separable, and are often separate in the software offerings, together they form the system that allows engineers to understand the

3 Worse, still, when provenance cannot be tracked, we will have governance problems (compliance, ethics, legal). For example, if we cannot prove that the data that we trained on is owned by us or licensed for this use, we are open to claims that we misused it. If we cannot demonstrate the chain of connection that created a dataset, we cannot show compliance with privacy rules and laws.

models in production, how they are built, who built them, and what features they are built upon.

Just as with feature stores, everyone has a rudimentary form of a model management system, but if it amounts to "whatever configuration files and scripts are in the lead model developer's home directory," it may be appropriate to check whether that level of flexibility is still appropriate for the needs of the organization. There are reasonable ways to get started with model management systems so that the burden is low. It does not need to be complicated but can be a key source of information, tying serving all the way back through training to storage. Without this data, it's not always possible to figure out what is going wrong in production.

Orchestration

Orchestration is the process by which we coordinate and keep track of all the other parts of the training system. This typically includes scheduling and configuring the various jobs associated with training the model as well and tracking when training is complete. Orchestration is often provided by a system that is tightly coupled with our ML framework and job/process scheduling system, but does not have to be.

For orchestration here, think of Apache Airflow as an example. Many systems are technically workflow orchestration systems but are focused on building data analytics pipelines (such as Apache Beam or Spark, or Google Cloud Dataflow). These typically come with significant assumptions about the structure of your tasks, have additional integrations, and have many restrictions built in. Note that Kubernetes is not a pipeline orchestration system: Kubernetes has a means of orchestrating containers and tasks that run in them, but generally does not by itself provide the kinds of semantics that help us specify how data moves through a pipeline.

Job/process/resource scheduling system

Everyone who runs ML training pipelines in a distributed environment will have a way of starting processes, keeping track of them, and noticing when they finish or stop. Some people are fortunate enough to work at an organization that provides centralized services, either locally or on a cloud provider, for scheduling jobs and tasks. Otherwise, it is best to use one of the popular compute resource management systems, either open source or commercial.

Examples of resource scheduling and management systems include software such as the previously mentioned Kubernetes, although it also includes many other features such as setting up networking among containers and handling requests to and from containers. More generally and more traditionally, Docker could be regarded as a resource scheduling system by providing a means of configuring and distributing virtual machine (VM) images to VMs.

ML framework

The *ML framework* is where the algorithmic action is. The point of ML training is to transform the input data into a representation of that data, called a *model*. The ML framework we use will provide an API to build the model that we need and will take care of all of the boilerplate code to read the features and convert them into the data structures appropriate for the model. ML frameworks are typically fairly low level and, although they are much discussed and debated, are ultimately quite a small part of the overall ML loop in an organization.

Quality Evaluation

The ML model development process can be thought of as continuous partial failure followed by modest success. It is essentially unheard of for the first model that anyone tries to be the best, or even a reasonably adequate, solution to a particular ML problem. It necessarily follows, therefore, that one of the essential elements of a model-training environment is a systematic way to evaluate the model that we just trained.

At some level, model quality evaluation has to be extremely specific to the purposes of a particular model. Vision models correctly categorize pictures. Language models interpret and predict text. At the most basic level, a model quality evaluation system offers a means of performing a quality evaluation, usually authored by the model developer, and storing the results in a way that they can be compared to previous versions of the same model. The operational role of such a system is ultimately to be sufficiently reliable that it can be an automatic gate to prevent "bad" models from being sent to our production serving environment.

Evaluation starts with factors as simple as verifying that we are loading the right version of the right model and ensuring that the model loads in our model server. Evaluation also must include performance aspects to make sure that the model can be served in the memory and computational resources that we have available. But also, we have to care about how the model performs on requests that we believe to be representative of the requests we will get. For significantly more detail on this topic, see Chapter 5.

Monitoring

Monitoring distributed data processing pipelines is difficult. The kinds of straightforward things that a production engineer might care about, such as whether the pipeline is making sufficiently fast progress working through the data, are quite difficult to produce accurately and meaningfully in a distributed system. Looking at the oldest unprocessed data might not be meaningful because there could be a single old bit of data that's stuck in a system that is otherwise done processing everything.

Likewise, looking at the data processing rate might not be useful by itself if some kinds of data are markedly more expensive to process than others.

This book has an entire monitoring chapter (Chapter 9). Harder questions will be tackled there. For this section, the single most important metric to track and alert on is training system throughput. If we have a meaningful long-term trend of how quickly our training system is able to process training data under a variety of conditions, we should be able to set thresholds to alert us when things are not going well.

General Reliability Principles

Given this simple, but workable, overall architecture, let's look at how this training system should work. If we keep several general principles in mind during the construction and operationalization of the system, things will generally go much better.

Most Failures Will Not Be ML Failures

ML training systems are complex data processing pipelines that happen to be tremendously sensitive to the data that they are processing. They are also most commonly distributed across many computers, although in the simplest case they might be on a single computer. This is not a base state likely to lead to long-term reliability, and production engineers generally look at this data sensitivity for the most common failures. However, when experienced practitioners look at the experienced failures in ML systems over time, they find that most of the failures are not ML specific.[4] They are software and systems failures that occur commonly in this kind of distributed system. These failures often have impact and detection challenges that are ML specific, but the underlying causes are most commonly not ML specific.

Amusing examples of these failures include such straightforward things as "the training system lost permission to read the data so the model trained on nothing," and "the version that we copied to serving wasn't the version we thought it was." Most of them are of the form of incorrectly monitored and managed data pipelines. For many more examples, see Chapters 11 and 15.

To make ML training systems reliable, look to systems and software errors and mitigate those first. Look at software versioning and deployment, permissions and data access requirements, data updating policies and data organization, replication systems, and verification. Essentially, do all of the work to make a general distributed system reliable before beginning any ML-specific work.

4 For example, see "How ML Breaks: A Decade of Breaks for One Large ML Pipeline" (*https://oreil.ly/Y1tk8*) by Daniel Papasian and Todd Underwood.

Models Will Be Retrained

Perhaps this section should be titled something even stronger: models must be *retrained*. Some model developers will train a model from a dataset once, check the results, deploy the model into production, and claim to be done. They will note that if the dataset isn't changing and the purpose of the model is achieved, the model is good enough and there is no good reason to ever train it again.

Do not believe this. Eventually, whether in a day or a year, that model developer or their successor will get a new idea and want to train a different version of the same model. Perhaps a better dataset covering similar cases will be identified or created, and then the model developers will want to train on that one. Perhaps just for disaster-recovery reasons you'd like to prove that if you delete every copy of the model by mistake, you can re-create it. You might simply want to verify that the toolchain for training and validation is intact.

For all of these reasons, assume every model will be retrained and plan accordingly—store configs and version them, store snapshots, and keep versioned data and metadata. This approach has tremendous value: most of the debates about so-called "offline" and "online" models are actually debates about retraining models in the presence of new data. By creating a hard production requirement that models can be retrained, the technical environment is much of the way to facilitating periodic retraining of all models (including rapid retraining).[5]

Assume every model will be retrained.

Models Will Have Multiple Versions (at the Same Time!)

Models are almost always developed in cohorts, most obviously because we will want to train different versions with different hyperparameters. One common approach is to have a named model with multiple minor changes to it. Sometimes those changes arrive in a rapid cluster at the beginning, and other times they arrive over time. But just as models will be retrained, it is true that they will be changed and developed

5 This recommendation has one interesting exception, but one that will not apply to the vast majority of practitioners: huge language models. Multiple very large language models are being trained by large ML-centric organizations in order to provide answers to complex queries across a variety of languages and data types. These models are so expensive to train that the production model for them is explicitly to train them once and use them (either directly or via transfer learning) "forever." Of course, if the cost of training these models is significantly reduced or other algorithmic advances arise, these organizations may find themselves training new versions of these models anyway.

over time. In many environments, we will want to serve two or more versions of the same model at the same time in order to determine how the different versions of the model work for different conditions (for those familiar with traditional web development for user experience, this is essentially A/B testing for models).

Hosting simultaneous versions of the same model requires specific infrastructure. We need to use our model management infrastructure to keep track of model metadata (including things like the model family, model name, model version, and model creation date). We also need to have a system to route a subset of lookups to one version of the model versus another version.

Good Models Will Become Bad

We should assume that the models we produce will be hard to reproduce and will have subtle and large reliability problems in the future. Even when a model works well when we launch it, we have to assume that either the model or the world might change in some difficult-to-predict way that causes us enormous trouble in future years. Make backup plans.

The very first backup plan is to make a non-ML (or at least "simpler-ML") fallback path or "fail-safe" implementation for our model. This is going to be a heuristic or algorithm or default that ensures that at least some basic functionality is provided by our application when the ML model is unable to provide sophisticated predictions, categorizations, and insights. Common algorithms that accomplish this goal are simplistic and extremely general but at least slightly better than nothing. One example we've mentioned earlier is that for recommendations on the *yarnit.ai* storefront, we might simply default to showing popular items when we don't have a customized recommendation model available.

This approach has a tremendous problem, however: it limits how good you can let your ML models be. If the models become so much better than the heuristics or defaults, you will come to depend upon them so thoroughly that no backup will be sufficient to accomplish the same goals. Depending on defaults and heuristics is a completely appropriate path for most organizations that are early in the ML adoption lifecycle. But it is a dependency that you should wean yourself off of if you'd like to actually take advantage of ML in your organization.

The other backup plan is to keep multiple versions of the same model and plan to revert to an older one if you need to. This will cover cases where a newer version of the model is significantly worse for some reason, but it will not help when the world as a whole has changed and therefore all versions of this model are not very good.

Ultimately, the second backup plan, combined with the ability to serve multiple models at the same time and quickly develop new variations of existing models, provides a path to understanding and resolving future model quality problems when the world has changed in a way that makes the model perform poorly. It is important for production traditionalists to note that no fixed or defensible barrier exists between model development and production in this case. Model quality is both a production engineering problem and a model development problem (which might occur urgently in production).

Data Will Be Unavailable

Some of the data used to train new models will not be available when we try to read it. Data storage systems, especially distributed data storage systems, have failure modes that include small amounts of actual data loss, but much higher amounts of data unavailability. This is a problem worth thinking through in advance of its occurrence because it will definitely occur.

Most ML training datasets are already sampled from another, larger dataset, or simply a subset of all possible data by virtue of when or how we started collecting the data in the first place. For example, if we train on a dataset of customer purchase behavior at *yarnit.ai*, this dataset is already incomplete from the start, in at least two obvious ways.

First, there was some date before which we were not collecting this data (or some date before which we choose not to use the data for whatever reason). Second, this is really only customer purchase behavior on our site and does not include any customer purchase behavior of similar products on any other sites. This is unsurprising because our competitors don't share data with us, but it does mean that we're already seeing only a subset of what is almost certainly relevant training data. For very high-volume systems (web-browsing logs, for example), many organizations subsample this data before training automatically as well, simply to reduce the cost of data processing and model training.

Given that our training data is already subsampled, probably in several ways, when we lose data, we should answer this question: is the loss of data biased in some way? If we were to drop one out of every 1,000 training records in a completely random way, this is almost certainly safe to ignore for the model. On the other hand, losing all of the data from people in Spain, or from people who shop in the mornings, or from the day before a large knitting conference—these are not ignorable. They are likely to create new biases in the data that we train on versus the data that we do not.[6]

6 Statisticians refer to these various properties of the data as *missing completely at random* (the propensity for the data point to be missing is completely random); *missing at random*, or *MAR* (the propensity of the data to be missing is not related to the data but is related to another variable—a truly unfortunate name for a statistical term); and *not missing at random* (the likelihood of the data to be missing is correlated with something in the data). In this case, we're describing MAR data because the propensity for any given data point to be missing is correlated with another variable (in this case, geography or time of day, for example).

Some training systems will try to pre-solve the problem of missing data for the entire system in advance of it occurring. This will work only if all of your models have a similar set of constraints and goals. This is because the impact of missing data is something that matters to each model and can be understood only in the context of its impact on each model.

Missing data can also have security properties that are worth considering. Some systems, especially those designed to prevent fraud and abuse, will be under constant observation and attack from outside malicious parties. Attacks on these systems consist of trying different kinds of behaviors to determine the response of the system and taking advantage of gaps or weaknesses that appear. In these cases, training system reliability teams need to be certain that there is no way for an outside party to systematically bias which particular time periods are skipped during training. It is not at all unheard of for attackers to find ways to, for example, create very large volumes of duplicate transactions for short periods of time in order to overwhelm data processing systems and try to hit a high-rate discard heuristic in the system. This is the kind of scenario that everyone working on data loss scenarios needs to consider in advance.

Models Should Be Improvable

Models will change over time, not just by adding new data. They will also see larger, structural changes. The requirements to change come from several directions. Sometimes we will add a feature to the application or implementation that provides new data but also requires new features of the model. Sometimes our user behavior will change substantially enough that we need to alter the model to accommodate. Procedurally, the most challenging change to model training in the training system we're describing here is adding a completely new feature.

Features Will Be Added and Changed

Most production ML training systems will have some form of a feature store to organize the ML training data. (Feature stores offer many advantages and are discussed in more detail in Chapter 4.) From the perspective of a training system, what we need to note is that a significant part of model development over time is often adding new features to the model. This happens when a model developer has a new idea about some data that might, in combination with our existing data, usefully improve the model.

Adding features will require a change to the feature store schema that is implementation specific, but it might also benefit from a process to "backfill" the feature store by reprocessing raw logs or data from the past to add the new feature for prior examples. For instance, if we decide that the local weather in the city that we believe our customers to be shopping from is a salient way to predict what they might buy, we'll have to add `customer_temperature` and `customer_precipitation` columns to the feature store.[7] We might also reprocess browsing and purchasing data for the last year to add these two columns in the past so that we can validate our assumption that this is an important signal. Adding new columns to the feature store and changing to schema and content of data in the past are both activities that can significantly impact reliability of all our models in the training system if the changes are not managed and coordinated carefully.

Models Can Train Too Fast

ML production engineers are sometimes surprised to learn that in some learning systems, models can train too fast.[8] This can depend a bit on the exact ML algorithm, model structure, and parallelism of the system in which it's being implemented. But it is entirely possible to have a model that, when trained too quickly, produces garbage results, but when trained more slowly, produces much more accurate results.

Here is one way this can happen: there is a distributed representation of the state of the model used by a distributed set of learning processes. The learning processes read new data, consult the state of the model, and then update the state of part of the model to reflect the piece of data they just read.[9] As long as there are either locks (slow!) or no updates to the particular key we are updating (unlikely at scale) in the middle of that process, everything is fine.

The problem is that multiple race conditions can exist, where two or more learner tasks consult the model, read some data, and queue and update to the model at the same time. One really common occurrence then is that the updates can stack on top of each other, and that portion of the model can move too far in a certain direction. The next time a bunch of learner tasks consult the model, they find it skewed in one direction by a lot, compared to the data that they are reading, so they queue up changes to move it substantially in the other direction. Over time, this part of the model (and other parts of the model) can diverge from the correct value rather than converge.

---

7 Adding these features might have significant privacy implications. These are discussed briefly in Chapter 4 and much more extensively in Chapter 6.

8 This is most common in model architectures that use gradient descent with significant amounts of parallelism in learning. But this is an extremely common setup for large ML training systems. One example of a model architecture that does not suffer from this problem is random forests.

9 An architecture in which the parameters of the model are stored is described well in "Scaling Distributed Learning with the Parameter Server" (*https://oreil.ly/dSXst*) by Mu Li et al. and in Chapter 12 of *Dive Into Deep Learning* by Joanne Quinn et al. (Corwin, 2019). Variants of this architecture have become the most common way that large-scale ML training stacks distribute their learning.

For distributed training setups, multiple race conditions is an extremely common source of failure.

Unfortunately for the discipline of ML production engineering, there is no simple way to determine when a model is being trained "too fast." There's a real-world test that is inconvenient and frustrating: if you train the same model faster and slower (typically, with more and fewer learner tasks), and the slower model is "better" in some set of quality metrics, then you might be training too fast.

The main approaches to mitigating this problem are to structurally limit the number of updates "in flight" by making the state of the model as seen by any learning processes closely synchronized with the state of the model as stored. This can be done by storing the current state of the model in very fast storage (RAM) and by limiting the rate at which multiple processes update the model. It is possible, of course, to use locking data structures for each key or each part of the model, but the performance penalties imposed by these are usually too high to seriously consider.

Resource Utilization Matters

This should be stated simply and clearly: ML training and serving is computationally expensive. One basic reason that we care about resource efficiency for ML training is that without an efficient implementation, ML may not make business sense. Consider that an ML model offers some business or organizational value proportional to the value that it provides, divided by the cost to create the model. While the biggest costs at the beginning are people and opportunity costs, as we collect more data, train more models, and use them more, computer infrastructure costs will grow to an increasingly large share of our expenditure. So it makes sense to pay attention to it early.

Specifically and concretely, utilization describes the following:

$$\frac{\text{portion of compute resources used}}{\text{total compute resources paid for}}$$

This is the converse of wastefulness and measures how well we're using the resources we pay for. In an increasingly cloud world, this is an important metric to track early.

Resource utilization is also a reliability issue. The more headroom we have to retrain models compared to the resources we have available, the more resilient the overall system will be. This is because we will be able to recover from outages more quickly. This is true if we're training the model on a single computer or on a huge cluster. Furthermore, utilization is also an innovation issue. The cheaper models are to train, the more ideas we will be able to explore in a given amount of time and budget. This markedly increases the likelihood that we will find a good idea among all of the bad ones. It is easy to lose track of this down here in the details, but we are not really here to train models—we're here to make some kind of difference in our organization and for our customers.

So it's clear we care about efficient use of our resources. Here are some simple ways we can make ML training systems work well in this respect:

Process data in batches
: When possible (algorithm-dependent), train on chunks of data at the same time.

Rebuild existing models
: Early-stage ML training systems often rebuild models from scratch on all of the data when new data arrives. This is simpler from a configuration and software perspective but ultimately can become enormously inefficient. This idea of incrementally updating models has a few other variants:

  - Use transfer learning to build a model by starting with an existing model.[10]
  - Use a multimodel architecture with a long-term model that is large but seldom retrained, and a smaller, short-term model that is cheaply updated frequently.
  - Use online learning, whereby the model is incrementally updated as each new data point arrives.

Simple steps such as these can significantly impact an organization's computational costs for training and retraining models.

Utilization != Efficiency

To know whether our ML efforts are useful, we have to measure the value of the process rather than the CPU cycles spent to deliver it. Efficiency measures the following:

$$\frac{\text{value produced}}{\text{cost}}$$

10 Transfer learning most generally involves taking learning from one task and applying it to a different, but related task. Most commonly in production ML environments, transfer learning involves starting learning with the snapshot of an already trained, earlier version of our model. We will either train only on new features, not included in the snapshot, or train only on new data that has appeared since the training of the snapshot. This can speed up learning significantly and thereby reduces costs significantly as well.

Cost can be calculated two ways, each of which provides a different view of our efforts. *Money-indexed cost* is the dollars we spend on resources. Here we just calculate the total amount of money spent on training models. This has the advantage of being a very real figure for most organizations. It has the disadvantage that changes in pricing of resources can make it hard to see changes that are due to modeling and system work versus exogenous changes from our resource providers. For example, if our cloud provider starts charging much more for a GPU that we currently use, our efficiency will go down through no change we made. This is important to know, of course, but it doesn't help us build a more efficient training system. In other words, money cost is the most important, long-term measure of efficiency but is ironically not always the best way to identify projects to improve efficiency.

Conversely, *resource-indexed cost* is measured in dollar-constant terms. One way to do this is to identify the most expensive and most constrained resource and use that as the sole element of the resource-indexed cost. For example, we might measure cost as *CPU seconds* or *GPU seconds*. This has the advantage that when we make our training system more efficient, we will be able to see it immediately, regardless of current pricing details.

This raises the difficult question of what, exactly, is the value of our ML efforts. Again, there are two kinds of value we might measure: *per model* and *overall*. At the per model level of granularity, *value* is less grandiose than we might expect. We don't need to measure the actual business impact of every single trained model. Instead, for simplicity, let's assume that our training system is worthwhile. In that case, we need a metric that helps us compare the value being created across different implementations training the same model. One that works well is something like

number of features trained

or even

number of examples processed

or possibly

number of experimental models trained

So for a per model, resource-indexed, cost-efficiency metric, we might have this:

$$\frac{\text{millions of examples}}{\text{GPU second}}$$

This will help us easily see efforts to make reading and training more efficient without requiring us to know anything at all about what the model actually does.

Conversely, *overall* value attempts to measure value across the entire program of ML model training, considering how much it costs us to add value to the organization as a whole. This will include the cost of the staff, the test models, and the production model training and serving. It should also attempt to measure the overall value of the model to our organization. Are our customers happier? Are we making more money? Overall efficiency of the ML training system is measured at a whole-program or whole-group basis and is measured over months rather than seconds.

Organizations that do not have a notion of efficiency will ultimately misallocate time and effort. It is far better to have a somewhat inaccurate measure that can be improved than to not measure efficiency at all.

Outages Include Recovery

This is somewhat obvious but still worth stating clearly: ML outages include the time it takes to recover from the outage. This has a huge and direct implication for monitoring, service levels, and incident response. For example, if a system can tolerate a 24-hour outage of your training system, but it takes you 18 hours to detect any problems and 12 hours to train a new model after the problems are detected, we cannot reliably stay within the 24 hours. Many people modeling production-engineering response to training-system outages utterly neglect to include model recovery time.

Common Training Reliability Problems

Now that you understand the basic architecture of a training system and have looked at general reliability principles about training systems, let's look at some specific scenarios where training systems fail. This section covers three of the most common reliability problems for ML training systems: data sensitivity, reproducibility, and capacity shortfalls. For each, we will describe the failure and then give a concrete example of how that might occur in the context of YarnIt, our fictional online knitting and crochet supply store.

Data Sensitivity

As has been repeatedly mentioned, ML training systems can be extremely sensitive to small changes in the input data and to changes in the distribution of that data. Specifically, we can have the same volume of training data but have significant gaps in the way that the data covers various subsets of the data. Think about a model that is trying to predict things about worldwide purchases but has data from only US and Canadian transactions. Or consider an image-categorization algorithm that has no pictures of cats but many pictures of dogs. In each of these scenarios, the model will have a biased view of reality by virtue of training on only a biased set of data. These gaps in training data coverage can be present from the very beginning or can occur over time as we experience gaps or shifts in the training data.

Lack of representativeness in the input data is one common source of bias in ML models; here, we are using *bias* in both the technical sense of the difference between the predicted value and the correct value in a model, but also in the social sense of being prejudiced against or damaging for a population in society. Strange distributions in the data can also cause a wide variety of other much more mundane problems. For some subtle and interesting cases, see Chapter 11, but for now let's consider a straightforward data sensitivity problem at YarnIt.

Example Data Problem at YarnIt

YarnIt uses an ML model to rank the results of end-user searches. Customers come to the website and type some words for a product they are looking for. We generate a simple and broad list of candidate products that might match that search and then rank them with an ML model designed to predict how likely each product is to be useful to the user who is doing this query right now.

The model will have features like "words in the product name," "product type," "price," "country of origin of the query," and "price sensitivity of the user." These will help us rank a set of candidate products for this user. And we retrain this model every day in order to ensure that we're correctly ranking new products and adapting to changes in purchase patterns.

In one case, our pricing team at YarnIt creates a series of promotions to sell off overstocked products. The modeling team wants to capture the pre-discount price and the sale price separately, as these might be different signals to user purchase behavior. But because of the change in data formatting, they mistakenly exclude all discounted purchases from the training set after they add the discounted price to the dataset. Until they notice this, the model will train entirely on full-price purchases. From the perspective of the ML system, discounted items are simply no longer ever purchased by anyone, ever. As a result, the model will eventually stop recommending the discounted products, since there is no longer any evidence from our logging, data, and training system that anyone is ever purchasing them! This kind of very small error in data handling during training can lead to significant errors in the model.

Reproducibility

ML training is often not strictly reproducible; it is almost impossible to use exactly the same binaries on exactly the same training data and produce exactly the same model, given the way most modern ML training frameworks work. Even more disconcerting, it may not even be possible to get approximately the same model. Note that while *reproducibility* in academic ML refers to reproducing the results in a published paper, here we are referring to something more straightforward and more concerning: reproducing our own results from this same model on this same dataset.

ML reproducibility challenges come from several sources, some of them fixable and others not. It is important to address the solvable problems first. Here are some of the most common causes of model irreproducibility:

Model configuration, including hyperparameters
: Small changes in the precise configuration of the model, especially in the hyperparameters chosen, can have big effects on the resulting model. The solution here is clear: use versioning for the model configurations, including the hyperparameters, and ensure that you're using exactly the same values.

Data differences
: As obvious as it may sound, most ML training feature storage systems are frequently updated, and it is difficult to guarantee that there are no changes at all to data between two runs of the same model. If you're having reproducibility challenges, eliminating the possibility of differences in the training data is a critical step.

Binary changes
: Even minor version updates to your ML training framework, learning binaries, or orchestration or scheduling system can result in changes to the resulting models. Hold these constant across training runs while you're debugging reproducibility problems.[11]

Aside from those fixable causes for irreproducibility, at least three causes are not easily fixed:

Random initializations
: Many ML algorithms and most ML systems use random initializations, random shuffling, or random starting-point selection as a core part of the way they work. This can contribute to differences across training runs. In some cases, this difference can be mitigated by using the same random seed across runs.

System parallelism
: In a distributed system (or even a single-computer training system with multiple threads), jobs will be scheduled on lots of processors, and they will learn in a somewhat different order each time. There will be ordering effects depending on which keys are updated in what order. Without sacrificing the throughput and speed advantages of distributed computing, there's no obvious way to avoid

11 The astute reader might note how terrifying this is. Another way to read this is "my models could change any time I happen to update TensorFlow or PyTorch, even for a new minor version." This is essentially true but not common, and the differences often aren't pronounced.

this. Note that some modern hardware accelerator architectures offer custom, high-speed interconnections among chips much faster than other networking technologies. NVIDIA's NVLink or the interconnections among Google's Cloud TPUs are examples of this. These interconnections reduce, but do not eliminate, the lag in propagating state among compute nodes.[12]

Data parallelism
: Just as the learning jobs are distributed, so is the data, assuming we have a lot of it. Most distributed data systems do not have strong ordering guarantees without imposing significant performance constraints. We have to assume that we will end up reading the training data in somewhat different order even if we do so from a limited number of training jobs.

Addressing these three causes is costly and challenging to the point of being almost impossible. Some level of inability to reproduce precisely the same model is a necessary feature of the ML training process.

Example Reproducibility Problem at YarnIt

At YarnIt, we retrain our search and recommendations models nightly to ensure that we regularly adjust them for changes in products and customer behavior. Typically, we take a snapshot of the previous day's model and then train the new events since then on top of that model. This is cheaper to do but ultimately does mean that each model is really dozens or hundreds of incremental training runs on top of a model trained quite some time ago.

Periodically, we have small changes to the training dataset over time. The most common changes are charges that are due to fraud. Detecting that a transaction is fraudulent may take up to several days, and by that point we may have already trained a new model with that transaction included as an authorized purchase. The most thorough way to fix that would be to recategorize the original transaction as fraud and then retrain every model that had ever included that transaction from an older snapshot. That would be extremely expensive to do every time we have a fraudulent transaction. We could conceivably end up retraining the last couple of weeks' models constantly. The other approach is to attempt to reverse the fraud from the model. This is complicated because there is no foolproof or exact way to revert a transaction

12 As long as the speed at which processors (whether CPUs or GPUs/accelerators) and their local memory operate is significantly higher than the speed at which we can access that state from across a network connection, there will always be lag in propagating that state. When processors update a portion of the model based on input data that they have learned from, there can always be other processors using an older version of those keys in the model.

in most ML models.[13] We can approximate the change by treating the fraud detected as a new negative event, but the resulting model won't be quite the same.[14]

This is all for the models that are currently in production. At the same time, model developers at YarnIt are constantly developing new models to try to improve their ability to predict and rank. When they develop a new model, they train it from scratch on all the data with the new model structure and then compare it to the existing model to see if it is materially better or worse. It may be obvious where this is going: the problem is that if we retrain the *current* production model from scratch on the current data, that model may well be significantly different from the current production model that is in production (which was trained iteratively on the same data over time). The fraudulent transactions listed previously will just never be trained on rather than be trained on, left for a while, and then deleted later. The situation is actually even less deterministic than that: even if we train the exact same model on the exact same data with no changes, we might have nontrivial differences, with one model trained all at once and another trained incrementally over several updates.[15]

This kind of really unnerving problem is why model quality must be jointly owned by model, infrastructure, and production engineers together. The only real reliability solution to this problem is to treat each model as it is trained as a slightly different variant of the Platonic ideal of that model and fully renounce the idea of equality between trained models, even when they are the same model configuration trained by the same computers on the same data twice in a row. This, of course, may tend to massively increase the cost and complexity of regression testing. If we absolutely need them to be more stable (note that this is not "the same" since we cannot achieve that

13 A large and growing set of work exists on the topic of deleting data from ML models. Readers should consult some of this research to understand more about the various approaches and consequences of deleting previously learned data from a model. One paper summarizing some recent work on this topic is "Making AI Forget You: Data Deletion in Machine Learning" (*https://oreil.ly/GWShn*) by Antonio A. Ginart et al., but be aware that this is an active area of work.

14 Chapter 6 covers some cases where we want to delete private data from an existing model trained on that data. The short version is that if the data is truly private and is included in our model, unless we used differential privacy during model construction and provide careful guarantees on how the model can be queried, we probably have to retrain the model from scratch. Indeed, we have to do this every single time someone requests that their data be removed. This, alone, is a powerful argument for ensuring that our models do not include private data.

15 Details of why this is the case are really specific to the model and ML framework, and beyond the scope of this book. But it often boils down to nondeterminism in the ML framework exacerbated by nondeterminism in the parallel processing of data. Reproducing this nondeterminism in your own environment is tremendously educational and more than a tiny bit terrifying. And yes, this footnote did just encourage readers to reproduce irreproducibility.

in most cases), then we may have to start thinking about ensembles of copies of the same model so that we minimize the change over time.[16]

Compute Resource Capacity

Just as it is a common cause of outages in non-ML systems, lack of sufficient capacity to train is a common cause of outages for ML systems. The basic capacity that we need to train a new model includes the following:

I/O capacity
: This is capacity at the feature store so that we can read the input data quickly.

Compute capacity
: This is the CPU or accelerator of the training jobs so that we can learn from the input data. This requires a pretty significant number of compute operations.

Memory read/write capacity
: The state of the model at any given time is stored somewhere, but most commonly in RAM, so when the training system updates the state, the system requires memory bandwidth to do so.

One of the tricky and troubling aspects of ML training system capacity problems is that changes in the distribution of input data, and not just its size, can create compute and storage capacity problems. Planning for capacity problems in ML training systems requires thoughtful architectures as well as careful and consistent monitoring.

Example Capacity Problem at YarnIt

YarnIt updates many models each day. These are typically trained during the lowest usage period for the website, which is whatever is nighttime for the largest number of users, and are expected to be updated before the start of the busy period the following day. Timing the training in this way gives us the possibility to reuse some resources between the online serving and the ML training system. At the very least, we will need to read the logs produced by the serving system, since the models that YarnIt trains daily read the searches, purchases, and browsing history from the website the day before.

As with most ML models, some types of events are less computationally complicated to process than others. Some of the input data in our feature store requires connections to other data sources in order to complete the input for some training operations. For example, when we show purchases for products listed by our partners

16 *Ensemble models* are just models that are collections of other models. Their most common use is to combine multiple very different models for a single purpose. In this case, we would combine multiple copies of the same model.

rather than by YarnIt directly, we need to look up details about that partner in order to continue to build models that accurately predict customer preferences about those products from that partner. If, for whatever reason, the portion of our purchases from partners increases over time, we might see a significant capacity shortfall in the ability to read from the partner information datasets. Furthermore, this might appear as if we have run out of compute capacity, when actually the CPUs are all waiting on responses from the partner data storage system.

Additionally, some models might be more important than others, and we probably want a system for prioritization of training jobs in those cases where we are resource constrained and need to focus more resources on those important models. This commonly occurs after an outage. Imagine we have a 48-hour outage of some part of the training system. At that point, we have stale models representing our best view of the world over two days ago. Since we were down for so long, it is reasonable to expect that we will take time to catch up, even using all of the machine resources that we have available. In this case, knowing which models are most important to refresh quickly is extremely useful.

Structural Reliability

Some of the reliability problems for an ML training system come not from the code or the implementation of the training system, but instead from the broader context in which these are implemented. These challenges are sometimes invisible to systems and reliability engineers because they do not show up in the models or the systems. These challenges show up in the organization and the people.

Organizational Challenges

Many organizations adding ML capabilities start by hiring someone to develop a model. Only later do they add ML systems and reliability engineers. This is reasonable to a point, but to be fully productive, model developers need a stable, efficient, reliable, and well-instrumented environment to run in. While the industry has relatively few people who have experience as production engineers or SREs on ML systems, it turns out that almost all the problems with ML systems are distributed systems problems. Anyone who has built and cared for a distributed system of similar scale should be able to be an effective production engineer on our ML system with some time and experience.

That will be enough for us to get started adding ML to our organization. But if we learned anything from the preceding failure examples, it is that some are extremely straightforward but others really do involve understanding the basics of how the models are structured and how the learning algorithms work. To be successful over the long term, we do not need ML production engineers who are experts in ML, but we do need people who are actively interested in it and are committed to learning more details about how it works. We will not be able to simply delegate all model quality problems to the modeling team.

Finally, we will also have a seniority and visibility problem. ML teams are more likely to get more senior attention than many other similarly sized or scoped teams. This is at least in part because when ML works, it is applied to some of the most valuable parts of our business: making money, making customers happy, and so on. When we fail at those things, senior leaders notice. ML engineers across the whole ecosystem need to learn to be comfortable communicating at a more senior level of the organization and with nontechnical leaders who have an interest in their work, which can have serious reputational and legal consequences when it goes wrong. This is uncomfortable for some of these engineers, but managers building ML teams should prepare them for this eventuality.

For a more in-depth discussion about organizational considerations beyond just the training system, see Chapters 13 and 14.

Ethics and Fairness Considerations

ML can be powerful but also can cause powerful damage. If no one in our organization is responsible for ensuring that we're using ML properly, we are likely to eventually run into trouble. The ML training system is one place where we can have visibility into problems (model quality monitoring) and can enforce governance standards.

For organizations that are newer to implementing ML, the model developers and ML training system engineers may be jointly responsible for implementing minimal privacy, fairness, and ethics checks. At a minimum, these must ensure that we are compliant with local laws regarding data privacy and use in every jurisdiction in which we operate. They must also ensure that datasets are fairly created and curated and that models are checked for the most common kinds of bias.

One common and effective approach is for an organization to adopt a set of Responsible AI principles and then, over time, build the systems and organizational capacity to ensure that those principles are consistently and successfully applied to all uses of ML at the organization. Think about how to be consistent at the model level (Chapter 5), policy level (Chapter 6), but also apply principles to data (Chapter 4), monitoring (Chapter 9), and incident response (Chapter 11).

Conclusion

Although ML training system implementers will still need to make many choices, this chapter should provide a clear sense of the context, structure, and consequences of those choices. We have outlined the major components of a training system as well as many of the practical reliability principles that affect our use of those systems. With this perspective on how trained models are created, we can now turn our attention to the following steps in the ML lifecycle.

CHAPTER 8
Serving

You've made a model; now you have to get it out there into the world and start predicting things. This is a process often known as *serving the model*. That's a common shorthand for "creating a structure to ensure our system can ask the model to make predictions on new examples, and return those predictions to the people or systems that need them" (so you can see why the shorthand was invented).

In our *yarnit.ai* online store example, we can imagine that our team has just created a model that is great at predicting the likelihood that a given user will purchase a given product. We need to have a way for the model to share its predictions with our overall system. But how, exactly, should we set this up?

We have a range of possibilities, each with different architectures and trade-offs. They are sufficiently different in approach that it might not be obvious looking at the list that these are all attempts to solve the same problem: how can we integrate our predictions with the overall system? We could do any of the following:

- Load the model into 1,000 servers in Des Moines, Iowa, and feed all incoming traffic to these servers.
- Precompute the model's predictions for the 100,000,000 most commonly seen combinations of yarn products and user queries using a big offline batch-processing job. Write those to a shared database once a day that is read by our system, and use a default score of $p = 0.01$ for anything not in that list.
- Create a JavaScript version of the model and load it into the web page so that predictions are made in the user's browser.
- Create a mobile app that has the model embedded into it so that predictions are made on the user's mobile device.

- Have different versions of the model with different trade-offs of computational cost and accuracy. Create a tiered system in which versions of the model are available in the cloud, using different hardware with different costs. Send the easy queries to a cheaper (less accurate) model and send the more difficult queries to a more expensive (more accurate) model.

This chapter is devoted to helping us map out the criteria for selecting from choices like this. Along the way, we will also discuss critical practicalities like ensuring that the feature pipeline used in serving is compatible with that used in training, and strategies for updating a model in serving.

Key Questions for Model Serving

There are a lot of ways that we can think about creating structures around a model that support serving, each with very different sets of trade-offs. To help navigate this space, it's useful to think through some specific questions about the needs for our system.

What Will Be the Load to Our Model?

The first thing to understand about our serving environment is the level of traffic that our model will be asked to handle—often referred to as *queries per second* (*QPS*) when queries are made on demand. A model serving predictions to millions of daily users may be asked to handle tens of thousands of queries per second. A model that runs an audio recognizer that listens for a *wake word* on a mobile device, like "Hey YarnIt," may run at a few QPS. A model that predicts housing prices for a real estate service might not be served on demand at all, and may instead be run as part of a large batch-processing pipeline.

A few basic strategies can address large traffic load. The first is to replicate the model across many machines and run these in parallel—perhaps using a cloud-based platform to allow a combination of traffic distribution and easy scaling up as demand grows. The second is to use more powerful hardware, such as hardware accelerators like GPUs or other specialized chips. These often require batching requests together to maximize efficiency, because the chips are so powerful that they can be bottlenecked more on input and output rather than on computing the model predictions themselves.[1]

1 In some cases, however, your choices about how to arrange computation are fixed and cannot be changed. Models that must be served on a device are one such example. Say we are deploying a model within a mobile app that uses image recognition to identify knitting patterns for sweaters from pictures taken with a mobile camera phone. We might choose to implement that image recognition directly on the mobile device, and by avoiding sending pictures to servers elsewhere, we'll improve latency and reliability, and potentially even privacy—though for mobile devices, ML computation is generally battery-expensive.

We could also tune the computation cost of the model itself, perhaps by using fewer features, or a deep learning model with fewer layers or parameters, or approaches such as quantization and sparsification to make the internal mathematical operations less expensive. Model cascades can also be effective at cost reduction—this is where a cheap model is used to make first-guess decisions on easy examples, and only the more difficult examples are sent to a more expensive model.

What Are the Prediction Latency Needs of Our Model?

Prediction latency is the time between the moment we make a request and the moment we get an answer back. Acceptable prediction latency can vary dramatically among applications, and is a major determiner of serving architecture choices.

For an online web store like *yarnit.ai*, we might have a total time budget of only half a second between the time that the user types in a query like "merino wool yarn" and the time they expect to see a full page of suggested products. Factoring in network delays and other processing necessary to build and load the page, this might mean that we have only a few milliseconds for the model to make all of its predictions on candidate products. Other very low-latency applications might include models that are used in high-frequency trading platforms, or that do real-time guidance of autonomous vehicles.

On the other end of the spectrum, we might have a model that is being used to determine the optimal spot to drill for oil, or that is trying to guide the design of protein sequences to be used to create new antibody treatments. For applications like these, latency is not a major concern because using these predictions (such as actually creating an oil rig or actually testing a candidate protein sequence in the wet lab) is likely to take weeks or months. Other application modalities have implicit delays built in. For example, an email spam-filtering model might not need to have millisecond response time if a user checks their inbox only every morning.

Taken together, latency and traffic load define the overall computational needs of our ML system. If prediction latency is too high, we can mitigate the issue by using more powerful hardware, or by making our model less expensive to compute. However, it is important to note that parallelization by creating a larger number of model replicas is usually not a solution to prediction latency, as the end-to-end time it takes to send a single example through the model isn't affected by simply having more versions of the model available.

Real systems often produce a distribution of latency values, due to network effects and overall system load. It can be useful to look at tail latency, such as the worst few percent, rather than average latency, so that we do not miss noticing if a percentage of requests are getting dropped.[2]

Where Does the Model Need to Live?

In our modern world, defined as it is by flows of information and concepts like virtual machines and cloud computing, it can be easy to forget that computers are physical devices, and that a model needs to be stored on a physical device in a specific location. We need to determine the home (or homes) for our model, and this choice has significant implications on the overall serving system architecture. Here are some options to consider.

On a local machine

Although this is not really a production-level solution, in some cases a model developer may have a model running on their local machine, perhaps invoking small batch jobs to process data when needed. This is not recommended for anything beyond small-scale prototyping or bespoke uses. Even in these cases, it is easy to come to rely on this in early stages and create more trouble than expected when we need to migrate to a production-level environment.

On servers owned or managed by our organization

If our organization owns or operates its own servers, we likely will run our models on this same platform. This may be especially important when specific privacy or security concerns are in place. It may also be the right option if latency is a hypercritical concern, or if specialty hardware is needed to run our models. However, this choice can limit flexibility in terms of ability to scale up or down, and will likely require special attention to monitoring.

In the cloud

Serving our models by using a cloud-based provider can allow for easily scaling our overall computational footprint up or down, and may also allow us to choose from several hardware options. This may be done in two ways. The first, running model servers on our own virtual servers and controlling how many of them we use, is essentially indistinguishable from the preceding option of using servers owned or managed by our organization. In this case, it might be slightly easier to scale up or

2 *Tail latency* refers to the longest latencies of the total distribution of latencies observed when querying a model. If we query a model many times and order the latency it takes to get a response from shortest to longest, we might find a distribution for which the median response time is quite fast. But in some cases we have a long tail of much, much longer responses. This is the tail, and the durations are the tail latencies. See "The Tail at Scale" (*https://research.google/pubs/pub40801*) by Jeffrey Dean and Luiz André Barroso for more.

scale down the number of servers, but the management overhead is otherwise similar. Here we're more interested in the second case: using a managed inference service.

In a managed inference service, some monitoring needs may be automatically addressed—although we will still likely need to independently verify and monitor overall model quality and predictive performance. Round-trip latency will likely be higher because of network costs. Depending on the geographical location of the actual datacenters, these costs may be higher or lower, and we may be able to mitigate some of these issues by using datacenters in multiple major geographical locations if we will be fielding requests globally. Privacy and security needs are also highlighted here, as we will be sending information across the network and will need to ensure that appropriate safeguards are in place. Finally, in addition to privacy and security concerns, we may have governance reasons for being cautious about using particular cloud providers: some online activities are regulated by national governments in a way that requires certain data to be kept in particular jurisdictions. Make sure you know about these factors before making a serving layout plan.

On-device

Today's world is filled with computational devices that are part of our daily lives. Everything from mobile phones to smart watches, digital assistants, automobiles, thermostats, printers, home security systems, and even exercise equipment have a surprising amount of computational capacity, and developers are finding ML applications in nearly all of them. When a model is needed in these settings, it is much more likely that it will need to be stored on the device itself, because the alternative is to access a model in the cloud, which requires constant network connection and may also have complicated privacy concerns. These settings of serving "on the edge" typically have strict constraints on model size, because memory is limited, as well as the amount of power that may be consumed by model predictions.

Updating the model in such settings typically requires a push across the network, and is unlikely to happen for all such devices in a timely fashion; some devices may never receive any updates at all. Because of the difficulty of making fixes to push updates, testing and verification take on a whole new level of importance in these settings. In some critical use cases, such as a model that continually needs to scan input audio for certain commands, it may even be necessary to encode the model at the hardware level rather than the software levels. This can yield huge efficiency improvements, but at the cost of making updates more difficult—or even impossible.

What Are the Hardware Needs for Our Model?

In recent years, a range of computational hardware and chip options has emerged that has enabled dramatic improvements in serving efficiency for various model types. Understanding these options is important for informing our overall serving architecture.

The main thing to know about deep models in serving is that they rely on dense matrix multiplications, which basically means that we need to do a lot of multiplication and addition operations in a way that is compute intensive, but that is also highly predictable in terms of memory access patterns.[3] The little multiplication and addition operations that make up one dense matrix multiplication operation parallelize beautifully. This means that traditional CPUs will struggle to perform well. At the time of writing, a typical CPU has around eight cores, each with one or at most a small number of algorithmic logic units (ALUs), which are the pieces of the chip that know how to do multiplication and addition operations. Thus, CPUs can typically parallelize only a handful of these operations at once, and their strengths in handling branching, memory access, and a wide variety of computational operations don't really come into play. This makes running inference for deep learning models slow on CPUs.

A much better choice for serving deep learning models are chips called *hardware accelerators*. The most common ones are GPUs because these chips were first developed for processing graphics, which also rely on doing fast dense matrix multiplications.[4] The main insight in a GPU is that if a few ALUs are good, thousands must be better. GPUs are thus great at the special-purpose task of dense matrix multiplications, but typically are not well suited to other tasks.[5]

Of course, GPUs have drawbacks, the most obvious of which is that these are specialized hardware. This typically means that either we need to invest organizationally in serving deep models using GPUs, or we are using a cloud service that supplies GPUs (and may charge a premium accordingly), or that we are serving in an on-device setting where a GPU is locally available.

The other main drawback of GPUs is that they're not well suited to operations not involving large amounts of dense matrix multiplications. Sparse models are one such example. Sparse models are most useful when we need to use only a small number

3 Think millions or billions of individual arithmetic operations for one prediction from a deep neural network.

4 While GPUs are by far the most common type of ML hardware accelerator used in training and serving, many other specialized accelerator architectures are designed specifically for ML. Companies like Google, Apple, Facebook, Amazon, Qualcomm, and Samsung Electronics all have ML accelerator products or projects. This space is changing rapidly.

5 Indeed, GPUs are *so* good at computation that they are often bottlenecked not on their ability to do the matrix multiplications, but instead on bandwidth for getting data in and out of the chip. Batching requests together to amortize the input and output costs can be an extremely effective strategy, in many cases allowing us to process hundreds of requests with the same wall-clock latency as a single request. The only problem with batching is that we may be slower waiting for enough requests to come in to create a batch of sufficient size, but in environments with high load, this is not usually an issue.

of important pieces of information out of a large universe of possibilities, such as the specific words that show up in a given sentence or search query out of the large universe of all possible words. With appropriate modeling, sparsity can be used in these settings to dramatically reduce computational cost, but GPUs can't easily benefit from this and CPUs may be much more appropriate. Sparse models may include nondeep methods such as sparse linear models or random forests. They can also appear in deep learning models as sparse embeddings, which can be thought of as a learned input adapter that converts sparse input data (such as text) into a dense representation that is more easily used within a deep model.

How Will the Serving Model Be Stored, Loaded, Versioned, and Updated?

As a physical object, our serving model has a specific size that needs to be stored. A model that is serving offline in an environment might be stored on disk and loaded in by specific binaries in batch jobs whenever a new set of predictions needs to be made. The main storage requirements are thus the disk space needed to keep the model, as well as the I/O capacity to load the model from disk, and the RAM needed to load the model into memory for use—and of these costs, the RAM is likely more expensive or more limited in capacity.

A model that is used in live online serving needs to be stored in RAM in dedicated machines, and for high-throughput services in latency-critical settings, copies of this model likely will be stored and served in many replica machines in parallel. As we discuss in Chapter 10, most models will need to be updated eventually by retraining on new data, and some are updated weekly, daily, hourly, or even more frequently. This means that we will need to swap the version of a model currently used in serving on a given machine with a new version.

If we want to avoid production outages while this happens, we have two main strategies. The first is to allocate twice as much RAM for the serving jobs, so that the new version of the model can be loaded into the machine while the old one is still serving, and then hot-swapping which one is used once the new version is fully ready. This works well but is wasteful of RAM for the majority of the time when a model is not being loaded or swapped. The second is to overprovision in terms of the number of replica machines by a certain percentage and then to progressively take a proportion (e.g., 10%) offline in turn to update the model. This more gradual approach also allows for more graceful error checking and canarying.

It is also important to remember that if we want our system to support A/B testing, which most developers will want to use, then it will be important to create an architecture that allows both an A and a B version of the model to be served—and indeed, developers may want to have many kinds of Bs running in A/B tests at the same time. Deciding exactly how many versions will be supported and at what capacity is an important architectural choice that requires balancing resourcing, system complexity, and organizational requirements together.

What Will Our Feature Pipeline for Serving Look Like?

Features need to be processed at serving time as well as at training time. Any feature processing or other data manipulation that is done to our data at training time will almost certainly need to be repeated for all examples sent to our model at serving time, and the computational requirements for this may be considerable. In some cases, this is as simple as converting the raw pixel values of an image into a dense vector to be fed to an image model. In more typical production settings, it may require joining several sources of information together in real time.

For example, for our *yarnit.ai* store, we might need to supply a product recommendation model with the following:

- Tokenized normalized text from a user query, drawn from the search box entry
- Information about past purchase history, drawn from a stored database of user information
- Information about product prices and descriptions, drawn from a stored product database
- Information about geography, language, and time of day, drawn from a localization system

Each kind of information comes from a different source, and may have different opportunities for precomputation or reuse from query to query or session to session. In many cases, this means that the actual code used to turn these pieces of information into features for our ML model to use may be different at serving time from the code used for similar tasks at training time. This distinction is one of the main sources of classic *training-serving skew* errors and bugs, which are notoriously difficult to detect and debug. For a much more in-depth discussion of this kind of skew, and others, see Chapter 9.

One of the promises of modern feature stores is that they handle both training and serving together in a single logical package. The reality of this promise may differ by system and use case, so it is well worth ensuring robust monitoring in any case.

It is also worth noting that creating features for our model to use at serving time is a key source of latency, and in many systems will be the dominating factor. This means that the serving feature pipeline is far from an afterthought, and is indeed often the most production-critical part of the entire serving stack.

Model Serving Architectures

With the preceding questions in mind, we will now examine four broad categories of serving architectures. Obviously, each needs to be tailored to specific use cases, and some serving systems may use a combination of approaches. With that said, we observe that most of the architecture and deployment approaches fall into the following four broad categories:

- Offline
- Online
- Model as a service
- Serving at the edge

Wc look at each in detail now.

Offline Serving (Batch Inference)

Offline serving is often the simplest and fastest architecture to implement. The application serving the end user is not exposed to the models directly. Models are trained ahead of time, often referred to as *batch inference*.

Batch inference is a way to avoid the problem of hosting a model to be reachable for predictions on demand when you don't need that. It works by loading the model and executing its predictions offline against a predefined set of input data. As a result, the model's predictions are stored as a simple dataset, perhaps in a database or a *.csv* file or another resource that stores data. Once these predictions are needed, the problem is identical to any other problem for which static data resources need to be loaded from data storage. In essence, by computing the model predictions offline, you convert on-demand model predictions into a more standard problem of simple data lookup (Figure 8-1).

For example, the *popularity* of each product on *yarnit.ai* for a given subset of users can be computed offline—perhaps at a convenient low-load time, if doing so is expensive in some way—and used as a sort function helper when displaying arbitrary items at any point as required to render the page.

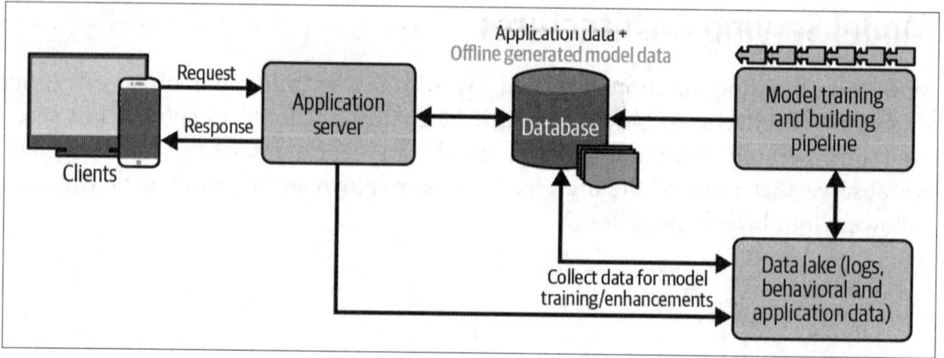

Figure 8-1. Offline model serving via data store

If the use case is less demanding, we might even be able to avoid the complexity of storing and serving model predictions via a database and write the predictions to a flat file, or in-memory data structure, and use them directly within the application (Figure 8-2). As an example, our web store could use a *search query intent classifier* (specific product versus broad category) to help the query engine rewrite the query for retrieving the search results efficiently. (You could, of course, build an approximation to the same structure by, say, indexing yarn by fiber content as a hash containing wool, cotton, acrylic, blends, etc., and/or a reverse hash.)

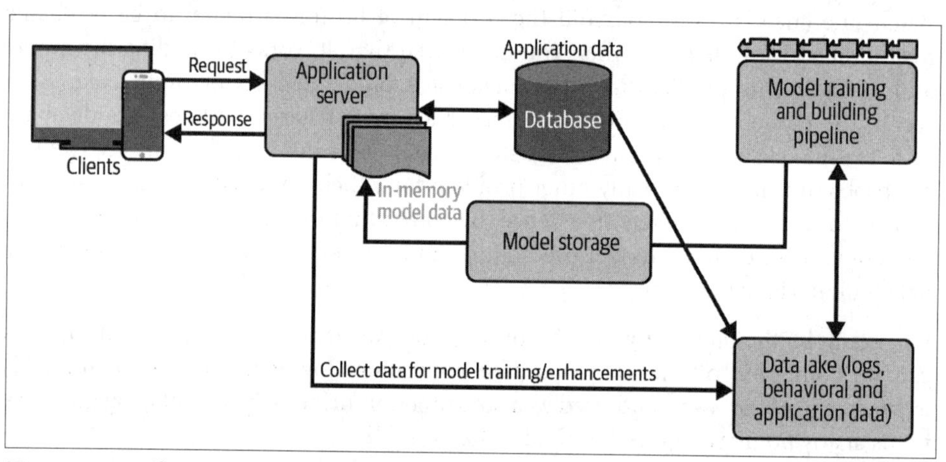

Figure 8-2. Offline model serving via in-memory data structures

Advantages

The advantages of offline serving are as follows:

Less complicated
: This approach requires no special infrastructure. Often you can reuse something you already have, or start something small and simple. The runtime system has fewer moving parts.

Easy access
: Applications facilitating the use case can perform simple key-value lookups or SQL queries based on the data store.

Better performance
: Predictions are provided quickly, since they have been precomputed. This might be an overriding consideration for certain mission-critical applications.

Flexible
: By using separate tables or records based on an identifier, this approach provides a flexible and easy way to roll out and roll back various models.

Verification
: The ability to verify all model predictions before use is a significant benefit for establishing correct operation.

Disadvantages

The disadvantages of offline serving are listed here:

Availability of data (training)
: Training data needs to be available ahead of time. Hence, model enhancements will take longer to deploy into production systems. Also, a critical upstream data outage could lead to stale models, days' worth of delays, permanently lost data, and expensive backfill processes to "catch up" the offline job to a current state.

Availability of data (serving)
: Effectively, the serving data needs to be available ahead of time; for fully correct operation, the system needs to know in advance every possible query that will be made of it. This is simply impossible in many use cases.

Scaling
: Scaling is difficult, especially for use cases dependent on large datasets or large query spaces. For example, you can't handle a search query space with a long tail—i.e., many different queries, the bulk of which aren't commonly used—with high accuracy and low latency.

Capacity limits
: Storing multiple model outputs in memory or in application databases will have storage limitations and/or cause performance problems. This will impact the ability to run multiple A/B tests at the same time. This may not cause a real problem, provided that the database and query resource requirements scale at similar rates and provided we have enough resources to provision.

Less selectivity
: Since models and predictions are precomputed, we won't be able to influence the predictions by using online context.

Online Serving (Online Inference)

In contrast to the preceding approach, *online serving* does not rely on precomputed outputs from a fixed query space. Instead, we provide predictions in real time by ingesting/streaming samples of real-time data, generally from user activity. In our web store example, we could build a more personalized shopping experience by having the model constantly learn the real-time user behavior by using the current context along with historical information to make the predictions. The current context might include location, views/impressions on precomputed recommendations, recent search sessions, items viewed, or items added to and removed from the basket.

Because all of this activity can be taken into account at prediction time, this allows significantly more flexibility in how to respond. Applications powered by inferences generated by offline models plus training the supplemental models for additional parameters in real time (Figure 8-3) provides huge benefits and significant business impacts.

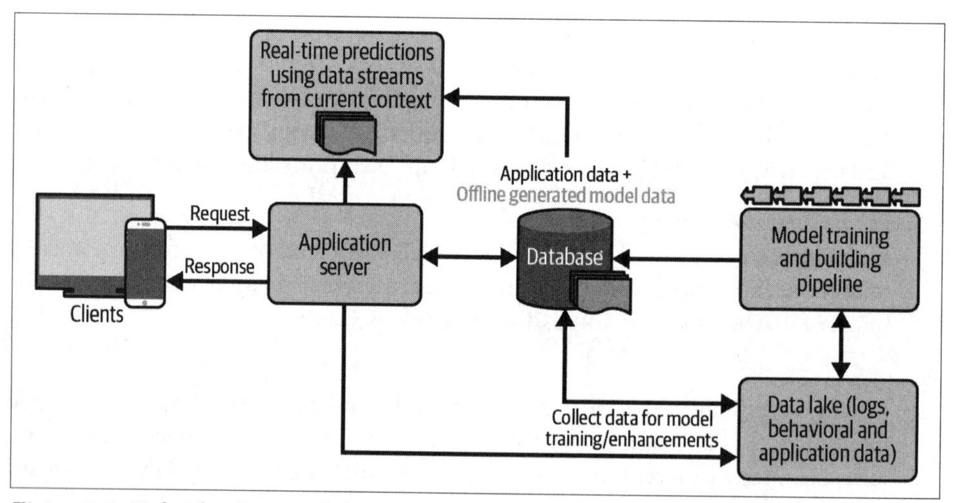

Figure 8-3. Hybrid online model serving in combination with predictions generated offline

Advantages

Advantages of online serving include the following:

Adaptability
: Online models learn as they go, and so greatly reduce the cadence with which model retraining and redeployment are required. Instead of adapting to concept drift at deployment time, the model adapts to concept drift at inference time, improving the performance of the model for customers.

Amenable to supplemental models
: Instead of training and changing one global model, we can tune more situation-specific models with a small subset of real-time data (for example, user- or location-specific models).

Disadvantages

Here are some disadvantages of the online-serving approach:

Latency budget required
: The model needs access to all relevant features. It will need quick access to new queries so that it can convert them into features and in turn look up relevant features stored elsewhere. If all the data we need for a single training example can't be sent to the server as part of the payload on the API call, we need to grab that data from somewhere else in milliseconds. Typically, that means using an in-memory store of some kind (for example, Redis).

Deployment complexities
: As the predictions are made in real time, rolling out the model changes is highly challenging, especially in a container-orchestration environment like Kubernetes.

Scalability constrained
: Since a model can and will change from time to time, it's not horizontally scalable. Instead, we might need to build a cluster of single-model instances that can consume new data as quickly as possible, and return the sets of learned parameters as part of the API response.

Higher oversight requirements
: This approach needs more advanced monitoring and adjustment/rollback mechanisms in place, since real-time changes could well include fraudulent behaviors caused by the bad actors in the ecosystem, and they could interact with, or influence, model behavior in some way.

Higher management requirements
: In addition to strong monitoring and rollback mechanisms, doing this correctly requires nontrivial expertise and fine-tuning to get right, both for data science and product engineering. Therefore, this approach is probably worth it for only

critical line-of-business applications, usually with high monetary impact for the business.

Serving models online is more powerful when it's combined with the model-as-a-service approach we discuss in the following section. We note in passing that real-time predictions can be served either synchronously or asynchronously. While synchronous mode is more straightforward and simpler to reason about, asynchronous mode gives us a lot more flexibility to handle the way results are passed around, and enable approaches like sending predictions via push *or* pull mechanisms, depending on the application and the end client (browser, app, device, internal service, etc.).

Model as a Service

The *Model-as-a-Service (MaaS)* approach is similar to software as a service and inherently favors a microservice architecture. With MaaS, models are stored in a dedicated cluster and served results via well-defined APIs. Regardless of the transport or serialization methods (e.g., gRPC or REST),[6] because models are served as a microservice, they're relatively isolated from the main application (Figure 8-4). This is therefore the most flexible and scalable deployment/serving strategy, since no in-process interaction or tight coupling are necessarily required.

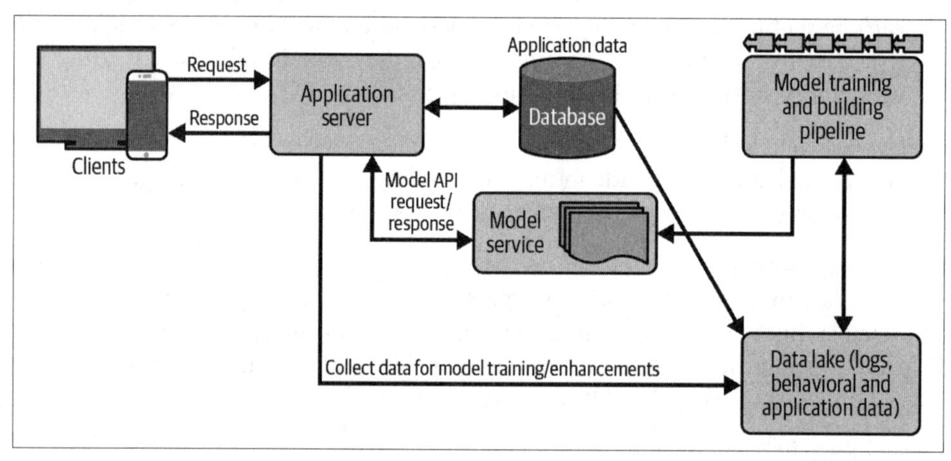

Figure 8-4. Models served as a separate microservice

Given the wide popularity of X-as-a-service approaches throughout the industry, we will focus on this particular method more than others, and will examine in detail various aspects of serving model predictions via APIs later in the chapter.

6 gRPC (*https://grpc.io*) is an open source RPC system initially developed by Google. Representational State Transfer (REST) is a widely used pattern for APIs that developers follow when they create web APIs.

Advantages

The following are advantages of MaaS:

Leveraging context
: By definition in the MaaS context, we have the ability to serve predictions in real time by using real-time context and new features.

Separation of concerns
: A separate service approach allows ML engineering to make model adjustments in a stabler way, and apply well-known techniques for managing operational problems. Most models of the MaaS type can be organized in a stateless way without any shared configuration dependencies. In these cases, adding a new model-serving capacity is as simple as adding new instances to the serving architecture, also known as *horizontal scaling*.

Deployment isolation
: As per any development architecture in which RPCs are the sole method of communication, the choice of technical stack could vary between application and model service layers, allowing respective teams to develop very differently if required. Independent deployment cycles could follow too, and make it a little easier to deploy versions on different timescales, or multiple environments: QA, staging, canaries, etc.

Version management
: Versioning is easy to proliferate, since we can store multiple versions of the models in the same cluster and point to them as required; this is extremely convenient for A/B testing, for example. The version-identifying information about the model being used can often be designed as a part of the service's response data as well. Among other benefits, this allows for rolling redeployments because stakeholder systems can rely on a model identifier to track, route, and collate any event data that may be generated as a result of using the ML model, such as tracking which model was used to serve a particular result in an A/B test.

Centralization facilitates monitoring
: Because the model architecture is centralized, it's comparatively easier to monitor system health, capacity/throughout, latency, and resource consumption, as well as per model business metrics like impressions, clicks, conversions, and so on. If we design architectural components that wrap inputs/outputs and standardize the process of identifying models and loading them from configs, many of the SRE "golden four" types of observability metrics can be obtained "for free" just by plugging into other predefined tools that provide these for other general microservices.

Disadvantages

MaaS disadvantages are as follows:

Management overhead
: When you climb aboard the microservice train, getting off is difficult, and a lot of overhead is required to stay onboard safely and well. However, this overhead does at least have the advantage of being somewhat well documented and understood.

Organizational compliance
: When we reply on the standard framework for deploying microservices, we might initially get a lot of things "for free," such as log aggregation, metrics scraping and dashboarding, tracking metadata for containers and compute usage, and managed delivery software that converts a code build or release into a real deployment. But we will also get change requests to comply with privacy, security standards, authentication, auditing, resource limitations, and various migrations.

Latency budgets required
: In any kind of microservice architecture that effectively externalizes your call stack, latency becomes a critical and unignorable constraint. Since user-perceived latency needs to be kept within reasonably tight constraints (subsecond, ideally), this imposes performance-related constraints on all the other systems you'll communicate with. It also potentially creates an organizational blind spot around user-perceived performance, since (by default in siloed enterprises) no one team will own that performance as a whole. As a result, the choice of underlying data stores, languages, and organizational structure and patterns becomes important.

Distributed availability
: Architectures built on distributed microservices must be able to tolerate partial failures robustly. The calling service must have reasonable fallbacks when the model service is down.

Serving at the Edge

A slightly less commonly understood serving architecture is used when a model is deployed onto edge devices (Figure 8-5). An *edge device* might be anything from an Internet of Things (IoT) doorbell to self-driving vehicles, or anything in between. Today, the bulk of edge devices with an internet connection are modern smartphones.

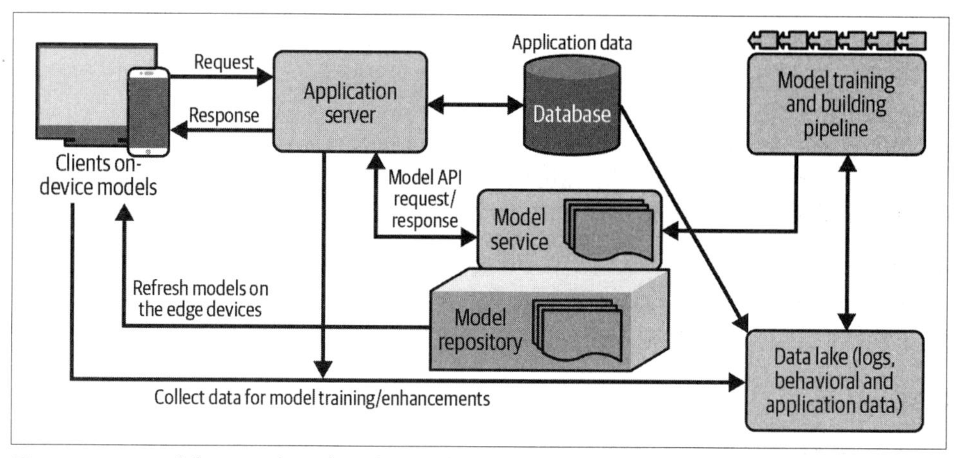

Figure 8-5. Models served at the edge and as a separate microservice on the server

Usually these models don't exist on their own: a server-side supplemental model of some kind helps fill the gaps. It's also common that most edge applications primarily rely on on-device inferences. That might change in the future, with emerging techniques like federated/collaborative learning.[7] The closeness to the user is a great advantage for some applications, but we often face severe resource limits in this architecture.

Advantages

Serving at the edge has these advantages:

Low latency
: Putting the model on the device allows it to be quicker. Near-instantaneous response (and no risk of dropped packets, etc.) to predict things can be absolutely critical for some applications: high latency or jitter in self-driving vehicles could cause accidents, injuries, or even deaths. Running models on edge devices is essentially compulsory here.

More-efficient network usage
: The more queries you can answer locally, the fewer you have to send over the network.

Improved privacy and security
: Making inferences locally means that the user data and the predictions made on that data are much harder to compromise. This is really useful for personalized

7 Federated learning is an approach that trains a model across multiple disconnected edge devices. Read more at TensorFlow (*https://oreil.ly/dYqeC*).

search, or recommendations that might require PII such as user profile, location, or transaction history.

More reliable
When the network connection is not consistently stable, being able to execute certain operations locally that were previously executed remotely becomes desirable and sometimes even necessary.

Energy efficiency
One key design requirement of edge devices is energy efficiency. In certain cases, local computing consumes less energy than network transmission.

Disadvantages

The following are disadvantages of serving at the edge:

Resource constraints (specialization)
With limited computing power, edge devices can perform only a few tasks. Non-edge infrastructure should still handle training, building, and serving of large models, while edge devices can perform local inferences with smaller models.

Resource constraints (accuracy)
ML models can consume lots of RAM and be computationally expensive; fitting these on memory-constrained edge devices can be difficult or impossible. The good news is a lot of research is ongoing to find alternative ways to address this; for example, parameter-efficient neural networks like SqueezeNet and MobileNet are both attempts to keep the models small and efficient without sacrificing too much accuracy.

Device heterogeneity (device-specific programming languages)
Coming up with a way to ensure that edge serving and on-device training happen precisely the same on both iOS and Android, for example, is a significant challenge. Doing it efficiently within the context of mobile development best practices also involves the intersections of two highly domain-specific groups of people (ML engineers and mobile engineers), which can create organizational strain on scarce shared specialty teams or prevent embracing standardized team models for full-stack development. A similar set of interactions exists whenever software is deployed into public-facing use as a service. For example, an accounting service available on the web will require that software engineers expert in building accounting systems deal with production engineers experienced at running software in production. The difference here is mostly of degree: ML engineers and mobile engineers come from extremely different worlds and technical contexts and are unlikely to communicate well without effort.

Device software versions are in the user's control

Unless you use a backend-for-frontend proxy service design pattern to route various calls off-device into a server-side backend, the owner of the edge device controls the software update cycles. We might push out a critical improvement to an on-device ML model in an iOS app, but that doesn't mean that millions of existing users have to update the version on their iPhones. They might wait around and continue using the outdated version for as long as they would like. Because any ML model deployed to the edge device might need to robustly keep operating and have its prediction and on-device learning setup continue working for a long time, it's a huge architectural commitment that might carry a lot of future-looking tech debt and legacy support with it, and should be chosen carefully.

One of the important attributes you need to track when serving ML models in production is versioning. Feedback loop data, backups, disaster recovery, and performance measurement all rely on it. We discuss these ideas in more detail in Chapter 9. In particular, we will look at suggested measurements in two sections: serving and SLOs.

Choosing an Architecture

Having talked about the various architecture options, we now need to choose the right one! Depending on the use case, that could be a complex affair; the differences between the model lifecycles, formats, and so on are one axis of consideration, never mind the vast implementation landscape that exists.

Our recommended approach is to first consider the *amount* of data and *speed* of the data required for your application: if extremely low latency is the priority, use offline/in-memory serving. Otherwise, use MaaS, except when you're running on an edge device, in which case serving at the edge is (obviously) the most appropriate.

The rest of this chapter focuses on MaaS, since it's more flexible, and pedagogically better since it suffers from fewer constraints.

Model API Design

Production-scale ML models are usually built using a wide variety of programming languages, toolkits, frameworks, and custom-built software. When trying to integrate with other production systems, such differences limit their *accessibility*, since ML and software engineers may have to learn a new programming language or write a parser for a new data format, and their *interoperability*, requiring data format converters and multiple language platforms.

One way of improving the accessibility and interoperability is to provide an abstracted interface via web services. Resource-oriented architectures (ROAs) in the REST style appear well suited to this task, given the natural alignment of REST's design philosophy with the desire to hide implementation-specific details.[8] Partially in support of this view, we've seen rapid growth in the area of ML web services in recent years: for example, Google Prediction/Vision APIs, Microsoft Azure Machine Learning, and many more.

Most service-oriented architecture (SOA) best practices apply to ML model/inference APIs too.[9] But you'll want to take note of the following points for models:

Data science versus engineering skills
: Many organizations have a pure data science team, with little or no experience running services in production. To gain all the benefits of DevOps, however, you will want to empower the data science team to take full ownership of releasing its models to production. Instead of "handing over" models to another team, they will collaborate with operations teams and co-own that process from start to finish.

Representations and models
: Even the slightest change in distribution of a feature may cause models to drift. For complex-enough models, creating this representation may mean numerous data pipelines, databases, and even upstream models. Handling this relationship is nontrivial for many ML teams.

Scale/performance characteristics
: In general, the *predict* part of the pipeline is purely compute-bound, something that is rather unique in a service environment. In many cases, the *representation* part of the workflow is more I/O bound, especially when we need to enrich the input, by loading data/features, or retrieve the image/video we're trying to predict on.

We believe the overwhelming factor that drives many design patterns in inference service design is—perhaps surprisingly—*organizational support and skill set*. A fundamental tension exists between requiring a data science team to fully own all end-to-end components of a production deployment and fully separating production

8 Resource-oriented architectures (as compared to service-oriented architecture) extend the REST pattern for web API building. A resource is an entity that has a state that can be assigned to a uniform resource locator (URL). See "An Overview of Resource-Oriented Architectures" (*https://oreil.ly/qzVwx*) by Joydip Kanjilal for an overview.

9 Similarly, service-oriented architecture is an approach whereby an application is decomposed into a series of services. It's a somewhat overused term that often means different things to different people in the industry (as is reflected in "Service-Oriented Architecture" (*https://oreil.ly/e5GzU*) by Cesar de la Torre et al.)

concerns away from the data science team so it may focus fully on its domain specialization of model training and model optimization.

Too far in either direction may prove unhealthy. If the data science team is asked to own too much, or doesn't have close collaborative partnership with operations support teams, it can become overwhelmed dealing with production concerns for which it has no training. If the data science team owns too little, it may be disconnected from the constraints or realities of the production system its models must fit into, and will be unable to remediate errors, assist in critical bug fixes, or contribute to architectural planning.

So, when we are ready to deploy models in production, we actually deploy two different things: the model itself and the APIs that go and query the model to fetch the predictions for a given input. Those two things also generate a lot of telemetry and a lot of information that'll later be used to help us monitor the models in production, try to detect drift or other anomalies, and feed back into the training phase of the ML lifecycle.

Testing

Testing model APIs, before deploying and serving in production, is extremely critical because models can be have a significant memory footprint and require significant computational resources to provide fast answers. Data scientists and ML engineers need to work closely with the software and QA engineers, and product and business teams, to estimate API usage. At a minimum, we need to perform the following tests:

- Functional testing (e.g., expected output for given input)
- Statistical testing (e.g., test the API on 1,000 unseen requests, and the distribution of the predicted class should match the trained distribution)
- Error handling (e.g., data type validation in the request)
- Load testing (e.g., *n* simultaneous users calling *x* times/second)
- End-to-end testing (e.g., validate that all the subsystems are working and/or logging as expected)

Serving for Accuracy or Resilience?

When serving ML models, a performance increase doesn't always mean business growth. Monitoring and correlating the model metrics with the business key performance indicators (KPIs) help bridge the gap between performance analysis and business impact, integrating the whole organization to function more efficiently toward a common goal. It is important to view every improvement in the ML pipeline through business KPIs; this helps in quantifying which factors matter the most.

Model performance is an assessment of the model's ability to perform a task accurately, not only with sample data but also with actual user data in real time in a production setup. It is necessary to evaluate performance to spot any erroneous predictions like drift in detection, bias, and increased data inconsistency. Detection is followed by mitigation of these errors by debugging, based on its behavior to ensure that the deployed model is making accurate predictions at the user's end and is resilient to data fluctuations. ML model metrics are measured and evaluated based on the type of model that the users are served by (for example, binary classification, linear regression, etc.), to yield a statistical report that enlists all the KPIs and becomes the basis of model performance.

Even though improvements in these metrics, such as minimizing log loss or improving recall, will lead to better statistical performance for the model, we find that business owners tend to care less about these statistical metrics and more about business KPIs. We will be looking for KPIs that provide a detailed view of how well a particular organization is performing, and create an analytical basis for optimized decision making. In our *yarnit.ai* web store example, a couple of main KPIs could be as follows:

Page views per visit
: This measures the average number of pages a user visits during a single site visit. A high value might indicate an unsatisfactory user experience due to the enormous digging the user had to do to reach what they want. Alternatively, a very low value might indicate boredom or frustration with the site and point to abandonment.

Returning customer order
: This measures the orders of an existing customer, and is essential for keeping track of brand value and growth.

A resilient model, while not the best model with respect to data science measures like accuracy or AUC, will perform well on a wide range of datasets beyond just the training set. It will also perform better for a longer period of time, as it's more robust and less overfitted. This means that we don't need to constantly monitor and retrain the model, which can disrupt model use in production and potentially even create losses for the organization. While no single KPI measures model resilience, here are a few ways we can evaluate the resiliency of models:

- Smaller standard deviations in a cross-validation run
- Similar error rates for longer times in production models
- Less discrepancy between error rates of test and validation datasets
- How much the model is impacted by input drift

We discuss more details about model quality and evaluation in Chapter 5, and API/system-level KPIs like latencies and resource utilization in Chapter 9.

Scaling

We've exposed the models via API endpoints so they can deliver value to the business and customers. This is good, but it's just the beginning. If all goes well, the model endpoints might see significantly higher workloads in the near future. If the organization starts to serve many more users, these increased demands can quickly bring down the ML services/infrastructure.

ML models deployed as API endpoints need to respond to such changes in demand. The number of API instances serving the models should increase when requests rise. When workload decreases, the number of instances should be reduced so that we don't end up in a state of underutilization of resources in the cluster, and we could potentially save a significant amount of operational expenses. This is similar to the autoscaling in any cloud computing environment in modern software architectures. Caching can also be efficient in ML environments, just as in traditional software architectures. Let's discuss these briefly.

Autoscaling

Autoscaling dynamically adjusts the number of instances provisioned for a model in response to changes in the workload. Autoscaling works by monitoring a target metric (e.g., CPU or memory usage) and comparing it to a target value we monitor for. Additionally, we can configure the minimum and maximum scaling capacity and a cool-down period to control scaling behavior and price. In our *yarnit.ai* web store example, the per language spelling-correction module used for powering search use cases can be scaled independently from scaling the personalized recommendations module for sending periodic emails recommending new/similar products based on customer purchase history.

Caching

Consider the problem of predicting categories and subcategories within our *yarnit.ai* online store. A user might search for "cable needles," and we might predict their intended shopping area is Equipment → Needles coming from an internal taxonomy of our store category layout. In a case like this, rather than repeatedly invoke the expensive ML model each time a repeat query like "cable needles" is encountered, we could leverage a cache.

For simple cases that have a small number of queries in the cache, this can usually be solved with a simple in-memory cache, possibly defined directly in the application logic or in the model's API server. But if we are dealing with a huge number of customer queries to fit in the cache, we may want to expand our cache into a separate API/service that can be independently scaled and monitored.

Disaster Recovery

ML serving via MaaS has all the same failure-recovery requirements as other software as a service (SaaS) platforms: surviving the loss of a datacenter and diversifying infrastructure risks, avoiding vendor lock-in, rolling back bad code changes quickly, and ensuring good circuit breaking to avoid contributing to failure cascades. Separate from these standard service failure considerations, the deep reliance of ML systems on training and data pipelines (whether online or offline) creates additional requirements, including accommodating data schema changes and database upgrades, onboarding new data sources, durable recovery of stateful data resources (like the state of online learning, or the state of on-device retraining in an edge serving use case after an app crashes), and graceful failure in the face of missing data or upstream data ETL job outages—to name but a few.

Data is constantly changing and growing in data warehouses, data lakes, and streaming data sources: adding new and/or enhancing existing features in the product/service creates new telemetry, a new data source may be added to supplement a new model, an existing database goes through a migration, someone accidentally begins initializing a counter at 1 instead of 0 in the last version of the model, and the list can go on. Any one of such changes brings more challenges to ML systems in production.

We discussed the challenges around data availability in Chapter 7. Without proper care for failure recovery, ML models that experience unexplained data changes or data disruptions may need to be taken out of production and iterated offline, sometimes for months or longer. During the early stages of architecture review, be sure to ask many questions about how the system will react to unusual data changes and how the system can be made robust to allow it to continue operating in production. Additionally, we will inevitably want to expand a successful model's scope or optimize a poorly performing model by adding additional data features. It is critical to factor in this data extensibility as an early architectural consideration to avoid failure scenarios where we are blocked from being able to ingest a new feature for the model because of the logistics of accommodating the new data in production.

Also for high availability, we may want to run the model API clusters in multiple datacenters and/or availability zones/regions in the cloud computing world. This will allow us to quickly route the traffic when an outage occurs in a specific cluster.

Such deployment architecture decisions are fundamentally driven by the SLOs of the organization.[10] We discuss SLOs in more detail in Chapter 9.

Just as with application data, we need to have backup strategies in place to constantly take snapshots of the current model data and use the last known good copies when needed. These backups could be used offline for further analysis and could potentially feed into training pipelines to enhance the existing models by deriving new features.

Ethics and Fairness Considerations

The general topic of fairness and ethics (along with privacy) is covered in depth in Chapter 6. This is a broad area that can be overwhelming for system implementers to consider. We strongly encourage you to read that chapter for a general introduction along with some concrete suggestions.

We should consider the following specific points for serving, however:

Organizational support and transparency
: When it comes to ethics and fairness while serving the ML models in production, we need to establish checks and balances as part of the development and deployment framework and be transparent with both internal stakeholders and customers about the data being collected and how it will be used.

Minimize privacy attack surface
: When we process a request through the model APIs, request and response schemas should try to avoid or at least minimize the need for user personal, demographic information. If it's part of the request, we need to make sure that data is not logged anywhere while serving the predictions. Even for serving personalized predictions, organizations that are extremely committed to ethics and privacy often interact with serving infrastructure with short-lived user identifiers/tokens instead of tracking the unique identifiers like user ID, device ID, and so on.

Secure endpoints
: Along with data privacy, especially when dealing with PII, product/business owners and ML/software engineers should invest more time and resources to secure the model API endpoints even though they are accessible only within the internal network (i.e., user requests are first processed by the application server before calling model APIs).

Everyone's responsibility
: Fairness and ethics are a responsibility for everyone, not just ethicists, and it is critical that implementers and users of an ML serving system be educated about

10 SLOs are thoroughly introduced in *Site Reliability Engineering: How Google Runs Production Systems*, and even more thoroughly covered in *Implementing Service Level Objectives* by Alex Hidalgo (O'Reilly, 2020).

these topics. Governance of these critical issues is not just the domain of ML engineers, and must be informed holistically by other members of the organization, including legal counsel, governance and risk management, operations and budget planning, and all members of the engineering team.

Conclusion

Serving reliably is hard. Making the models available to millions of users with millisecond latencies and 99.99% uptime is extremely challenging. Setting up the backend infrastructure so the right people can be notified when something goes wrong and then figuring out what went wrong is also hard. But we can successfully tackle that complexity in multiple ways, including asking the right questions about your system at the start, picking the right architecture, and paying specific attention to the APIs you might implement.

Serving isn't a one-time activity either, of course. Once we're serving, we then need to monitor and measure success (and availability) constantly. There are multiple ways to measure the ML model and product impact over the business, including input from key stakeholders, customers, and employees, and actual ROI as measured in revenue, or some other organizationally relevant metric. Hints on this, and other topics in deployment, logging, debugging, and experimentation are to be found in Chapter 9, while there is a much more complete coverage of measuring models in Chapter 5.

CHAPTER 9

Monitoring and Observability for Models

By Niall Murphy and Aparna Dhinakaran
Contributors/Reviewers: Ely Spears, Lina Weichbrodt, Tammy Le
Diagrams: Joel Bowman

Managing production systems is somewhere between an art and a science. Add the complexities of ML to this hybrid discipline, and it looks less like a science and more like an art. What we do today is very much a frontier, rather than a well-defined space. Despite that, this chapter outlines what we know about how to monitor, observe, and alert for ML production systems, and makes suggestions for developing the practice within your own organization.

What Is Production Monitoring and Why Do It?

This chapter is about how to monitor systems that are doing ML, rather than using ML to monitor systems. The latter is sometimes called *AIOps*; we are focusing on the former.

With that out of the way, let's talk about production monitoring generically, without the complexities of ML, so we can make things easier to understand—and where better to begin than with a definition? *Monitoring*, at the most basic level, provides data about how your systems are performing; that data is made storable, accessible, and displayable in some reasonable way. *Observability* is an attribute of software, meaning that when correctly written, the emitted monitoring data—usually extended or expanded in some way, with labeling or tagging—can be used to correctly infer the behavior of the system.[1]

1 You can't have observability without monitoring, but you can have the reverse—coarse-level detection without any ability to inspect in greater detail. This is, however, not the direction of travel of the industry.

Why would you care? It turns out there are lots of reasons. Most urgently, monitoring allows you to figure out whether your systems are actually working. If you are buying and reading this book of your own accord, you probably already understand how important that is. No less a luminary than Andrew Clay Shafer, cofounder of the DevOps movement, wrote (*https://sre.google/workbook/foreword-II*), "If the systems are down, the software has no value." If you don't accept this is important, or if you understand the arguments but don't believe them, we encourage you to read James Turnbull's *The Art of Monitoring* (2016). For the purposes of the rest of this chapter, though, we assume you understand that you need to monitor (and alert on) the state of systems, and what is up for discussion is how best to do that.

Of course, the situation has more nuance than that. For a start, systems don't usually behave as a Boolean, either entirely up or entirely down; generally, they can be performing anywhere on a spectrum from superb to very badly. Monitoring obviously needs to be able to handle this situation and represent the reality correctly.

Monitoring is hugely important in and of itself, but an offshoot of monitoring is absolutely crucial: alerting. A useful simplification is that when things go wrong, humans are alerted to fix them, and for the purposes of this paragraph, *alerting* is therefore both defining the conditions for "things going wrong," and being able to reliably notify the responsible folks that something isn't right—e.g., paging. This is a key technique in helping to "defend the user experience."

Less urgently, but still vital, is monitoring for long-term trend analysis, capacity planning, and general understanding of your service envelope. You use this kind of monitoring data to answer questions like these: Is my service cost-effective? Does it have any unobvious performance cliffs? Is there data distribution drift? How does service latency, for example, relate to user behavior on the weekend versus the working week? All these questions and more cannot really be answered well without monitoring and observability.

What Does It Look Like?

As we've alluded to, to do monitoring, you must have a *monitoring system* as well as systems to be monitored (here, called the *target systems*). Today, target systems emit *metrics*—a series, typically of numbers, with an identifying name—which are then collected by the monitoring system and transformed in various ways, often via *aggregation* (producing a sum or a rate across multiple instances or machines) or *decoration* (adding, say, event details onto the same data). These aggregated metrics are used for system analysis, debugging, and the alerting we mentioned previously.

A concrete example is a web server with a metric of the total number of requests it received; this metric has a name—say, in this case, `server.requests_total`. (Of course, it could be any request/response architecture, like an ML model!) The monitoring system will obtain these metrics, usually via *push* or *pull*, which refers to whether the metrics get pulled from the target systems or get pushed from them. These metrics are then collated, stored, and perhaps processed in some way, generally as a *time series*. Different monitoring systems will make different choices about how to receive, store, process, and so on, but the data is generally *queryable* and often (very often) there's a graphical way to plot the monitoring data so we can take advantage of our visual comparison hardware (eyes, retinas, optic nerves, and so on) to figure out what's actually happening.

By extension, an observable system uses these foundational ideas but goes a step further: instead of just getting a counter for the total number of requests, you get *labeled*[2] data for that metric, and indeed most metrics. Specifically, *labeled data* means that you don't just get a counter; you get subdivisions, or slices, of that metric. So, for example, you don't just have `server.requests_total`; you have `server.requests_total{lang=en}`, which means, "For all requests made where the client requested the page be rendered in English, what is the total number of requests?" Of course, not just `{lang=en}` either—also `{lang=fr}`, `{lang=pt}`, `{lang=es}`, `{lang=zh}`, and so on. A fully observable system allows slicing and dicing of such data on *extremely* fine-grained boundaries, such that it is possible to construct queries to look at the past 12 days of queries in Romanian that resulted in a HTTP 404 return code after 1200 ms of latency.[3]

Monitoring in general has many subtleties, particularly around how aggregation is done and how results are used, but it's a reasonable high-level picture—for *non*-ML system monitoring at least. When you add ML systems as *target systems* to this picture, you get not just all the issues mentioned previously, but also the special concerns of ML; Figure 9-1 may help illustrate.

2 Note that this use of "labeled" is distinct from the use of the term "labeled data" in supervised ML. Here, the labels are more like arbitrary key-value pairs associated with a time series.

3 Of course, you don't get this level of detail for free. Product developers have to write code to maintain labeled metrics and export them correctly, and you need a system capable of analysis and display. But it's worth it.

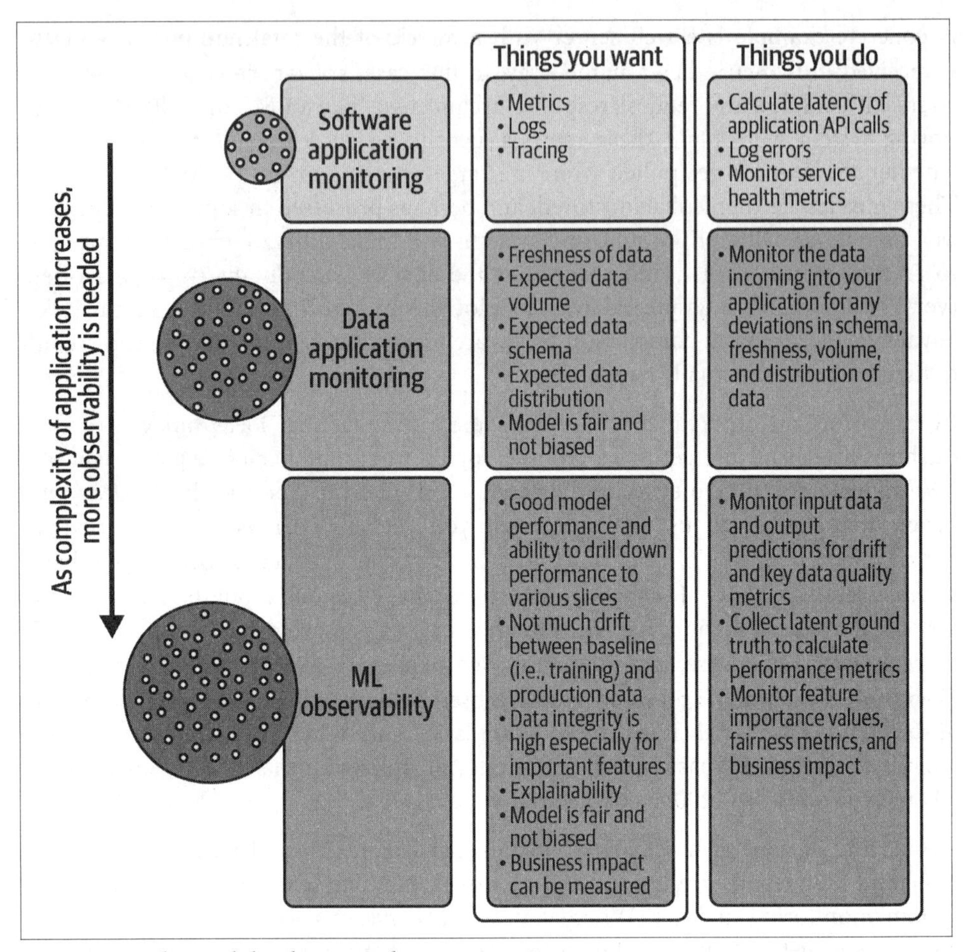

Figure 9-1. Observability layers and system requirements

The Concerns That ML Brings to Monitoring

One important concern is not necessarily the task of monitoring ML itself, but the *perception* of the act of monitoring by the model development community. What do we mean?

Well, we believe the model development community understands very well that software has inputs and outputs, and should be observed in order to figure out what's going on. (The whole act of model development can be looked on as the process of metric extraction, control, and optimization, for a start!) What is sometimes missing, though, is awareness and engagement with what happens *after* a model is developed and goes out into production. What we see as the mindset problem with monitoring ML also partially derives from how the word is used—semantically, *monitoring* can mean inspection activities applying to model development, or it can mean continual observation of systems in production. In actuality, the term is used in both contexts.

To put it another way, many model developers don't realize that exactly the same requirements for inspectability *could* and *should* apply when a model is running in production as they do in development. This gap is particularly true if your background is in using metrics for optimization, but not for *detection*. Detection turns out to be a hugely crucial use case, and the activity of monitoring should apply across the whole-model lifecycle.

This is not just a question of perception, though. The reality is that ML already struggles with explainability—particularly at the time of execution in production. This is partially because of the nature of ML, partially a function of the way models are developed today, partially the nature of production operation, and partially a reflection of the fact that tools for inspectability are generally aimed just at model development. All of these combine to make monitoring ML more difficult.

Reasons for Continual ML Observability—in Production

Observability data from your models is absolutely fundamental to business—both tactical operations and strategic insights. We have mentioned, and much has been written about, the negative consequences of not having monitoring and observability, but positive consequences arise too.

One example we like to use is the connection between latency and online sales. In 2008, Amazon discovered that each additional 100 ms of latency lost 1% of sales, and also the converse—so, the faster the better.[4] Similar results have been confirmed by Akamai Technologies, Google, Zalando, and others. We assert that without observability, there would be no way to have discovered this effect, and certainly no way to know for sure that you were either making it better or worse!

Ultimately, observability data *is* business outcome data. In the era of ML, this happily allows you not just to detect and respond to outages, but also to understand incredibly important things that are happening to your business. Ignore it at your peril.

Problems with ML Production Monitoring

ML model development is still in its infancy. The tools are immature, conceptual frameworks are underdeveloped, and discipline is in short supply, as everyone

4 *Discovered* is maybe not quite the right word. For more details, see the Digital Realty blog post "The Cost of Latency" (*https://oreil.ly/qawrq*) by Farhan Khan, or the 2018 Zalando study "Loading Time Matters" (*https://oreil.ly/UCN69*) by Christoph Luetke Schelhowe et al.

scrambles to get some kind of model—any kind of model!—off the ground as soon as possible and solving real problems. The pressure to ship is real and has real effects.

In particular, model development, which is inherently hard because it involves reconciling a wide array of conflicting concerns, gets harder because that urgency forces developers and data science folks to focus on those hard problems and ignore the wider picture. That wider picture often involves questions around monitoring and observability.

This leads to two important observations we would make about the difference between model development and production serving. One of these is a generic observation about all production environments, which just happens to be particularly complex and difficult in the ML world. The other is a specific situation about model development that is currently broadly true but may not be so forever; nonetheless, it is worth mentioning as a foundation to what follows.

Difficulties of Development Versus Serving

The first problem is that effectively simulating production in development is extremely hard, even in separate environments dedicated to that task (like test, staging, and so on.) This is not just because of the wide variety of possible serving architectures (model pools, shared libraries, edge devices, etc., with the associated infrastructure that you might or might not be running on) but also because in development you often invoke prediction methods directly or with a relatively small amount of code between you and the model for velocity reasons. Running in production also generally means you don't have the ability to manipulate input, logging level, processing, and so on, arbitrarily, leading to huge difficulties in debugging, reproducing problematic configurations, etc. Finally—and crucially—the data you have in testing is not necessarily distributed like the data the model encounters in production, and as always for ML, data distribution really matters.

The second problem is a little different. In conventional software delivery, the industry has a good handle on work practices that are known to improve throughput, reliability, and developmental *velocity*. The most important of these are probably the grouped concepts of continuous integration / continuous deployment (CI/CD), unit tests, small changes, and a collection of other techniques probably best described in *Accelerate* by Nicole Forsgren et al. (IT Revolution Press, 2018).[5] Unfortunately, today we are missing this equivalent of CI/CD for model development, and cannot yet say we have converged onto a good set of (telemetry-related, or otherwise) tools

5 CI/CD is too complicated to describe in detail here but basically means delivering software in a continuous, reliable stream. In addition, to be clear, just because we know what works under certain circumstances doesn't mean that the industry as a whole does it consistently—just that we have good, evidence-based reasons to believe it.

for model training and validation. We expect this will improve over time as existing tools (such as MLflow and Kubeflow) gain traction, vendors incorporate more of these concerns into their platforms, and the mindset of holistic or whole-lifecycle monitoring gains more acceptance.

An Important Note About Skew

Skew is used broadly to describe a variety of data problems. Among common uses of the term are biased distribution shifts in the underlying data, outliers (especially unexpected outliers), semantic violations in the interpretation of data, and missing feature values (especially including values missing in some features but not others). Skew also commonly refers to a failure of correspondence between two variables or data streams that are intended to be synchronized. Note that the way this is used in ML *systems* only rarely includes the use in statistics (biased distribution shifts in the underlying data).

The skew most likely to cause production problems is *training-serving skew*, which describes any difference between the performance of your model in training and in serving. Common causes are changes in feature definition between training and serving, changes or gaps in the data itself, or sometimes even some kind of a feedback loop between the algorithm and the task. Skew of this kind and other kinds is a common cause of avoidable outages in ML systems, and all of the simple causes of it should be targets for monitoring.

Different models and ML systems are subject to different kinds of skew. Astute readers will therefore note a problem: the techniques for monitoring and detecting skew are model specific (or at least specific to a particular model architecture, configuration, and purpose). There's no meaningful "just monitor all manner of skew for all models" function that anyone knows how to do yet (though it's certainly an intriguing idea). In other words, we can detect a difference between training and serving, but not know the underlying causes that difference has, or whether that difference is meaningful. We cannot detect differences in coverage of particular datasets without knowing what constitutes the domain for those datasets and what the expected coverage is.

In terms of impact on monitoring best practices, this implies that the monitoring system has to be general-purpose and flexible, but that individual model and model-family monitoring has to be implemented by production engineers and ML engineers working closely together. This is an area that many people are working on and is expected to improve in the future.

A Mindset Change Is Required

Though we have many technical challenges today, the organizational and cultural ones that act against holistic monitoring are arguably the most relevant ones here. In particular, model developers don't generally think in terms of detection of issues *post-deployment*, but instead think in terms of modeling KPI performance *pre*-deployment—and modeling KPIs are not necessarily directly connected to business KPIs![6]

This obviously presents a problem for whole-lifecycle monitoring, since both pre- *and* post-deployment turn out to be important, and like successful software generally, post-deployment often lasts longer than you think. Teams focused on developing and deploying models quickly are often impatient about rigorous delivery frameworks, as though these delivery frameworks will prevent them from organizing training and serving in whatever way is most convenient for them. Which, of course, they do—to some extent. But they do so by providing a set of monitoring and management guarantees in production that would otherwise be difficult to achieve and deployed on only an ad hoc basis.

If we accept this framing, the most important thing we should do is try to have a reasonable, flexible solution for maintaining the broadest useful picture of model behavior throughout its entire lifecycle, and make it adaptable for your own situation. Since the special case tools that today are used for model development (TensorBoard, Weights & Biases, and so on) don't usually naturally translate into production itself, the particular monitoring system in use, and so on, at the moment we will necessarily have to make some of this up ourselves. Given that, the overall goal for this chapter is to recommend a whole-lifecycle approach to monitoring, and in particular, suggest a default set of things to monitor *other* than the specific business metrics the model is intended to improve, since they are already well understood.

Best Practices for ML Model Monitoring

Let's start off with a few framing assumptions: for the purposes of this chapter, model development is generally done in a loop. We select data, train on it, build a model, do basic testing/validation, tweak, retrain, eventually release to production, learn how the model behaves, and begin the cycle again with ideas for improvement.

Monitoring in serving can be divided into model, data, and service (also known as *infrastructure*). Separation like this is useful because we don't have to handle every detail at every level, though we acknowledge that some crossover exists.

---

6 Indeed, we often see that modeling KPIs are actively hard to link to business KPIs, and teams can end up doing a series of A/B tests geared not toward safe rollout, but toward understanding the degree of coupling between online business metrics and offline modeling metrics.

Explainability—understanding what led the model to classify as it did, predict as it did, and so on—is a huge topic and likely to get more and more important as ML plays a larger role in more industries. A detailed explanation of explainability and an overview of best practices is beyond the scope of this chapter, and indeed this book. A handful of the ethical principles are addressed in Chapter 6, and most readers should review that.

From the more practical perspective of model monitoring only, however, explainability is important to understand in preproduction and, increasingly, production phases. The particularities of explainability vary according to model type, phase, business strategies, and so on. But the in-production cases are usually where ML is being used in safety-critical or socially significant ways, and where there might be a legal or ethical interest in understanding what led to the outcome. The primary objective here is to find as many ways as possible to "smooth out" the difference between model development and serving.

Generic Pre-serving Model Recommendations

We talk about this more in Chapter 3, but from a monitoring point of view, it's most important to keep in mind the business goal attached to the development of the model, and connect its KPIs to exported metrics for monitoring purposes. A model for which you have no business insight but plenty of infrastructural insight would be close to useless, and similarly, a model that you understood to work well in development but for which you had no insight into production behavior would arguably be potentially more harmful than useless.

The most important recommendation is that your business KPIs should correlate with model metrics; you should be able to trace these continuously from development to production. For example, if you are a ride-hailing app company, and you are predicting estimated time of arrival (ETA) for the ride, pickup location is something you'll want clearly available during the whole period. When monitoring for data integrity, the most important features of the model should therefore be given high priority.

Explainability and monitoring

As we've mentioned, explainability is a big area, but in particular for monitoring it presents some problems. Like debugging or observability in general, full explainability generally involves more resources (and is slower, more costly, and so on) in production than in development. Yet you often want explainability most urgently in a production context.

People responsible for ML models want explainability for a few reasons: establishing which features should be prioritized for data integrity, investigating a specific prediction or specific slices of predictions, and responding to business requirements for explainability generally. Since it's so expensive, and since business people may not have all the background required to understand it, being effective here often amounts to having a conversation with them about what they really care about. Often you can respond to their concerns (and potentially even build special monitoring solutions) without having to get into highly specific modeling details, which is a conversation that can sometimes distract from the essentials.

However, in some cases explainability is essential for troubleshooting. Let's take a lending use case: say a prediction is rejected because the request has a unique value for one feature that is not commonly seen—in this case, an application date of February 29. Application date is typically not one of the top 10 most important features across all of our predictions, but if for some reason our training dataset has only a few applications on February 29 and those end up being poor risks in the intervening years, we can imagine a model that uses the application date to heavily influence the decision to reject.

In such cases, explainability can surface what drove that individual prediction's output. This could be implemented via regularly logged summaries of predictions, exposing internal mechanisms via LIME and SHAP (essentially surrogate or duplicate models that use a conceptually analogous approach to differential cryptanalysis)[7], or another more customized approach. You generally find such explanations are not just used by model developers, but also risk and compliance teams, and can help nontechnical users understand systematic issues with models. Unfortunately, this is mostly too expensive to do in production.

Training and Retraining

In many ways, the training phase is the easiest to handle—from a classic monitoring point of view, anyway. The most important thing for monitoring training is keeping a holistic view, with the most important metric being how long it takes from starting training to producing a (hopefully working) model.

Having said that, other factors are important too—in particular, understanding when to *retrain*—i.e., to build a new model, in the hope or expectation that it might fix a problem for us. Though model rollback is the most common tactic used to resolve production problems, from time to time retraining is used—in this context, you can think of it as being like roll-forward for models: i.e., replacing what's in production with the latest version of everything.

7 We see an industry trend toward SHAP as the gold standard for model interpretability; LIME is also popular. See "Idea Behind LIME and SHAP" (*https://oreil.ly/yd9zo*) by Ashutosh Nayak for more details on SHAP and LIME.

Retraining is generally used when rollback hasn't worked or can't work (see "Fallbacks in validation" on page 203), or if roll-forward is easier for your infrastructure than rollback.[8] Equally, retraining has drawbacks in two circumstances. In the first, retraining would execute over the exact same data you used to train the old model (in which case, you would broadly expect the same behavior, since it's the same input). In the second, retraining takes so long that you can't use it to tactically resolve an outage. (This is one reason you might want to just run retraining automatically and periodically, if it's feasible for you to spend resources on this.)

For the first issue, you can *sometimes* "hack" the situation a little by changing the data you're training over—either by using an entirely new corpus, replacing some bad data, adding missing data, and so on. (Of course, significant variables for the training process here include changing the range of data we are training over, and whether or not we change any weightings, but hopefully deciding those is relatively quick.)[9]

Training is important in a monitoring context not only because of retraining, but also because this is where most developers establish their *baselines*, and these are used widely thereafter as a comparison point.[10]

Concrete recommendations

Next, we outline a list of things to monitor to get base-level coverage. Things that we consider minimal and mandatory, we highlight in italics; if you have to start somewhere, start there. Of course, you could put a *lot* of effort into monitoring, and perhaps in some cases it's worth it, but in some it's not. So we recommend you think carefully about the cost/benefit trade-off before implementing all of the following checks. (If you are looking for a more advanced guide to deciding what's important, we recommend you look at writing an SLO for your training pipelines and models—see "SLOs in ML monitoring" on page 216 for more details.)

8 In the annals of reliability literature, *roll-forward* is generally considered quite risky and to be seriously considered only when there is genuinely no possibility of a safe and clean rollback. Change is change, and even going back to an older version of a model represents some risk, but doing so may tend to minimize that risk by reducing the number of *untested* changes.

9 Be warned that sometimes if you press the retraining button, things get worse. You could press it at a time when a radically different model might be built than only a short time ago and thereby produce something equally broken. Think of the difference between training over December 24 and December 26 in the Anglosphere, for example. There's also no guarantee the new version is better, so unless your validation process is significantly automated, you won't just pay the CPU cost; you'll pay staff time costs as well.

10 Andrej Karpathy's blog on the software development lifecycle (SDLC) in an ML context (*https://oreil.ly/dtqF7*) and the Google Cloud pages on continuous ML (*https://oreil.ly/IKU5a*) are also very much worth reading in this context.

- Input data
  - *How large is the input dataset compared to the expected size?* Compare the training dataset to either the last time we trained this model or use another exogenous metric that can indicate rough size. If the dataset has shrunk by 50% unexpectedly, that's usually a bad sign.
  - *Are you comparing raw input data and feature data?* (Some problems emerge only when combining fields into features.)
  - *What are the youngest and oldest pieces of input data? Do they conform to expectations?* For extra credit, start to look at the histogram of the distribution of ages. In a distributed training situation, sometimes there can be a small number of very old pieces of data while everything else has been processed cleanly and effectively.
  - *What is the cardinality* (in this case, total number of elements or examples), and *how is the data distributed?* Is the distribution very different from the expected distribution—either compared to the previous time we trained on this data, or compared to other generated expected distributions?
  - In a batch-processing scenario, *can you enumerate the data and ensure it's finalized?* (Can you guarantee that all of the events from a particular day—say, yesterday—have arrived and include all relevant fields? Do you have any outliers from the day before that, or from today?)
  - In a streaming processing scenario, *what is the rate of arrival of incoming data?* Do you process on receipt of each "bundle" or after a fixed size has been collected (in which case, you should track how often processing is invoked)?
  - If access to the data is mediated, *what are the rates of access? Are there a large number of failures* (particularly authentication-related ones)?
  - If the data is copied from somewhere else to go into, say, a feature store, and the model is built from a feature store, *does that copying happen correctly?*
- Processing
  - *How many processing jobs are running? When did they last run? Did they complete by the time they should have? What was the rate of restarts and the successful job completion rate? Is there a backlog of unprocessed units? What is the distribution, in age, of the unprocessed units?*
  - *What is the processing rate* (measured, say, in input data elements processed per second?) *What is the 50th percentile, 90th percentile, and 99th percentile of processing rate?* (Pay particular attention to long-running shards—i.e., if you split your work across many processing jobs, you often see one of them running long—see the previous comment about input distributions.) How does that compare to the previous iteration? Or for more accurate comparisons,

how do you ensure that seasonality effects are properly accounted for in defining the previous iteration?[11]

  - *How much, in total, of the three axes of CPU, memory, and I/O have been consumed to produce the model?* (This can be nontrivial to determine, but is crucial for figuring out bottlenecks—in particular, if you add more of any resource, would you know if your system would get faster?)

- Holistic view
  - *How long did the entire run—from marshaling input data to producing the model—take, both in absolute duration and comparative duration?*
  - What is the *size of the output model*? If there is more than one file, is every file present that should be?
  - *Can the model be successfully loaded and make simple predictions?*
  - For those who do testing in a separate environment, *does the model pass the testing process?*
  - *What is the time to get the model into production and serving queries?*
  - For those who do testing in production, *does the model pass exposure to users?* Are the business or use metrics that you track in production affected in an unexpected way by the new model?
  - *Do you understand what the largest contribution to getting models into production is?* Is it manual action or automatic action? (You might have to add decoration or annotation information to achieve this: for example, a way to annotate a particular period as being a time of heavy network load.)

Finally, we note that some of these suggestions might also usefully apply to subcomponents of your training pipeline. It's not uncommon to have, for example, various data landing zones that have some very simple checks applied to them before they get copied for fuller processing elsewhere (think yarn supplier delivery manifests, for example, often sent over relatively inflexible electronic fund transfer, or EFT, processes). Being aware if the size of those dropped by 50% would be useful. Similar arguments might apply to chains of feature preprocessing prior to feature-store assumption, and so on. We can't describe every possible architecture in advance here, but we can say that business risk analysis can help you to understand where best to place your scarce resources.

11 For example, using week-on-week to avoid comparing (say) a Saturday to a Friday, which often have very different shopping patterns. In general, being aware of these patterns is useful for monitoring work.

Model Validation (Before Rollout)

Models are generally designed to achieve a certain business impact, also known as *improving a metric*. *Business validation* can therefore be understood as attempting to understand the business impact of a model,[12] for example, looking at how it affects profit and loss (also sometimes known as the *profit/loss curve*), or using a confusion matrix, which tries to understand where the model classified no and the answer should have been yes, and so on. Of course, we always need to beware of user behavior being noisy, but the primary goal is to find out whether the model improves a certain metric from a baseline.

The secondary goal is trying to figure out whether the new model is better than what we already have. To do that, we not only have to test versus historical data, which we are probably doing anyway, but also compare two models against each other.[13] We can choose from at least two good approaches:

- Test in preproduction environments (sometimes called a *sandbox*), and run the models either in parallel or serially, depending on your capacity, to compare behaviors.
- Test in production, with the model getting a small subset of real user traffic (typically, between 1% and 5%), sometimes called a *canary test*.

The second approach also has the good effect of exposing any problem with the model reacting to user traffic before a full rollout. However, this does require you to have some way of engineering traffic such that a particular version gets only a subset of traffic—not all infrastructure supports this feature. But if you have it, it's really good for allowing safe transitions from old to new models, and overlaps nicely with A/B testing. (General model updates can also fit well within such a framework, and indeed contemporary software deployment makes a lot of use of this approach.)

A hybrid approach is to send some data to the model locally, and the same data to the model running in production (though with this approach, the local model is not enabled to take full production traffic; it is just taking your test traffic). This helps expose feature code differences between modeling and serving, configuration differences, and so on. Finally, some folks send production traffic to a model, but don't serve the results to end users, which tests many components of the serving path

12 Chapter 5 covers this topic in more detail, but it is necessary to repeat some of the elements, and emphasize a slightly different set, here.

13 Of course, that raises the possibility that, given performance may be a multidimensional assessment, there might be no clear winner. In that case, there are good arguments for doing two opposite things: (1) preferring the model currently running on the basis that it changes the least and is well understood, and (2) preferring a new model on the basis that something built over more recent data is probably going to survive change better and will be less painful to transition from. Only you can judge which is best for your circumstances.

without exposing users to risk. This is called *shadowing*, and provides another place on the spectrum to balance correctness and risk.

Fallbacks in validation

For all kinds of reasons, it's highly advisable to have a *fallback plan*, which is a set of steps you take if the new model fails, the rollout fails, or even the old model is found to have some weird behavior in a particular subset of circumstances. There are two main approaches (as seen in "Emergency Response Must Be Done in Real Time" on page 235):

Use an older version of the model (roll back)
: This implies it's a good idea to keep the actual binaries of these around, versioned correctly, as well as monitoring the actual versions in production, since stressing your training infrastructure to build a model from old data at a time of production outage is generally the opposite of a good idea.

Fall back to a simpler algorithmic or even hardcoded approach
: For example, when attempting to supply ranked recommendations for product purchases, instead of broken recommendations, just display the top 10 most popular products on the site—though it won't be right for anyone, it won't be too wrong either (in most cases)!

Be very careful in your rollback, since many subtle problems wait in the long grass, waiting to strike: *schema changes*, *format changes*, or changes in *semantics*—whereby a crucial database, data source, or feature store format changes between releases—are all examples of areas that are easy to overlook in rollbacks. This can sometimes make it effectively impossible or impractical to do so unless the whole suite of dependencies is rebuilt, but might also make it hard to roll forward to a new version unless the actual error is fully understood. In these situations, algorithmic fallbacks can be life-saving. However, algorithmic fallbacks also can themselves fail because of categorical errors.

A nuance here is that even if the schema stays the same, feature transformations applied to one version of the model may not be the same as what's applied to another version. For example, how to handle a missing value for a ride-hailing app pickup location in one version of the model might be to simply default to where the rider was last seen, and in another version of the model might be the closest previously saved pickup location (e.g. home, office, etc.). These differences in the way the features are transformed might differ among versions and would also complicate rollback. Even worse would be if you roll back to one feature transformation on one "side" of the system and stay with the new one on the other "side."

Call to action

We don't yet have anything like the CI/CD workflow used in the infrastructure-as-code (IaC) community today that is commonly available for model development.[14] Different companies and even different teams within the same company could solve the problems differently. That doesn't mean you can't do checks, or can't do them automatically—homegrown approaches have, of course, some utility—but today's reality involves manual, peer-based review of both code and data, and detailed validation requiring PhDs in statistics.

For what it's worth, this situation imposes serious friction for ML development as it stands. The industry actively needs to work toward making as much of this as automatic as possible, and this work is too important to be left to the platforms. We therefore call on the industry to move toward a state where IaC approaches are used as widely in ML development as CI/CD and IaC enjoy today in product development. Even if we don't get that far, being in a situation where everything after manual validation moves the training artifacts safely and automatically to production would be a significant improvement than what we have on average across industry today.

Concrete recommendations

Commonly used numerical KPIs are measured to assess an ML model's performance. These are also covered in some detail and with a deeper theoretical foundation in Chapter 5, and with specific attention to serving use cases in Chapter 8. We summarize these KPIs here for continuity and ease of access:

Accuracy
: The fraction of predictions for which the model is correct.

Precision
: The ratio of true positives to total positives.

Recall
: The ratio of predicted positives to total positives.[15]

---

14 IaC is the act of managing computer infrastructure as if it was code—which is to say, with statements in files, version control, release processes, and so on.

15 For those with a search engine background, by way of an example, if you search for a particular term, *precision* indicates the number of documents you retrieve that contain the term in a relevant way, divided by the total number of documents retrieved, and *recall* indicates the number of relevant documents retrieved, divided by the total number of relevant documents. So *high precision* means you don't give the user irrelevant documents, and *high recall* means you give them almost every relevant document. The converses are, alas, both easy to intuit the definition of, and easy to generate.

Precision and recall (PR) curve
: The space of tradeoffs between precision and recall at different decision thresholds.

Log loss
: See Chapter 5 for more detail.

ROC curve and AUC
: A threshold-independent measure of model quality.

Many other possible mathematical measures could be used, some of which you might recognize from early statistics classes (e.g., *p*-values and coefficient effect sizes) and some of which are a little more involved (e.g., posterior simulations). On the whole, though, if you understand the preceding list in terms of your model, and understand your output distribution *shape*, you have a very good handle on what's going on.

Serving

Monitoring ML models while they are serving in production has all of the difficulties of observing things in production, combined with the numerous challenges of figuring out *why* things are happening that are peculiar to ML. Nonetheless, we can start with a set of simple questions:

- What are good metrics to measure my model in serving? (In some sense, you want the more relevant ones; arbitrary metrics have a habit of growing without limit until signal/noise is degraded.)
- Is my model performing as expected? If it's not performing as well, why, and how can I resolve the issue?

To choose good metrics for measuring model performance, we need to properly understand how a model can fail. Three components are needed to make a model work successfully:

Model
: The model making predictions.

Data
: The data that is flowing in and out of the model. This includes the features that the model uses to make a prediction and the prediction itself.

Service
: The service that actually renders the model. This typically involves the deployment of the model and serving of inferences—infrastructure, in the broadest sense.

Each of these three components needs to be working as expected in order for the model to be successful. A failure in any can have an impact on the overall business KPIs that the model was designed to improve. Therefore, we need to measure *all* of these to have a complete picture. In the next sections, we examine them in more detail.

Model

Starting with the model itself, we again note that measuring model performance *in* serving is a lot trickier than measuring performance *prior to* serving. Given those difficulties, we prefer to use the same metrics to evaluate the model in serving as we do in training/validation, since that will give us more confidence that we're actually measuring in some sense "the same thing." Of course, doing this requires being able to match the prediction with the corresponding observed reality.

For example, let's assume a ride-hailing company has a model predicting the ETA of a car for a customer. While evaluating this model in validation, you care a lot about minimizing the error (usually referred to as *root mean squared error*, or *RMSE*[16]) so that the model's prediction is close to the actual ETA: in this context, it turns out that customers tend to get more upset if you overpromise and underdeliver than the other way around. So to calculate this same metric in a production environment, there must be some process that can reconcile the predicted ETA with the actual ETA (i.e., calculate the error).

In practice, a few scenarios describe the delay with which the real results (known as *actuals*) arrive, and how we cope with that.[17]

Case 1: Real-time actuals. The ideal ML deployment scenario—and often the only one taught in the classroom—occurs when you get fast actionable and fast performance information back on the model as soon as you deploy your model to production.

Many industries are lucky enough to have this ideal scenario. Probably the most famous is digital advertising: a model attempts to predict which ad a consumer is most likely to engage with. Almost immediately after the prediction is made, the ground truth—whether they clicked or not—is determined. A similar example is food delivery; as soon as the pizza has arrived at the hungry customer's house, you have real measurements you can compare your model predictions with, and pizza delivery as a business has a strong time limit baked into it (as well as other ingredients)—

16 Much more detail is available on RMSE in many other places; the C3 AI Glossary page (*https://oreil.ly/KqpdH*) is probably a good place to start.

17 The engineering challenge here is how to join these sets of data, which can often be done with session IDs, user journey tokens, or something similar.

take too long, and the customer typically no longer wants it.[18] Ultimately, a strong detection KPI is what you want, if you can get it.

The key thing this quick feedback loop enables is the ability to measure the effectiveness of your models essentially *instantaneously* (or at least very quickly); of course, once you have this latent ground truth linked back to your prediction event, no matter how long it took to get it, model performance metrics can easily be calculated and tracked.[19] Tracking such metrics on a regular cadence allows you to make certain that performance has not degraded drastically from when a model was trained, or when it was initially promoted to production.

However, many real-world environments change the way you get access to ground-truth data, as well as the tools you have at your disposal to monitor your models.

Case 2: Delayed actuals. In this case, you don't get the benefit of the fast real-time feedback outlined previously. Indeed, these kinds of situations are arguably more common in business generally than case 1, and certainly a lot more awkward to deal with. Imagine you are trying to predict the likelihood of physical infrastructure failing—say, bridges collapsing—or use ML to perform predictions in the real estate market. Both of these scenarios have durations that could plausibly be measured in years, or even decades. This makes it tricky to ensure that a model developed to help you make higher-quality decisions is behaving as expected.

This delay in receiving ground truth, as well as being very long, might also be *unbounded*. Take, for example, a fintech company trying to classify which credit card transactions are fraudulent. You likely won't know whether a transaction is truly fraudulent until you get a customer card loss report or charge dispute. This can happen a couple of days, weeks, or months after the transaction cleared—or in the scenario where the transaction goes undetected by the customer, might *never* happen.

Ultimately, you can still make the first approach work if you get "enough" semi-real-time data, the data arrives reliably enough, and so on, but if you can't make that approach work, teams may need to turn to proxy metrics. *Proxy metrics* are alternative signals that are correlated with the ground truth that you're trying to approximate, but are selected specifically because they arrive more quickly.

In the case of bridge failures, we might look at results of bridge inspections, maintenance schedules, whether a bridge is in an area prone to flooding, and the age of a bridge. For real-estate purchase-price predictions, a common technique is to look at prices for similar houses, but with as few components changed as possible: for

18 Though YarnIt has been experimenting with drop-ship capabilities for delivering its products, the feedback loop is still limited by the physical delivery cycle and product review time.

19 The best model metric to use primarily depends on the type of model and the distribution of the data it's predicting over. This should typically match up with the metrics used in training/validation.

example, the same number of bedrooms and bathrooms but in a different area, or different number of bedrooms but in the same area, and so on. For construction timescales of months or years, even a messy composite proxy like this can be better than nothing.

Ultimately, proxy metrics serve as a powerful tool in the face of delayed ground truth since, if you can't get a strongly correlated metric in real time, you can at least get a more weakly correlated one; often that's good enough. Don't forget the mathematical requirement that your proxy metric has statistical significance, however—and proxy metrics may even change relevance over time, so they need to be continually reevaluated.

Case 3: Biased actuals. One important thing to note is that not all ground truth is created equal. For example, if we are a fintech company looking at creditworthiness in loan contexts, the central problem with predictions is that declining a loan means you no longer have any information about whether the applicant could have paid you back. In other words, only the people you decide to give a loan to will result in outcomes that you can use to train future models on—a kind of selection bias that could allow bias of other kinds to creep in. As a result, we will never know whether someone the model predicted will default could have actually paid the loan back in full. As described more fully in Chapter 6, it is therefore critical to assess our data for potential biases, blindspots, and areas of under-representation.

Case 4: No/few actuals. In some ML applications, getting back actuals (responses) in a reasonable time window is just not possible. This could be for a variety of reasons, including these:

- Manual intervention is required to verify the model's predictions.
- There is no way to attribute the response to the prediction.
- The time window of getting actuals is so delayed it cannot meaningfully inform the modeler that the model's performance should be looked at.

For example, many image-classification applications require a human in the loop to manually verify that the images were classified correctly. This might require sampling so that only a valuable segment of the predictions is manually verified to improve the model's future training dataset.

So what does a team do if it doesn't get back actuals? In these scenarios, it's not uncommon for ML teams to once again use proxy metrics to give signals of model performance. A weaker correlation may be better than nothing in these extreme cases. It is also common while testing new versions of models in production to A/B test models and compare their impact on product metrics. Monitoring data of the model becomes even more important to know if the model has deviated from what is expected.

Other approaches. Quite apart from how a model might handle the relationship with actuals, it is possible to use generic measurements of the behavior of a model that can be useful for figuring out whether things have gone truly awry. Our top selection here is the share of "useless" responses—i.e., empty, incomplete, or with subpar fallback—versus "good enough" responses, though another critical one is our old friend, data distribution.

Troubleshooting model performance metrics. Let's assume that model performance metrics can be calculated. Inevitably, at some point your model will not meet performance expectations. The hard question to answer is *why*. The most common causes are typically undersampling in training data, drift, and data integrity issues impacting the quality of data the model uses to make predictions. A best practice is to look beyond averages and investigate various *slices* of predictions, i.e., a specific subsegment of predictions, such as everyone in California.

Imagine that your model is predicting likelihood of fraud. You expect the volume of false negatives (your model missed catching a real fraud transaction) to be less than 0.01%. Let's say you suddenly see false negatives jump to 2%. A natural way to proceed is to see whether this behavior is localized in any particular region or merchant. By doing so, you can surface where the missed classifications happen, and what slices of data to upsample when retraining the model. Of course, understanding the worst-performing slices of the model can provide feedback to model builders for ways to improve model performance.

Data

A key component of ML monitoring is monitoring the inputs and outputs of the model. As features are added or dropped, monitoring must be adapted to the schema of the model. We have two common ways to monitor data—drift detection and data quality checks. Drift is better for catching slow changes to the distribution of the data, while data quality checks are better for catching sudden, large changes in the data.

Drift. Drift measures change in distribution over time. Models do not perform equally well on every possible input: they are highly dependent on the data they were trained on, perform well when they see data that resembles that, and perform less well when they don't. Especially in hyper-growth businesses where data is constantly evolving, accounting for drift is important to ensure that your models stay relevant. As a result, measuring the distribution of your data in various dimensions, if only via histogram-style methods, is crucial to understanding what's going on in production. Some models are resilient to minor changes in input distributions, but accepting infinite resilience does not exist; at some point data distributions will stray far from what the model saw in training, and performance on the task at hand will suffer. This kind of drift is known as *feature drift*, or *data drift*. Conversely, when model outputs deviate from the established baseline, this is known as *model drift*, or *prediction drift*.

It would be great if the only things that could change were the inputs to your model, but unfortunately, that's not the case. Assuming your model is deterministic, and nothing in your feature pipelines has changed, it should give the same results if it sees the same inputs.

While this is reassuring, what would happen if the distribution of the correct answers, the actuals, change? Even if your model is making the same predictions as yesterday, it can make mistakes today! This drift in actuals can cause a regression in your model's performance and is commonly referred to as *concept drift*.

Measuring drift. Understanding the differences in two distributions is important, but is a sophisticated topic that we cannot go into in detail at this point. So if this material is foreign to you, don't fret. Just either use it as it is, or go look things up and learn more. (Either way, you'll probably end up coming across some of these terms if you engage with data distributions at all, never mind actively monitoring them.)

As we talked about previously, we measure drift by comparing the distributions of the inputs, outputs, and actuals between two distributions. The two distributions are typically the production data and the baseline distribution. Commonly used baseline distributions are training datasets, test datasets, or a prior window of time in production. But how do you quantify the distance between these distributions?

Several ways to do this are popular. The *population stability index* (*PSI*), commonly used across banking and fintechs, relies on bucketizing data for one metric, calculating the percentages residing in the buckets for each distribution, and comparing the log of the divided percentages. This detects shifts within buckets or between buckets. The *Kullback-Leibler divergence* (*KL divergence*) is similar to PSI, but asymmetric so it can detect distribution order switching. The *Wasserstein distance* (also known as *earth mover's distance*) calculates the amount of work required to move one distribution over to another, which is useful for recognizing the amount of shift between buckets—i.e., if things spill from one bucket to the next, the Wasserstein score will be lower than if a more radical jump occurs from one end of the distribution to the other far end.

While each of these distribution distance measures differ in the way they compute distance, they are all doing the same thing: providing a way to quantify how different two distributions are. Especially when actuals are not available, drift is used in the real world to identify changes in model predictions, features, and actuals.

Troubleshooting drift. In many ML use cases where performance metrics cannot be calculated directly, drift often becomes the primary way to monitor changes in model predictions. When examining why predictions have started the drift, we usually start with looking at which inputs have also drifted, and we can couple feature drift and feature importance together to determine which features might be more strongly correlated with the change. In turn, that helps us figure out which features might need to be resampled from baseline or could potentially need a data quality fix.

Data quality

While drift is focused on *slow* failures, data quality monitoring for models is focused on the *hard* failures. Models rely on the input features coming in to make a prediction, and these input features can come from a variety of data sources. Different types of data can be monitored for data quality issues—categorical fields, numerical fields, as well as unstructured data types such as embeddings, and so on. Here, we will dive into common strategies for monitoring structured categorical and numerical data.

Categorical data. *Categorical data* is a stream of selections of a single value from a limited but larger collection of values. Think of categories like the type of pet someone owns: dog, cat, bird, and so on. You might not think much can go wrong here, but a sudden shift in the distribution of categories is always possible, either because of user behavior (this year's hot Christmas pet is, for example, reindeer) or other failures. Let's say, for example, your hypothetical model predicting which pet food to buy for your pet supply store sees data suggesting that people own only cats now. This might cause your model to purchase only cat food, and all your potential customers with dogs will have to go to the pet supply store down the street instead.

In addition to a sudden cardinality shift in your categorical data—i.e., literally how many are counted in each category—your data stream might start returning values that are not valid for the category. This is, quite simply, a bug in your data stream, and a violation of the contract (semantic expectations) you have set up between the data and the model. This could happen for a variety of reasons: your data source being unreliable, your data processing code going awry, an upstream schema change, etc. At this point, whatever comes out of your model is undefined behavior, and you need to make sure to protect yourself against type mismatches like this in categorical data streams.

Examples might include the following:

- You were expecting a string for a feature and suddenly received floats.
- You are case-sensitive for a feature (i.e., state inputs) and were expecting lowercase values, but are now receiving uppercase values (e.g., *ca* versus *CA*).
- You are receiving data from a third-party vendor, and the order of their schema has an off-by-one error. You are now seeing each feature receive another feature's values.

An unfortunately common—and tricky-to-handle—situation is missing data. This can arise for any number of reasons to do with infrastructure, application, storage, and network failures, or simple bugs. The real question is how to handle it. In a training context, sometimes just discarding the row will allow things to proceed and not be too bad; in a production context, you can throw an error (but not a fatal one; otherwise, that turns an intermittently flaky storage service into fleet-wide death, which is not to be recommended). While these techniques help you compensate for this problem, it's not really a sustainable solution: if you have hundreds, thousands, or tens of thousands of data streams used to compute one feature vector for your model, the chance that one of these streams is missing can be very high!

Though this is more to do with robustness than monitoring, it is possible to compensate for missing values in categorical data in a number of ways, a process commonly referred to as *imputation*. You could choose the most common category that you have historically seen in your data,[20] or you could use the values that are present to predict what this missing value likely is. The complexity of your solution to this problem is entirely up to your application scenario, but it's important to know that no solution is perfect.

Numerical data. A numerical data stream is also pretty self-explanatory. *Numerical data* is data generally represented by floats (though sometimes integers), such as the amount of money in your bank account, or the temperature in Fahrenheit or Celsius.

To start off our analysis of what can go wrong with numerical streams, one thing you sometimes see is an *outlier*, a value that exists far out of the range of historical values. Outliers are potentially dangerous to your model, as they could make your model mispredict in spectacular fashion.

Type mismatch can also affect numerical data. It's possible that a particular data stream where you're expecting a temperature reading instead gives you (say) a categorical data point, and you have to handle this appropriately. It's possible that the default behavior may be to cast this categorical value to a number which, though now valid, has entirely lost its semantic meaning and is now an error in your data that is incredibly hard to track down. Another possibility is something that maybe doesn't change type but changes semantics: maybe you're tracking a total counter (with some arbitrarily large or small wrap), and it suddenly gets changed to a delta. Depending on the data, of course, this might go undetected for quite a while!

20 This strengthens the argument for your application keeping distribution state, or at least bucket counters. Other techniques for robustness actually use *dropping* data in order to improve robustness (which is really a synonym for avoiding overfitting).

Lastly, just as with categorical data, numerical data also suffers from the same missing data problems, but the ordering and span implied by a numerical sequence gives us more options for imputation compared to the pure categorical situation. For example, we can take an average, median, or other aggregate distribution metric to impute a missing numeric value such as yarn weight (in grams); see Figure 9-2.

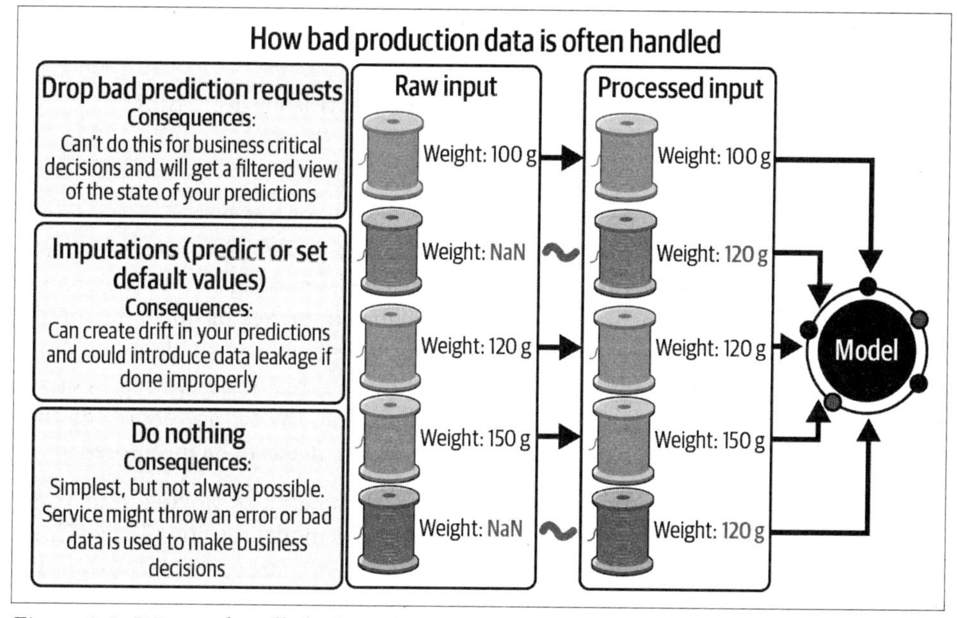

Figure 9-2. Ways to handle bad production data

Measuring data quality. It's not surprising to ML practitioners today that many models rely on very large numbers of features to perform their task. With training set sizes exploding into the hundreds of millions and even billions, models with feature vector lengths in the tens and hundreds of thousands are not uncommon.

This leads us to a major challenge that practitioners face today. To support these incredibly large feature vectors, teams have poured larger and larger data streams into feature generation. The reality is that this data schema will inevitably change often as the team experiments to improve the model. It's common to add a feature, drop a feature, or change how it is computed/processed, and platforms such as feature stores are becoming widespread for tackling precisely this management overhead. Concretely, here are the most important checks to do across your features:

- Categorical data
  - Cardinality: Has the number of unique values changed?
  - Missing values: Is this feature missing data?

— Type mismatch: Has the data type changed?
— Volume of data: Has the volume of data seen for this feature changed?

- Numerical data
  — Out-of-range violations: Has the value gone outside an appropriate range?
  — Missing values: Is this feature missing data?
  — Type mismatch: Has the data type changed?
  — Moving averages: Have the feature values been increasing/decreasing?

Service

For an ML system to be successful, you need to understand not just the data going in and out of the ML system, and the performance of the model itself, but the overall service performance in rendering or serving the model—making its predictions or classifications available. Even if the model performance improves business outcomes and data integrity is maintained, but it takes several minutes for a single prediction, it might not be performant enough to be deployed in a real-time serving system. Similar to other software services deployed to production, the service deploying the model into production and serving the model's inferences needs to be monitored.

Numerous options exist for serving a model into production: deploying your own APIs/microservices, using an open source framework for model serving (TensorFlow serving, PyTorch serving, Kubeflow serving, and so on), or using a third-party service. Regardless of what is used for model serving, it is important (especially in real-time services) to monitor the prediction latency of the service because the expected prediction should happen immediately after the request is sent. There are two ways to do this:

Model level
: Reduce the time it takes for the model to make a prediction.

Serving level
: Reduce the time the system takes to service the prediction when it receives the request. This is not just about the model, but also gathering the input features (sometimes precomputing or caching them), and quickly catching the predictions to serve.

Optimizing performance of the model. To optimize the model for lower prediction latencies, the best approach is to reduce the complexity of the model. Some examples of reducing complexity could be reducing the number of layers in a neural network, reducing levels in decision trees, or reducing any irrelevant/unused part of the model. Architecture makes a difference too: for example, bidirectional encoder representations for transformers (BERT) is slower than feed-forward approaches, and tree-based models are faster than deep learning.

In some cases, this might be a direct trade-off to the model efficacy. For example, if there are more levels in a decision tree, more-complex relationships can be captured from the data and therefore increase the overall effectiveness of the model. However, fewer levels in a decision tree can reduce prediction latency. Balancing the efficacy of the model (accuracy, precision, AUC, and so on) with its required operational constraints is important to strive for in any model to be deployed. This becomes especially relevant for models that are embedded on mobile or devices.

Optimizing performance of the service. To optimize service performance, here are suggestions for areas you could monitor and improve.

First, let's consider *input feature lookup*. Before the model can even make a prediction, all of the input features must be gathered, and this is often accomplished by the service layer of the ML system. Some features will be passed in by the caller, while others might be collected from a datastore or calculated in real time. For example, a model predicting the likelihood of a customer responding to an ad might take in the historical purchase information of this customer. The customer wouldn't provide this when they view the page themselves, but the model service would query a feature store or a real-time database to fetch this information. Gathering input features can generally be classified into two groups:

Static features
: Features that are less likely to change quickly and can be stored or calculated ahead of time. For example, historical purchase patterns or preferences for a customer can be calculated ahead of time.

Real-time calculated features
: Features that require being calculated over a dynamic time window. For example, when predicting ETAs for food delivery, this might require knowing how many other orders have been made in the last hour.

In practice, we typically have a mix of user- or application-provided features, static features, and real-time calculated features. Monitoring the lookup and transformations needed for these features is important to trace where the latency is coming from in the ML system.

Next, let's consider *precomputing predictions*. In some use cases, it is possible to reduce prediction latency by precomputing predictions, storing them, and serving them using a low-latency read datastore. For example, a streaming service might store ahead of time the most popular recommendations for a new user of their service. This type of offline batch-scoring job can vastly reduce latencies in the serving environment because the brunt of the work has been done before the model has even been called.

Other Things to Consider

Even with all of the preceding considerations, we have by no means exhausted the list of potential concerns you might have with monitoring and observability. Here are some other areas you might want to keep in mind for your monitoring journey.

SLOs in ML monitoring. SLOs are a popular and still growing technique used to explicitly decide in advance an appropriate level of reliability for systems—in some sense, determining what the user experience will be—and deciding whether to do feature work or reliability work, depending on which side of the threshold we are on at a given moment. For example, if we have decided the customer should have 99.9% availability, and we move below that at some point, this triggers fixing broken systems until we are trending at 99.9% again. That way, the union set of ML engineers, the overall product team, and the SRE team (if any) cooperate to "bend the curve" of the user's experienced reliability toward the optimal, decided-on level.

This is a very simple, indeed simplistic, example, and the reality is considerably more complex in multiple key areas. Even though ML is complex, you can still get some value out of doing SLOs. But watch out for the following subtleties:

- As we have already established, you cannot determine the suitability of a model for production by just looking at the model. Therefore, SLOs have a strong case for being scoped to cover data as well (and data distribution, freshness, and so on).
- Availability for request-response serving systems, like the frontend web server for *yarnit.ai*, is a relatively clear-cut issue—it is either serving correct content within a certain acceptable latency, or it isn't. What does availability for an ML system mean when the things being served might have, for example, classifier confidences ranging from 0.0 to 1.0, and any individual result or set of results might be completely, correctly, and justifiably less than any arbitrary quality threshold? At the very least, SLOs for ML therefore have to encompass the idea of quality or confidence thresholds.
- The best way to do this is, perhaps a little surprisingly, not to focus on the ML system performance, but instead to focus on the business objective that the ML system in question is supposed to deliver. Or in other words, the system as a whole can suffer degradations of various—perhaps temporary—kinds, if the overall user experience is still within the defined limits. For more of this topic, see Chapter 8 of *Implementing Service Level Objectives* (*https://oreil.ly/hZ76d*).

- For environments with lots of models, or lots of model churn, a promising approach is a self-service infrastructure so ML engineers can define and enforce per model or per class of model SLOs (generally by comparison to a golden dataset); SREs could develop, offer, and support such a service, thus helping the overall SLO approach scale for everyone. (The same approach can be extended to separate use cases for a single model, too, as long as such tagging can be done deterministically.)
- Finally, as is clear from Chapter 11, ML is *entangled* across the business, and it may not be practical to specify SLOs for an ML system without also specifying them for all or most adjacent systems. (Indeed, you'll have to, for anything other than the smallest systems.) Additionally, since inspectability or explainability is often difficult to obtain, reacting quickly enough to defend a sufficiently demanding SLO (say, 99.5% or more) is not going to be something a human-driven process can achieve on its own: this will be a question of automatic systems acting in concert with the humans. Automatic systems acting on their own in the non-ML case are often very useful, but we should move with caution in the ML case.

If you wish to get more concrete about establishing SLOs for your ML infrastructure, we have a recommendation for where to start. If you look at Table 9-1, you'll see our suggestions for what to base your SLOs on, given what you care about. The rows cover the context of what you're trying to measure, and the columns address the scope. So, for example, if you're worried about the overall system health of your ML training system, the top-left quadrant is where to go to find ideas on where to start. In essence, the table outlines behaviors, indicators, or example metrics to base SLOs on. Of course, this is a nonexhaustive list.

Table 9-1. Behaviors or metrics to monitor per ML context

	Training	Serving	Application (YarnIt)
Overall system health	Model build time (excluding snapshots) Number of concurrent trainings Number of failed build attempts Resource saturation (e.g., GPU or I/O throughput)	Latency (time to respond to request) Traffic (number of requests) Error count (number of failed requests) Saturation (exhaustion of any particular resource: CPU, RAM, I/O)	Mostly tied to user journey or session Purchase rate Login rate Page subcomponent failure rate (e.g., cart display) Cart abandonment

	Training	Serving	Application (YarnIt)
Generic ML signals / basic model health	Pretraining: Source data size (i.e., hasn't grown or shrunk a lot since last build) Training: Training time as a function of model type and input data size Training configurations (e.g., hyperparameters) Post-training: Model quality metrics (accuracy, precision, recall, other model-specific metrics)	Model serving latency (as a fraction of overall serving latency or compared over time) Model serving throughput Model serving error rate (time-out/ empty value) Serving resource use (RAM in particular) Age/version of model in serving	Model-specific metrics (visible on individual page load) Number of recommendations per page Estimated quality (predicted click-through) of each recommendation Similar metrics for search model
Domain-specific signals	Built model passes validation tests (not only that aggregate filters work, but results on golden set and held-out set tests are of equal or better quality). Recommendations for new products in the store are of similar or greater quality.	These metrics are often delayed. Offline signals match served predictions in aggregate and in relevant slices. Number of recommendations served for specific queries matches pre-serving expectations.	Session- or user-journey-specific metrics. Percentage of visits with purchases is not declining. Average sale per visit maintained. Session duration maintained.

Actually specifying an SLO is best left to other texts, but The Site Reliability Workbook (*https://oreil.ly/1lbVy*), edited by Betsy Beyer et al. (O'Reilly, 2018) has a reasonable example of a fully specified, exhaustively detailed SLO document (*https://sre.google/workbook/slo-document*); it is also possible to be have something as simple as "99% of this data is younger than two hours" and for that still to provide value.

Monitoring across services. As you scale out your ML infrastructure further, two of the major things that you'll probably find yourself doing, which probably won't be easy, are distributed tracing, and understanding the latency distribution of your systems. (Though we've been talking about distributions a lot in this chapter, we are specifically referring to monitoring in this section, rather than ML.)

Distributed tracing, strongly related to observability, enables you to trace the full path of a request as it goes between different services in a distributed architecture. Think of it as labeling a request with "give me a full report on the path this takes." Every system that handles the request should turn on some degree of extra visibility for it as it cascades throughout the system. Various commercial and free systems implement this—you could do worse than look at OpenTelemetry (*https://opentelemetry.io*) to get a handle on how it works.

Another crucial area is understanding distributions—in this case, not of ML data, but of latency. It's not necessarily intuitive, but in a sufficiently complicated architectural arrangement, even a surprisingly small proportion of requests that run slowly can end up affecting customer experience significantly.[21] Most organizations don't have the storage or compute power to track this for every request, so some kind of *request sampling* is required—i.e., to track detailed statistics for certain requests, but only a comparatively small proportion of them by default. You can achieve this in various ways, depending on your environment, but we advise you to enable these capabilities early, since the additional insight into customer experience and production behavior generally can be vital.

Finally, one issue that is typically pertinent to only very large organizations is how to scale to very large monitoring setups. Though we can't cover this in detail here, there are three main issues to consider. First is (1) how to run multiple monitoring entities without duplication of alerts, data, and so on. Usually this is related to (2) finding a way to divide the entities to be monitored in a reasonable way (this usually implies approximately equal partitioning of the monitoring targets). Finally, (3) we must monitor the monitors, to make sure they themselves are running. Different monitoring products support these features to a different extent—Prometheus, for example, can monitor itself and deduplicate alerts. Our experience suggests that deduplication is potentially the most important capability here—if your system can't perform that, the next time you have a serious outage, you potentially have *N* × *<number of pages you would receive>*, where *N* is the number of monitoring instances you have. That could be quite a large number.

Further advice on constructing and running sophisticated monitoring setups is available from a variety of sources, but one particularly good source is *Observability Engineering* (*https://oreil.ly/WFkOm*) by Charity Majors et al. (O'Reilly, 2022).

Fairness in monitoring. In this chapter, we have covered important metrics for monitoring the service health, model efficacy, and data integrity of the model. However, the topic of monitoring includes other things to consider that we will not be able to cover in this chapter. For example, our deployed models, almost by definition, have real-world effects: deciding who gets a loan, who gets a job, or what purchasing decisions we will make. In the increasingly automated world of such decision making, it is critical we do not codify systemic bias and discrimination through the models. Fairness is obviously a critical topic and is covered in depth in Chapter 6. From a monitoring point of view, our major concern is the requirement to facilitate such visibility into model decisions, while not facilitating inappropriate visibility—see the next section for more detail.

Privacy in monitoring. A special case of fairness is the question of privacy in monitoring. Almost by default, production monitoring dashboards will display the information that is fed to them. You can therefore understand immediately that if your

21 See "The Tail at Scale" (*https://research.google/pubs/pub40801*) by Jeffrey Dean and Luiz Andre Barroso.

monitoring stack exposes PII or other sensitive information, that will go right to the dashboard and be viewable to all who can access it at the dashboard level—even if they can't access it at the production level. This is clearly a violation of privacy expectations, if not some privacy-concerned legal frameworks themselves (though we are not lawyers).

Though every situation has nuances, in general we recommend that if you can't tolerate the risk of PII escaping, you *either* tighten dashboard access (and every intermediate point where that information is collected), or you control what happens to the PII on egress from the first point of aggregation—for example, names, addresses, and so on, are fully anonymized and the system has had a thorough privacy review. In general, though controlling access to a dashboard is simpler and easier, it's really not a strong control point. You almost certainly want (need!) to PII-proof the data at the first point of egress from the monitoring system, or even do it in a staged way, where different kinds of data can be processed or stripped at different ingress points to different systems—a multistage filtering approach. Again, see Chapter 6 for a much more complete treatment of this important issue.

Business impact. Another important category we don't cover is relating the model performance metrics to the business impact of the model. We measure model performance with statistical metrics such as AUC or log loss, but those don't incorporate the concrete impact that the business has experienced when the metrics degrade. The more representative the metrics are, the easier it will be to make that connection—choose carefully![22]

Dense data types (image, video, text documents, audio, and so on). Another important category we won't be covering is monitoring for dense data types, also sometimes called *unstructured data*—though since these formats are, in fact, highly structured, this is a bit of a misnomer! As ML increasingly uses images, video, and so on as inputs into its models, it's necessary to monitor data integrity for these nontabular data types too. There aren't commonly available approaches today, and we call on the industry to actively work toward supporting this. One growing approach is to monitor the *embedding* outputs of the data itself. ML practitioners use embeddings to map these items (e.g., movies, images) to lower-dimensional vectors where similar items are closer together. Monitoring these lower-dimensional vectors provides a proxy to monitoring dense data types.

22 The business impact of a model is ultimately the only reason it exists. Every other metric is secondary to the purpose for which the model was built. Therefore, direct monitoring of business impact is the holy grail of monitoring, however difficult it might be. Determining the correlation between available monitoring metrics and business value is almost certainly the most important thing for the organization to do in order to extract the most value out of its ML efforts in production.

High-Level Recommendations for Monitoring Strategy

"Serving" on page 205 made many detailed recommendations for how to start monitoring, but we'd like to cover a few high-level recommendations for your overall monitoring strategy here:

Actuals or not
: If your model is able to get back actuals in a near real-time way, monitoring model/KPI performance is the best signal, since that corresponds most closely to your concept of what the model is doing. If you don't have reasonable actuals, you'll have to build a picture of what's happening from a set of partial sources, including infrastructural elements as well as model performance. Look at the "Generic recommendations" item toward the end of this list for hints on how to do that.

Model performance metrics
: In general, you are best served by exposing the model performance statistics we talked about in "Model" on page 206. Doing this is strongly dependent on your local monitoring situation, ML platform usage, and so on, so we can't tell you how to do this in detail, but at the very least you should be tracking them. There's also some value in looking at how the model performance is working at a pure service level. Think of it as a simple request-response service; if the model is unable to make predictions/recommendations, if the model produces predictions with a drastically lower confidence, and if any algorithmic fallback path is being invoked more than expected, these are all good things to know (and you should therefore monitor them).

Data concerns (drift)
: It is vital to track the distribution of input data on an ongoing basis, so computing the various measures of distribution (PSI, KL divergence, Wasserstein distance, etc.) and also surfacing that in your monitoring system is vital. (The distribution of your *output* data—i.e., the predictions themselves, reconciled with reality—is covered previously.)

Data concerns (quality)
: Track missing data, type mismatches, data corpus size, and related attributes as discussed previously in this chapter. Some of the recommendations from the training section are also hugely relevant in the serving context too—not the duration of training itself, of course, but everything about the availability of data.

Service or infrastructure performance
: As a first approximation—i.e., you need a place to start—you can treat this as any serving system. From *Site Reliability Engineering* (*https://oreil.ly/Y1gel*), there are four *golden signals* you can use to look at the high-level state of your serving system: latency, traffic, errors, and saturation. If you establish good levels

for these—via, for example, Chapter 8 of *Implementing Service Level Objectives* (*https://oreil.ly/efwiE*)—it's quite hard for something to go seriously wrong in infrastructure and *not* have it be reflected in those numbers. In some sense, "goodput" is what you're looking for: questions are arriving at a reasonable rate, and getting answered correctly and quickly enough for the needs of the application.

Alerting and SLOs
: We haven't said much specifically about alerting in this chapter, since in general, once you're monitoring a metric and have established some kind of threshold, alerting is (supposed to be) a fairly mechanical achievement. But it's important to pay special attention to the questions of deciding what the right metrics to alert on are—definitely not every metric—and what makes for a good threshold. Otherwise, everyone gets drowned in alerts, sucking away valuable cognitive attention, almost all of which are basically irrelevant to the customer experience. For more on this, a good place to start might be Chapter 8 of *Implementing Service Level Objectives* (*https://oreil.ly/Ruf03*).

Though many of the conceptual underpinnings of request/response systems apply to ML serving systems, training is different, being primarily either batch or streaming oriented. One potential problem you'll often come across in monitoring training is the question of how to alert for model building being late. The major challenge here is that since the model building duration can be a large number of minutes, hours, or even days, it doesn't make much sense to alert on every fluctuation in training performance; otherwise, you'll be alerting too much.

The quantity that becomes important therefore is *catch-up time*. If your model takes roughly (say) 20 hours to build, and you have a requirement for a new model every 24 hours, what happens if you get 8 hours through building and then stall for 4 hours? Having consumed 12 wall-clock hours, with 12 wall-clock hours to go, if you resume training at your previous rate, you will have 12 hours of learning before you're finished. So you may already have blown your deadline, and predicting that 12 hours early would be a significant optimization. Therefore maintaining a notion of how long it would take to catch up, and alerting when current time + catch-up time is greater than threshold will bring attention to potentially stuck situations usefully quickly.

Generic recommendations
: An important monitoring and/or debugging technique is to log predictions (ID, value) to a table, or some otherwise easily searchable format—if you're concerned about speed, head-of-line blocking, and so on, then sampling is also a perfectly tractable approach (though you will then lose guaranteed explainability). For extra points, have a column for the *actual* value, as opposed to the prediction, so you can backfill and compare. In the world of microservices and/or distributed

tracing, it's also useful to get the client consumer of a prediction to log the prediction ID as well, so you have a chain of joinable events. (Sometimes it's OK to kick away this kind of scaffolding in the process of moving from development to prediction, but it's generally handy in both contexts.)

Explainability during serving
: The gold standard for explainability during serving is what's called *individual inference level explainability*—in other words, why was this specific transaction granted, or denied, or what led to the classifier making that particular decision in this particular context? As per the preceding "Generic recommendations" paragraph, success here really looks like being able to tie the specific state of the model at the time of specific prediction request with a specific state. In some industries, such as lending, models that predict whether to give an individual a loan may use explainability to surface why an individual was rejected. This is often communicated to the downstream users via reason codes.

Conclusion

We hope this chapter has provided a useful overview to monitoring your ML systems from birth to happy life in production. To reiterate: the main battle is to realize that you need to monitor at as high a level of fidelity as you can—for explainability, production debugging, and just generally knowing how your business is doing. Once you accept that, you can choose from multiple approaches to implementation, and aggregating the various "Concrete recommendations" sections of this chapter should be of use to you.

CHAPTER 10
Continuous ML

Up until now, our discussions of ML systems have sometimes centered on the idea that a model is something we train and then deploy, almost as though this is something that happens once and only once. A slightly deeper view is to draw a distinction between models that are trained once and deployed versus those that are trained in a more continuous fashion, which we will refer to as *continuous ML systems*. Typical continuous ML systems receive new data in a streaming or periodic batch fashion and use this to trigger training an updated model version that is pushed to serving.

Clearly, there are major differences from an MLOps perspective between a model that is trained once versus a model that is updated in a continuous manner. Moving to continuous ML raises the stakes for automated verification. It introduces the potential for headaches around feedback loops and model reactions to changes in the external world. Managing continuous data streams, responding to model failures and corruptions, and even seemingly trivial tasks like introducing new features for the model to train on all increase system complexity.

Indeed, on the surface, it may seem like creating a continuous ML system might be a terrible idea. After all, in doing so, we expose our system to a set of changes that is inherently unknowable in advance because of potential changes in the real external world, and thus may cause unexpected or undesired system behavior. If we remember that in ML, data is code, the idea of continuous ML is to accept the equivalent of a steady stream of new code that can change the behavior of our production system. The only reason to do this is when the benefits of a continuously updating system outweigh the costs. In many cases, the benefits are indeed considerable, because having a system that can learn and adapt to new emerging trends in the world can enable a level of overall system quality that could help us achieve complex business and product goals.

The goal of this chapter is to examine the areas where costs can accrue and problems may arise, with the aim of keeping the overall cost of continuous ML systems as low as possible while retaining the benefits. These include the following observations:

- External world events may influence our systems.
- Models may influence their own future training data through feedback loops.
- Temporal effects can arise at several timescales.
- Crisis response must be done in real time.
- New launches require staged ramp-ups and stable baselines.
- Models must be managed rather than shipped.

Each of these points summarizes a range of underlying complexities, which we will dive into in the bulk of this chapter. The technical challenges are not the end of the story, of course. In addition to the practical and technical challenges that continuous ML introduces into our ML development and deployment processes, it also creates organizational opportunities and complexities.

Models that are continuously improved need organizations capable of managing that continuous improvement.

We need frameworks for generating and tracking ideas for improvements to the model. We need a way to evaluate the performance of various versions of our models over long periods of time rather than focusing sharply on the point of time when we launch one model to replace another. We need to think of modeling as a long-lived, value-generating program that has costs and risks but also huge potential benefits, and discuss the organizational implications of these needs at the end of this chapter.

Anatomy of a Continuous ML System

Before we look in detail at the implications of continuous ML systems, let's take some time to take a pass through the typical ML production stack and see how things change in the continuous setting compared to the noncontinuous setting. At a high level, a continuous ML system regularly takes data in from the world in a steady stream, uses it to update the model, and then after appropriate validation, pushes out an updated version of the model to serve new data.

Training Examples

Rather than existing as a fixed set of immutable data, training data in a continuous ML system comes in a steady stream. This might include things like sets of recommended products from a set of possible yarn products, along with the query that generated these recommendations. In high-volume applications, the stream of training examples may resemble a firehose, with significant amounts of data being collected every second from all regions of the globe. Significant data engineering can be required to ensure that this stream of data is processed effectively and reliably.

Training Labels

The training labels for our examples also come from the world in a stream. Interestingly, the source of this stream may well be distinct from the stream of training examples itself. For example, say we wish to use whether the user purchased a given yarn product as the training label. We know which products a user is shown at query time, and can log those as they are sent to the user. However, purchase behavior cannot be known at query time—we have to wait for some time to see if they choose to buy—and this information may come from a purchase-handling system that resides in a completely different part of the overall service infrastructure. In other settings, we may see delays in training labels when those are provided by human experts.

Joining together the examples with their correct labels thus requires unavoidable delay, and likely involves some relatively sophisticated infrastructure to process efficiently and reliably. Indeed, this joining is production critical. Just imagine the headache that would be caused if the label information was unavailable because of an outage, and unlabeled examples were sent to the model for training[1].

Filtering Out Bad Data

Whenever we allow our models to learn directly from behavior in the world, we run the risk that the world will send along behaviors that we wish our model did not have to learn from. For example, spammers or scammers may try to interfere with our yarn product prediction model by issuing many fake queries without purchasing in an attempt to make it appear that some products are desired less by users than they actually are. Or some bad actors may try to make our helpful *yarnit.ai* chatbot learn rude behavior by entering offensive text repeatedly into the chat window. These attacks must be detected and dealt with. Less malicious but equally damaging forms of bad data may be caused by pipeline outages or bugs. In all cases, it is important to remove such forms of bad data from the pipeline before training, so that the model training is not impacted.

1 See "5. Ad Click Prediction: Databases Versus Reality" on page 353 for an example where this happens.

Effective removal of spammy or corrupted data is a difficult task. It requires automated methods for anomaly detection, and models whose primary purpose is to detect bad data. Often an arms race of sorts arises between the bad actors trying to inappropriately influence the model and ops teams trying to detect these attempts and filter them out. Effective organizations often have fully dedicated teams devoted just to the problem of filtering out bad data from a continuous ML pipeline.

Feature Stores and Data Management

In typical production ML systems, raw data is converted into features, which in addition to being useful for learning are also more compact for storage. Many production systems use a *feature store* for storing data in this way, which is essentially an augmented database that manages input streams, knows how to convert raw data into features, stores them efficiently, allows for sharing among projects, and supports both model training and serving.[2] For high-volume applications, it is often necessary to do some amount of sampling from the overall stream of data to reduce storage and processing costs. In many cases, this sampling will not be uniform. For example, we might wish to keep all of the (rare) positives but only a fraction of the (very common) negatives, which means we will need to track these sampling biases and incorporate them with appropriate weighting into training.

Even though the featurized data is more compact and generally more useful, it will almost always be necessary to also keep some amount of logged raw data as well. This is important to have, both for developing new features and for testing and verifying correctness of the feature extraction and transformation code paths.

Updating the Model

In continuous ML systems, it is often preferable to use a training methodology that allows for incremental updates. Training methods based on stochastic gradient descent (SGD) can be used without any modifications in continuous settings. (Recall that SGD forms the basis of most deep learning training platforms.) To use SGD in a continuous setting, we essentially just close our eyes and pretend that the stream of data shown to the model comes in a stochastic (random) order. If our data stream is actually in a relatively shuffled order, this would be totally fine.

In reality, a stream of data often has time-based correlations that are not really random, and then we have to worry about how much breaking this assumption hurts us in practice. The absolute worst way that our data could be nonrandom is if we had an upstream batch-processing job that ordered, say, all the positives to come in one batch, and then all the negatives to come in another. SGD approaches would fail

2 See "Feature store" on page 72 for an in-depth treatment of feature stores.

miserably on data like this, and we would need to create intermediate shuffling of the data to help put SGD on safer, more randomized ground.

Some pipelines enforce a strict policy of training on a given example only once, in the order that it appeared temporally in the stream. This policy simplifies many things, both from an infrastructure standpoint and from a model analysis and repeatability standpoint, and when data is plentiful has no real drawbacks. However, more data-starved applications may need to visit individual examples many times to converge to a good model, so this strategy of visiting each example exactly once in order cannot always be followed.

Pushing Updated Models to Serving

In most continuous ML settings, we refer to major changes to a model as a *launch*. Major changes might include changes to a model architecture, the addition or removal of certain features, a change in hyperparameter settings like learning rates, or other changes that would motivate us to fully reevaluate the performance of a model before launching it as our production model. Minor changes such as small modifications to internal model weights based on new incoming data are referred to as *updates*.

As the model is updated, we will periodically write out checkpoints saving the current state of the model. These checkpoints are then pushed out to serving, but they are also important for disaster recovery. One way to think about pushing a new model checkpoint out to serving is that it is actually a small, automated model launch, and if we push a new checkpoint four times an hour, then we are doing almost a hundred small, automated model launches a day.

All of the things that can go wrong with a major model launch can also go wrong when we push a new checkpoint to serving. The model may be corrupted somehow—perhaps by bugs, perhaps by having been trained on bad data. The model file itself may be flawed, perhaps corrupted by write errors, hardware bugs, or even (yes, really) cosmic rays. If the model is a deep learning model, there is a chance that in the most recent training step it has "exploded" and the internal model parameters contain `NaN`s, or that we run into the vanishing gradients problem, effectively halting further learning. If our checkpointing and pushing process is automated, there may be bugs in that system. If our checkpointing and pushing process is not automated, and relies instead on manual effort, then the system is probably not ready to be run in a continuous mode.

Of course, lots of things can go wrong with our system if we do *not* regularly push model updates based on new data, so the point is not to avoid updating models, but rather to point out that validation of model checkpoints is a critical step before they are pushed to serving. Typical strategies use staged validation. First we use tests that can be performed offline without impacting the production system, such as loading the checkpoint and scoring a set of golden set data in a sandbox environment. All of the offline evaluation methods discussed in Chapter 5 apply here. Then we load the new checkpoint into a *canary*—a single instance that we observe carefully to see if it fails—and allow it to serve a tiny amount of traffic, and then as long as monitoring holds, we slowly ramp up the amount of traffic served by the updated version until it is finally serving 100% of the data.

Observations About Continuous ML Systems

Now that you've gotten a bit familiar with the ways that continuous ML pipelines can differ from their noncontinuous cousins, we can dive into some of their unique characteristics and challenges.

External World Events May Influence Our Systems

When we look at the API for a widely used class or object, like a `vector` from the C++ standard template library or a Python `dictionary`, they typically do not include a stark line of documentation that reads, "WARNING: behavior undefined during the World Cup." Thank goodness they don't, and don't have to.

In contrast, continuous ML systems have—or should have—exactly this form of warning. It could read something like this:

Any change in input distributions to the model in production may cause erratic or unpredictable system behavior, because the theoretical guarantees for most ML models really hold for only the IID setting.[3]

The sources of such changes may be incredibly varied and unexpected. Sporting events, election nights, natural disasters, daylight savings, bad weather, good weather, traffic accidents, network outages, pandemics, new product releases—all of these are potential sources of changes to our data streams, and thus to our system behavior. In all cases, we are likely to have little or no warning about the events themselves, although monitoring strategies described in Chapter 9 might help you to think about what's required for prompt alerts.

What sorts of things can happen from an external event? Here's an example. For our yarn store, let's imagine the impact of having a major political figure appear on national television on a frigid day wearing hand-knit brown woolen mittens. Searches

3 In statistics and ML parlance, the IID assumption is that data is drawn independently from an identical distribution—that is, that our test sets and training sets are randomly drawn from the same sources in the same way. This is covered in much more detail in Chapter 5.

and purchases for "brown wool" spike suddenly. After a short delay, the model is updated on this new influx of search and purchase data, and learns to predict much higher values for brown wool products. Our model is trained with a form of SGD, which ends up overconfident and making scores extremely high for these products. Because of the sudden high scores, these products are shown to nearly all users, and the available stock is rapidly sold out. Once all stock is sold out, no more purchases are made, but nearly all searches are still showing brown wool products because of their high score from our model.

The next influx of data shows that no users are purchasing any products, and the model overcompensates, but because the `brown_wool` products have been shown on such a broad range of queries to such a broad range of users, the model now learns to give lower scores for nearly all products, resulting in no results or junk results for all user queries. This reinforces the trend that no users are purchasing anything, and the system spirals down, until our MLOps team identifies the issue, rolls the model back to a previous well-behaved version, and filters the abnormal data from our store of training data before re-enabling training.

Clearly, this example can be addressed with potential fixes and monitoring at multiple levels, but it illustrates the way that system dynamics can have consequences that are difficult to anticipate in advance.

One subtle danger of knowing that a wide variety of world events can cause unexpected system behavior is that we can end up explaining away changes in observed metrics or monitoring too quickly. Knowing that Argentina and Brazil are playing an important soccer (football) match today may cause us to assume that this is the root cause for an observed system instability, and miss rooting out a pipeline error or other system bug.

What would it look like to have a continuous ML system that was completely robust to distribution shifts? Basically, we would need to have a way to adaptively weight the training data so that its distribution did not depend on changes in the world. One way to do this would be to create a model to do *propensity scoring*, which shows how likely a given example is to occur in our training data at a given time. We would then need to weigh our training data by the inverse of this propensity score, so that rare examples are given greater weight. The propensity scores would need to be updated quickly enough that when a world event caused some examples to suddenly be much more likely than in the past, they were down-weighted accordingly. Most importantly, we would need to make sure that all examples had nonzero propensity scores to avoid divide-by-zero issues when doing inverse propensity scoring.

We need to avoid propensity scores that are too small, so that the inverse propensity weighting does not get blown up by a small number of examples with huge weights. This could be done by capping the weights, except that we would also need these scores to be statistically unbiased, and capping weights would cause bias. Instead, we could use extensive randomization to ensure that no example has too low a probability of being included, but this may well mean exposing users to random or irrelevant data or having our models suggest random actions that may be undesirable. All in, achieving a setup like this is possible in theory but extremely difficult in practice.

The reality is that we are likely to have to find ways to manage instability due to distribution shifts, rather than completely solve them.

Models Can Influence Their Own Training Data

One of the most important questions to answer for our continuous ML system is whether a *feedback loop* exists between a model and its training data. Clearly, all trained models are influenced by the stream of data that comes in for training, but some models also, in turn, influence the data that is collected.

To help understand the issues, consider that some model systems have no influence over the stream of data that is collected for retraining. Weather prediction models are a good example. No matter what the weather station might like us to believe, a prediction that tomorrow will be a nice sunny day has no actual influence on atmospheric conditions. Such systems are clean in the sense that they do not have feedback loops, and we can make changes to models without fearing that we might impact tomorrow's actual chance of rain.

Other models do influence the collection of their training data, especially when those models make recommendations to users or decide on actions that impact what they can learn about next. These create implicit feedback loops, and as anyone who has heard screeching feedback from a microphone knows, feedback loops can create unexpected and detrimental effects.

As an easy case, some systems that rely on feedback loops to learn about new trends might miss them entirely if they never have a mechanism that allows them to try a new thing in the first place. We might reflect on the experience of trying to get a child to try a new food for the first time as a way to think about this effect. As a more concrete example, consider a model that helps recommend wool products to show to users; the model may then be trained on the user response to those selected products, such as clicks, page views, or purchases. The model will receive feedback about the products that were selected to be shown to the user, but will *not* get feedback about products that were not selected.[4] It is easy to imagine that a new wool product, such as a new color of organic alpaca yarn, might be something that users would love to purchase, but for which the model has no previous data. In this case, the model might continue recommending previous nonorganic products and be oblivious to its omission.

Not discovering new things is bad, but even worse behaviors can happen. Imagine a stock market prediction model that is in charge of picking stocks to buy and sell based on market data. If an external entity mistakenly makes a large sale, a model might observe this and predict that the market is about to go down and most holdings should also be sold. If this sale is large enough, this will drive down the market, which may make the model even more aggressive about wanting to sell. Interestingly, other models—potentially from completely disjointed organizations—may see this signal in the market and also decide to sell, creating a reinforcing feedback loop that creates an overall market crash.

While the stock prediction scenario is an extreme case, it did, in fact, happen.[5] However, we don't need to exist in a broad market of competing models to experience these effects. For example, imagine that in our *yarnit.ai* store, we have one model that is in charge of recommending products to users, and one model that is in charge of determining when the user should be given discounts or coupons. If the product recommendation relies on purchase behavior as a signal in training, and the presence of discounts influences purchase behavior, then there is a feedback loop that links these two models, and changes or updates to one model can influence the other.

Feedback loops are another form of distribution shift, and the propensity-weighting approaches we've described can be helpful, although they are difficult to get completely right. The effect of feedback loops can be lessened to some degree by logging model version information along with other training data, and using this information as a feature in the model. This at least gives our model the opportunity to disambiguate whether a sudden change in observed data is due to a change in the real world—such as the holidays are over, and nobody wants to buy wool (or indeed much of anything) in early January—or a change in the model's learned state.

4 Those familiar with different subfields of ML will note that this is technically a contextual bandit setting, and thus subject to an explore versus exploit trade-off. The main idea in this setting is that when our system learns only about the things it intentionally selects, it is important to sometimes randomly change our selections to explore the world a little more and make sure we do not lock our system into a self-perpetuating loop of beliefs. Exploring too little leads to missing out on the very best; exploring too much is a waste of time and resources. Parallels to any of our own life choices are, of course, completely coincidental.

5 See *The Flash Crash: A New Deconstruction* (*https://oreil.ly/sAkYH*) for an in-depth treatment of the events that lead up to the flash crash, including consideration of the triggering incident as well as the facilitating market conditions. One point to take away from this discussion is just how hard it can be to do root-cause analysis on systems that involve multiple models interacting with one another.

Temporal Effects Can Arise at Several Timescales

We create continuous ML systems when we care about the ways that data changes over time. Some of these changes are deeply meaningful to the underlying product needs, like the introduction of a new form of synthetic wool or the creation of automated knitting machines suitable for home use. Incorporating data about these emergent trends as soon as possible would be important for our *yarnit.ai* store.

Other temporal effects are cyclical, with cycles occurring over at least three major timescales:

Seasonal
: Many continuous ML systems experience profound seasonal effects. For online commerce sites like *yarnit.ai*, dramatic changes and increases in purchasing behavior can occur as the winter holidays approach, followed by a sudden drop in early January. Warm weather months and cool weather months may have very different trends—and may also vary significantly by geographic region or even by Northern versus Southern Hemisphere. The most effective way to deal with seasonality effects is to make sure that our model has trained on data from more than one full year in the past—if we are fortunate enough to have it—and that time-of-year information is included as a feature signal in the training data.

Weekly
: Just as data can vary by season, it can also cycle on a weekly basis, based on day of week. Weekend days may have significantly more usage for some cases, like the portion of our *yarnit.ai* store targeted to hobbyists—or significantly less for others, like the portion of the yarn store that targets business-to-business sales. Locality is strongly tied in here, as weekend days may differ by country, and time-zone effects also matter strongly, as it may be Monday in Tokyo while it is still Sunday in San Francisco.

Daily
: Things start to get nontrivial when we look at the daily effects, based on time of day. At first glance, it is obvious that many systems will experience different data at different times of day—midnight behavior is likely different than early morning, or during the work day. It is also likely clear that locality is crucial here, because of time-zone effects.

The subtlety for daily cycles comes in when we consider that most continuous ML systems actually run continuously behind reality, because of the inherent need for delays in pipelines and data streams waiting for training labels—such as click or purchase behavior that may take some time for a user to decide on—as well as filtering out bad data, updating models, validating checkpoints, and pushing checkpoints to serving and ramping them up fully. Indeed, such delays could add up to 6 or even 12 hours. Therefore, our models may be operating very much out of phase with reality, serving a version of the model that thinks it is the middle of the night when, in fact, it is in the heart of the workday.

Fortunately, fixing such issues is relatively easy, by logging time-of-day information along with other training signals and using these as inputs to our models. But it highlights the importance of thinking through ways that the version of the model we happen to have loaded in serving may be stale or otherwise misinformed about the actual reality it is asked to work with at that moment.

Emergency Response Must Be Done in Real Time

By this point, it should be clear that while continuous ML systems can provide great value, they also have a broad range of system-level vulnerabilities. It can be argued that a continuous ML system inherits all of the production vulnerabilities of large, complex software systems—which carry plenty of reliability issucs on thcir own—and adds in a whole set of additional issues that can produce undefined or unreliable behavior. In most settings, we cannot rely on theoretical guarantees or proofs of correctness.

When such issues arise in continuous ML systems, they not only need to be fixed, but also need to be fixed (or mitigated) in real time. There are several reasons for this. First, our models are likely mission critical, and having a production fire that impacts our ability to serve useful predictions from our models may impact our organization minute to minute. Second, our model may have a feedback loop with itself, meaning that if we do not address issues quickly, the stream of input data may also be corrupted and require care to fix as well. Third, our model may be part of a larger ecosystem that is difficult to reset to a known state. This can occur when our model is in a feedback loop with other models, or when poor predictions from a model create lasting harm and are difficult to undo, such as an otherwise helpful *yarnit.ai* chatbot suddenly issuing rude curses at users.

Real-time crisis response requires first detecting issues quickly, which means that from an organizational standpoint, a good litmus test for determining whether we are ready for continuous ML is to examine the thoroughness and timeliness of our monitoring and alerting strategies. It can take time for the full effect of data imperfections to create downstream detrimental effects, because of pipeline delays and slow changes in systems that learn gradually on new data. This makes it especially important to have simple canary metrics that alert on changes to input distributions, rather than waiting for model outputs to change.

Monitoring or altering strategies will be helpful in real time only if they are supported by folks who have the responsibility to respond when the alerts fire. Well-functioning MLOps organizations set up specific service-level agreements about how quickly an alert must be responded to, and set up mechanisms like pager rotations to ensure that the alert is seen by the right person at the right time. Having a global team with teammates in several time zones, should you be in that fortunate position, can help tremendously with avoiding the need for folks to be woken up by an alert at 3 A.M.

Once we receive an alert, we need to have a well-documented playbook of responses that can be carried out by any member of the MLOps team, along with an escalation path for further action.

For continuous ML systems, we have a set of basic immediate responses to any given crisis. These are stop training, fall back, roll back, remove bad data, and roll through.

Not every crisis needs all of these steps, and the choice of which response is most appropriate depends on the severity of the issue and the speed with which we are able to diagnose and cure the root-cause issue. Let's look at each of these basic crisis response steps for continuous ML systems and then discuss factors for choosing a response strategy.

Stop training

It has been said that the First Rule of Holes is this: *when you find yourself in a hole, stop digging*. Similarly, when we find that our data stream is corrupted in some way, perhaps by a bad model, or an outage, or a code-level bug somewhere in the system, a useful response can be to stop model training and halt pushing any new model versions out to serving. This is a short-term response that at least helps ensure that problems will not get worse while we decide on a mitigation or fix. It makes sense to ensure that there is an easy way for MLOps folks to stop training on any model that is their responsibility. Automated systems are helpful here, but of course need to alert sufficiently so we do not discover that a model has silently stopped training three weeks ago.

It is always useful to have the equivalent of a Big Red Button that can be used to stop training manually in a detected emergency.

Fall back

In continuous ML systems, it is important to have a fallback strategy that can be used in place of our production model that provides acceptable (even if nonoptimal) results. This could be a far simpler model that does not train in a continuous fashion, or a lookup table of the most common responses, or even just a small function that returns the median prediction to all queries. The key thing is that if our continuous

ML system encounters sudden massive failures—what we might describe as "being on fire"—we have an ultra-reliable method that can be used as a temporary replacement without the larger product becoming completely unusable. Fallback strategies are typically less reliable in overall performance than our main model (otherwise, we would not use an ML model in the first place), so fallback strategies are very much intended to be short-term responses that allow for emergency responses to take place in other parts of the system.

Roll back

If our continuous ML system is currently in a crisis state, it makes sense to revert the system to a state from before the crisis and see if everything is OK. The root cause of our crisis may have come in through two basic areas: bad code or bad data.

If we believe that the root cause of our issue is bad code, from a recently introduced bug, then rolling back our production binaries to use a previously known-good version may fix the issue in the short term. Of course, any rollback to a previous production binary must be done in a staged ramp-up in case any new compatibility issues or other flaws exist that make the old version of the binary no longer usable. At any rate, it is important to keep on hand a set of fully compiled previous binaries so that rollback can be done quickly and efficiently when needed.

If we believe that the root cause of our issue is bad data that has caused the model to train itself into a bad state, it makes sense to roll back the model version to a previously known-good version. Again, it is important to keep checkpoints of our trained production model on hand so that we have a set of previous versions to choose from. For example, imagine that Black Friday sales in the US cause such a large increase in purchase requests from users to our *yarnit.ai* store that the fraud detection portion of the system starts to label all purchases as invalid, making it look to our model as though all products are extremely unlikely to be purchased. Rolling back to a version of the model that was checkpointed a week before the Black Friday date would at least allow the model to serve reasonable predictions while the rest of the larger system was fixed.

Remove bad data

When we have bad data in our system, we need to have an easy way to remove it so that our model will not be corrupted by it. In the preceding example, we would want to remove the data that was corrupted by the faulty fraud detection system. Otherwise, when we re-enable training to proceed, this data will be encountered by our rolled-back model as it moves forward in time through the training data, and it will be corrupted by the bad data again. Removing bad data is a useful strategy whenever we believe that the data itself is highly unrepresentative of typical data and is unlikely to give the model useful new information, and that the root cause of the bad data is temporary, due to an external world event or to bugs in our system that can be quickly fixed.

Roll through

If we have stopped training for our continuous ML system, at some point we need to cross our fingers and enable it to resume training. We typically do this after bad data has been removed and we are sure that any bugs have been addressed. However, if a crisis is detected due to an external world event, sometimes the best response is to just cross our fingers and roll through it, allowing the model to train on the atypical data and then to recover itself as the world event ends. Indeed, it is unfortunately true that this world has few days with no political event, major sporting event, or other major newsworthy disaster happening somewhere, and making sure that our model has enough exposure to atypical data like this from different global regions can be an important way to ensure that our model is generally robust.

Choosing a response strategy

How do we choose which events to stop, roll back, and remove, and which to roll through? To answer that question, we need to observe our model's response to similar historical events, which is most easily done when we have trained the model on historical data in sequential temporal order. Another important question to answer is whether the crisis-indicating metrics we are currently seeing are due to a bad model or to the atypical state of the world. In other words, is our model broken, or is it just being asked to handle much more difficult requests right now? One way to judge this is to observe offline metrics on golden set data, which should be recomputed for our model on a frequent periodic basis for this reason. If the model is actually corrupted, the golden set results may show a sharp decrease in performance, and rolling through is probably not the right approach.

Organizational considerations

When a crisis is currently going on, it can be a difficult time to learn new skills, to figure out roles within a team, or to decide on how to implement various response and mitigation strategies. Real-world firefighters regularly train together, refine best practices, and ensure that all of the infrastructure they need to respond to an alarm is in excellent condition and is ready to roll into action at a moment's notice. Similarly, we do not know exactly when our continuous ML systems will require crisis response, but we can confidently say it will happen and that we need to be well prepared. Creating an effective crisis response team is part of the cost of creating and maintaining a continuous ML system, and must be accounted for when we move in this direction. This is discussed in more detail in "Continuous Organizations" on page 242.

New Launches Require Staged Ramp-ups and Stable Baselines

When we have had a model running as part of our continuous ML system for a period of time, we will eventually want to launch a new version of that model that creates improvements in various ways. Maybe we want to use a larger version of the model for improved quality and now have serving capacity to handle it, or perhaps a model developer has created several new features that significantly improve predictive performance, or maybe we have discovered a more efficient model architecture that reduces serving costs in an important way. In cases like these, we need to explicitly launch the new version of the model to replace the old version.

Launching a new model most often involves some amount of uncertainty because of the limitations of offline testing and validation. As we describe in Chapter 11, offline testing and validation can give useful guidance on whether a new version of a model is likely to perform well in production, but often cannot give a complete picture. This is especially true when our continuous ML model is part of a feedback loop, because the data that we have previously trained on was most likely chosen by a previous model version, and evaluation on offline data is limited to data that has been collected based on actions or recommendations made by that previous model. We can imagine this situation as similar to that of a student driver, who is first evaluated by sitting in the passenger seat and being asked to give their opinion about the instructor's actions in driving the car. Just because they agree 100% with the actions of the instructor does not mean that they will not make some poor decisions when first given the opportunity to steer the car for themselves.

In this way, a new model launch requires some amount of *testing in production* as the final form of validation. We need to give our new model the ability to demonstrate that it is capable of being in the driver's seat. But that doesn't mean we just hand it the keys and expect everything to be perfect. Instead, we will most often use a staged ramp-up, first allowing the model to serve only a fraction of the overall data, and increasing that amount only as we observe good performance over time. This strategy is commonly known as an *A/B test*: we test out our new model A against the performance of our old model B, in a format that resembles a controlled scientific experiment and helps verify that the new model will show the appropriate performance on our final business metrics (which may be distinct from offline evaluation metrics like accuracy).

The difference between a model launch and an ideal A/B test is that in a scientific experiment, A and B are independent and do not influence each other. For example, if we run an experiment in a scientific setting to determine whether cotton sweaters (A) keep people as warm as natural wool sweaters (B), the people wearing wool sweaters are unlikely to make the people wearing cotton ones report feeling any warmer or colder. However, if the wool sweater folks are so warm and happy that they go make tea and hot soup for the cotton sweater folks, this would definitely ruin our experiment.

For A/B experiments comparing continuous ML models, it turns out that A and B may well influence each other when our models are part of a feedback loop. For example, imagine that our new model A does a great job of recommending organic wool products to *yarnit.ai* users, whereas our previous model B had never done so. An A/B experiment might initially show that the A model is much better in this regard, but then as training data is produced by A that includes many more organic wool recommendations and purchases, the B model (which is also continuously updating) may then also learn that these products are liked by users and begin to recommend them as well, making the two models appear identical over time. If these effects are more diffuse, it can be hard to say whether the benefits of A have disappeared because it was never actually better than B, or if B has itself improved.

We could try to fix this by restricting A and B to each train on only the data that they themselves serve. This strategy can work well when each model serves the same amount of data, such as 50% of the overall traffic each, but can make for flawed comparisons in other cases. If A is looking bad early on, is that because the model is bad, or because it has only 1% of training data while B has 99%?

Another strategy is to try to create a stable baseline of some sort, which can help serve as a reference point in comparisons so that we can figure out whether comparisons between A and B are changing because A is getting worse or because B is getting better—or indeed, if both are getting dramatically worse in lockstep. A stable baseline is a model C that is not influenced by either A or B, and is allowed to serve a certain amount of traffic so that we can use those results as a comparison. The basic idea is to then look at (A-C) against (B-C) rather than A against B directly, with the idea that this will allow us to see any changes more clearly.

The four general strategies for creating stable baselines have different advantages and disadvantages:

Fallback strategy as baseline
: When we have a reasonable fallback strategy that does not involve continuous retraining, this is useful not only for crisis response but also as a potential independent datapoint. This can work well if the quality of the fallback strategy is not too much worse than the main production model. If the difference is very large, however, statistical noise may overwhelm any comparisons between A and B using this as a reference point.

Stop trainer
: If we have a copy of our production model B and halt training on it, then by definition it will not be influenced by any future behavior of A or B. If we allow it to serve a small amount of traffic, this can provide a useful stable baseline C, with

the caveat that the overall performance of a "stop trainer" model will degrade slowly over time. It can be useful to run independent experiments to observe how much degradation can be expected and whether this strategy will be useful.

Delay trainer
: If we expect our overall launch process to take, say, two weeks, then a reasonable alternative can be to run a copy of our production model that is set to update continuously, but at a two-week delay. This has the advantage over the stop trainer in that the relative performance is unlikely to degrade, with the drawback that after it has been running for a length of time equal to its delay, it will begin to become influenced by A and B and will lose its utility. Thus, a two-week delay trainer model will become useless after two weeks.

Parallel universe model
: An approach that maintains strict independence from A and B but does not have a limited shelf life is a parallel universe model that is allowed to serve a small fraction of overall data and learns only on the data that it serves itself. A and B do not train on this data, keeping these data universes completely separate.

Why is this useful? Imagine that the act of putting B into production changes the overall ecosystem in some way. We could imagine stock prediction market models operating this way—perhaps pushing the overall market up or down in some special cases. In this case, both A and B might end up strongly increasing or decreasing their median predictions, but the difference A-B might be small and appear stable. Having this third point C allows us to detect whether the changes between models are isolated to differences in A and B themselves, or are due to a broader impact.

Parallel universe models often take time to stabilize after being set up because of the restricted amount of training data and the overall distribution shift. But after this initial period, they can provide a useful independent evaluation point—again, up to the limits of statistical noise when comparing evaluation metrics.

Models Must Be Managed Rather Than Shipped

Overall, model launches require particular care because at these times our systems are most vulnerable to crisis. If we are in a model launch process with A and B both serving roughly half the traffic, we have just doubled the potential sources of error and doubled the amount of work needed to address any emergency that may arise. Like crisis response, model launches are done best when they rely on well-communicated, well-practiced processes.

Some products are like brick walls: they take a lot of planning and effort to get right, but once completed are more or less done and require only occasional maintenance. The default state is that they just work. Continuous ML systems are at the opposite end of the spectrum and require daily attention. Problems that arise with a continuous ML model may be difficult to solve in a complete or permanent way. For example, if one of our overall product issues is that we would like to do a better job of recommending wool products to users in warm climates, this is one that is likely to require a variety of approaches, and the utility of those approaches may change over time as tastes and fashions adapt season to season and year to year.

In this way, a continuous ML system is something that requires regular management. Managing a model effectively requires daily access to metrics that report the model's predictive performance. This can be done through dashboards and related tooling that allow a model manager to understand how things look today, how trends might be changing, and where problems might be arising. Any dashboard's utility is upper bound by the amount of attention paid to it, so there needs to be a clear owner who regularly spends time with it. A useful metaphor is that it is useful to have a cup of coffee with our model every day, just to get to know it by working with a dashboard to understand how it is doing that day. And just as a people manager provides regular performance evaluation, a model owner should provide regular reports about model performance to upper levels of the organization to share knowledge and visibility.

When we learn something useful about our model, a strong best practice is to write it down in the form of a short writeup. A writeup, which can be just a few paragraphs accompanied by a screenshot of a dashboard or similar supporting evidence, can help build organizational knowledge, and is of the most benefit when the observation is accompanied by a short summary about what it means about model behavior. Such writeups have historically proven extremely useful, both for helping guide future model development and for understanding and debugging unexpected behaviors seen during crises.

Lastly, when we do encounter a crisis, it is important that organizationally we extract as much learning from the experience as possible by creating postmortem documents that describe in detail what happened, how the issue was diagnosed, what the damage was, what mitigation strategies were applied and how successful they were, and finally recommendations for how to make improvements to either reduce recurrence of the issue or enable more effective response in the future. Creating these postmortem documents is useful both in the short term to help identify fixes, but also in the long term as a repository of organizational knowledge and experience that can be referenced over time.

Continuous Organizations

At this point, it should be clear that an organization that is taking on a continuous ML system is committing to a long-term responsibility. Like a puppy, a continuous ML system requires daily attention, care, feeding, and training. Structuring our organizations to be well equipped to handle the responsibilities of managing a continuous ML system requires numerous structures to do well.

Determining evaluation strategies is a key leadership responsibility. Evaluation strategies allow us to assess the health and quality of our models, both in terms of long-term business goals and short-term decisions, such as whether to include or exclude a given feature from the model. As noted in Chapter 9, it can be tempting to reduce this problem to determining metrics, but a given metric (such as revenue, clicks, precision, recall, or even latency) is meaningless without a reference point, baseline, or distribution. Decision making in continuous ML settings often requires some amount of counterfactual reasoning, thinking through the impact of feedback loops, or wrestling with noise and uncertainty that makes effective decision making challenging. We can help reduce the difficulty of these challenges to some degree by having clearly defined and documented standards and processes in place for evaluation.

Organizationally, making investment decisions is similarly challenging. How much should we invest in more compute to create a larger and potentially more powerful model, and will that investment pay for itself in terms of improved product outcomes relative to opportunity cost? How should we trade off investing in model quality improvements versus ML system-level reliability? Organizationally, how do we direct the time and effort of our constrained human experts to best benefit the overall mission? These are fundamentally hard problems, in no small part because different parts of our organization may have different or even incompatible viewpoints on priorities.

We have two main strategies available for dealing with these kinds of issues. The first is to ensure that we have organizational leadership with enough scope and context to effectively weigh the differing needs of, for example, improving model monitoring and crisis response handling with that of improving model accuracy. Ensuring that the lessons learned from each part of the organization are well communicated across the entire organization can be one way to help different parts of an organization understand one another's challenges and pain points. This is best done by regularly sharing postmortems from incidents—and also by proactive "pre-mortem" discussions identifying potential weaknesses and failure modes. The second strategy is to invest in infrastructure that ensures that alerts and other warnings are propagated well in the full chains of producers and consumers. This can require a serious organizational commitment, but can also pay off over time in terms of reducing the human burden of verification within complex systems.

Organizationally, understanding that a continuous ML system relies on a steady stream of incoming data to determine system behavior makes clear that the data pipeline itself requires serious, dedicated oversight and management. In addition to simply ensuring that data is flowing and the pipeline is functioning well, key questions around which kinds of data to collect, how long to store data, and how our data pipelines should interact with upstream producers and downstream consumers are all critical strategic questions for organizational leaders to address. Deeper questions of privacy, ethics, data inclusivity, and fairness all also play important roles and must be part of the overall organizational strategy.

As we've noted, the launch process for ML systems and further improvements necessarily requires a staged ramp-up procedure in the continuous ML setting. A critical role of leadership is providing the oversight and review for the results of each stage, and making approval decisions to move to the next stage or determining that we are not yet ready to proceed or must even ramp down if things are not looking as intended. These decisions require high-level oversight because the consequences can be far reaching and interactions can occur with multiple producer or consumer systems, especially in the presence of feedback loops or other complex system-to-system interaction points. The process for assessing the wider impact of various launch stages and ensuring stability before proceeding must be well established and rigorously followed for a continuous ML organization to be effective in the long run.

Finally, when incidents or crisis moments occur, we need to have a process in place for responding effectively. For continuous ML systems, we have an advantage and a disadvantage in handling incidents. The advantage is that we almost always have a slightly (or somewhat) older version of the current model that we can roll back to in serving while we take a look at what went wrong. This can be incredibly helpful when we need a quick mitigation for something that went disastrously wrong with the current model. The disadvantage is that the entire system is constantly evolving, making it difficult to isolate root causes or breaking changes. Our model's change might have been motivated or required by other concurrent changes in the system (such as new data added, or new integrations using the model). In this context, troubleshooting exactly what has gone wrong can be much more difficult in a continuous ML environment. For concrete examples of this, see Chapter 11.

The most important prework for an organization running a continuous ML system is to pre-negotiate outage consequences and handling. This goes beyond just determining, in advance, who will fill which roles—although you need to do that too. ML production engineers shouldn't be determining the urgency of resolving a particular incident while in the middle of it, and model developers shouldn't be guessing at the costs and consequences of a given outage while it is ongoing. Each person should know their role, their authority, and their path to escalate a decision to someone else because these should be worked out in advance. But where possible, the whole organization should also have agreed upon some general service reliability standards. Examples might be the following:

- This model can be up to 12 hours old without serious consequences, even though we generally prefer that it be based on data that is no more than about 1 hour old.
- If this model has quality metrics worse than a particular known threshold, the approved fallback is to roll back to an old model.

- If the model is worse than an additional threshold that is predetermined, waking up the following business leaders is appropriate...

And so on. The general idea is to predetermine the parameters of various kinds of outages so that incident responders have the maximum ability to take action and minimum latency to a decision about how to respond. Many organizations tend not to work out these decisions in advance until they experience a few outages, but after that, it becomes quite reasonable to pre-decide what to do in the case of really bad ML problems.

Rethinking Noncontinuous ML Systems

We have talked about a range of issues for continuous ML systems in this chapter. But in addition to giving what we hope are useful mitigation strategies for creating robust and reliable systems even in the face of the difficulties, we would like to make a broader recommendation:

> All production ML systems should be treated as continuous ML systems.

We should think about every model that is a key part of a production system as one that is trained continuously, even if it is not actually updated on new data every minute, every hour, or even every day or week.

Why would we make such a recommendation? After all, continuous ML systems are full of complexity and vectors for failure. One reason is that if we apply the standards and best practices from continuous ML systems to all production-grade ML systems, we will definitely be ensuring that our technical infrastructure, model development, and MLOps or crisis response teams are set up to meet challenges as they arise. If we assume that our ML system is a continuous ML system and plan accordingly, we will be in a good spot.

Is this overkill? If a model is trained only once, applying the standards and best practices from continuous ML may be seen as a waste of resources. But in reality, no production models are trained only once. In our experience, we have seen that every production-level ML model will eventually be retrained or have a new version launched—maybe in a few months, or next year, as new data becomes available or models are developed. This can be done in an ad hoc fashion, every few weeks or months, but this irregular approach is likely to lead to failures and oversights. Our strong recommendation is that effective MLOps teams ensure that their models are updated on a regular schedule, be it daily, weekly, or monthly, so that validation procedures and checklists can become part of the organizational culture. From this standpoint, then, the recommendations for continuous ML systems are applicable to every ML system.

Conclusion

In this chapter, we have laid out sets of procedures and practices that can form the foundation of an organizational playbook for the ongoing care and oversight of continuous ML systems. These systems offer a remarkable range of benefits, enabling models that adapt to new data over time and allow for responsive learned systems that interact with users, marketplaces, environments, and the world.

It is obvious that any system that is both highly impactful and easily influenced requires considerable oversight. In our experience, the requirements for oversight in these cases go far beyond what can be expected of any individual, no matter how capable, and cannot be left to intuition or improvisation. Any organization managing a continuous ML system needs to think of this as an ongoing high-priority mission, with special care at times of launches or major updates, but also with continual monitoring and contingencies in place to allow fast response to emergent crises.

We covered six basic insights about continuous ML systems that we hope you will come away with:

- External world events may influence our systems.
- Models may influence their own future training data through feedback loops.
- Temporal effects can arise at several timescales.
- Crisis response must be done in real time.
- New launches require staged ramp-ups and stable baselines.
- Models must be managed rather than shipped.

And finally, we ended the chapter with the idea that all ML systems should likely best be thought of as continuous ML systems, as all models are eventually retrained, and having strong standards in place will benefit any organization in the long run.

CHAPTER 11
Incident Response

In this world, sometimes bad things happen, even to good data and systems. Disks fail. Files get corrupted. Machines break. Networks go down. API calls return errors. Data gets stuck or changes subtly. Models that were once accurate and representative models become less so. The world can also change around us: things that never, or almost never, previously happened can become commonplace; this itself has an impact on our models.

Much of this book is about building ML systems that prevent these things from happening, or when they happen—and they will—recognizing the situation correctly and mitigating it. Specifically, this chapter is about how to respond when bad, urgent things happen to ML systems. You may already be familiar with how teams handle systems going down or otherwise having a problem: this is known as *incident management*, and best practices exist for managing incidents that are common across lots of computer systems.[1]

We cover these generally applicable practices, but our focus is on how to manage outages for ML systems, and in particular how those outages and their management differ from other distributed computing system outages.

The main thing to remember is that ML systems have attributes that make resolving their incidents potentially very different from the incidents of non-ML production systems. The most important attribute in this context is their strong connection to real-world situations and user behavior. This means that we can see unintuitive effects where there is a disconnect among the ML system, the world, or the user behavior we are trying to model. We cover this in detail later, but the major thing

1 If you're looking for detailed coverage of general incident management, you may consider instead reviewing *Site Reliability Engineering* (*https://oreil.ly/cpdFY*) and the PagerDuty incident response handbook (*https://response.pagerduty.com*).

to understand now is that troubleshooting ML incidents can involve very much more of the organization than standard production incidents do, including finance, supplier and vendor management, PR, legal, and so on. ML incident resolution is not necessarily something that only engineering does.

A final serious point we would like to make here at the beginning is that, as with other aspects of ML systems, incident management has serious implications for ethics in general and very commonly for privacy. It is a mistake to focus on getting the system working first and worry about privacy afterward. Do not lose sight of this critical part of our work in this section. Privacy and ethics will make an appearance in several parts of the chapter and are addressed directly toward the end because by then we will be in a better place to draw some clear conclusions about how ML ethics principles interact with incident management.

Incident Management Basics

Three basic concepts for successful incident management are *knowing the state the incident is in*, *establishing the roles*, and *recording information for follow-up*. Many incidents are prolonged because of failures to identify what state the incident is in, and who is responsible for managing which aspects of it. If this continues for long enough, you have an *unmanaged incident*, which is the worst kind of incident.[2]

Indeed, if you've worked with incidents for long enough, you've probably seen one already, and it probably starts something like this: an engineer becomes aware of a problem; they troubleshoot the problem alone, hoping to figure out the cause; they fail to assess the impact of the problem on end users; and they don't communicate the state of the problem, either to other members of their team or the rest of the organization. The troubleshooting itself is typically disorganized and characterized by delays between actions, and assessing what happened after the actions. Once the initial troubleshooters realize the scope of the incident, even more delays may arise while they try to figure out which other teams need to be involved and send pages or alerts to track them down. If the problem continues indefinitely, other parts of the organization can notice that something is wrong and independently (sometimes counterproductively) take uncoordinated steps to resolve the problem.

The key idea here is to actually have a process—a well-rehearsed one—and to apply it reliably and methodically when something bad that's happened is worthy of being called an incident. Of course, creating a managed incident has a cost, and formalizing communications, behavior, and follow-up incurs overhead. So we don't do it for everything; not every `WARNING` in our logs warrants a couple of hours of meetings or phone calls. Being an effective on-call engineer requires developing a sense for what

2 See Chapter 14 of *Site Reliability Engineering* (*https://oreil.ly/cpdFY*) for more details.

is serious and what isn't, and smoothly engaging incident machinery when required. It is enormously helpful to have clearly defined guidelines ahead of time about when to declare an incident, how to manage it, and how to follow up after it.

Life of an Incident

Incidents have distinct phases of their existence. Although people of good will may differ on the specifics, incidents probably include states such as the following:

Pre-incident
: Architectural and structural decisions that set the conditions for the outage.

Trigger
: Something happens to create the user-facing impact.

Outage begins
: Our service is affected in a noticeable way by at least some users for at least some functions.

Detection
: The owners of the service become aware of the problem, either through automated monitoring notifying us or outside users complaining.

Troubleshooting
: We try to figure out what is going on and devise a means of fixing the problem.

Mitigation
: We identify the fastest and least risky steps to prevent at least the worst of the problems. This can range from something as mild as posting a notice that some things don't work right all the way to completely disabling our service.

Resolution
: We fix the underlying problem and the service returns to normal.

Follow-up
: We conduct a retrospective, learn what we can about the outage, identify a series of things we'd like to fix or other actions we would like to take, and then carry those out.

Computer system outages can roughly be described by these phases. We'll briefly cover the roles in a typical incident and then will try to understand what differs in handling an ML incident.

Incident Response Roles

Some companies have thousands of engineers working on systems infrastructure, and others might be lucky to have a single person. But whether your organization is large or small, the roles described in this section need to be filled.

It's important to note that not all of the roles must be filled by a separate person, since not all of the responsibilities are equally urgent, and not all incidents demand isolated focus. Also, your organization and your team has a particular size—not every team can fill every position directly. Furthermore, certain problems emerge only at scale: communication costs, in particular, tend to increase in larger organizations, often correlated with the complexity of infrastructure under management. Conversely, smaller engineering teams can suffer from tunnel vision and a lack of diversity of experience. Nothing in our guidance frees you of the necessity of adapting to the situation, and making the right choices—often by first making the wrong ones. But one critical fact is that you must plan ahead for the organizational capacity to properly support incident management duties. If they are a poorly staffed afterthought or you assume anyone can jump in when incidents occur with no structure, training, or spare time, the results can be quite bad.

The framework we are most familiar with for incident management derives from the US Federal Emergency Management Agency (FEMA) National Incident Management System (*https://oreil.ly/pFkaa*). In this framework, the minimum viable set of roles is typically as follows:

Incident commander
: A coordinator who has a clear understanding, at a high level, of the incident as a whole, and is responsible for assigning and monitoring the other roles.

Communications lead
: Responsible for outbound and inbound communication. The actual responsibilities for this role differ significantly based on the system but may include updating public documents for end users, contacting other internal services groups and asking for help, or answering queries from customer-facing support personnel.

Operations lead
: Approves, schedules, and records all production changes related to the outage (including stopping previously scheduled production changes on the same systems even if unrelated to the outage).

Planning lead
: Keeps track of longer-term items that should not be lost but do not impact immediate outage resolution. This includes recording work items to be fixed, storing logs to be analyzed, and scheduling time to review the incident in the

future. (Where applicable, the planning lead should also order dinner for the team.)

These roles are invariant, whether or not you are dealing with an ML incident. The things that *do* vary are listed here:

Detection
: ML systems are less deterministic than non-ML systems. As a result, it is harder to write monitoring rules to catch all incidents before a human user detects them.

Roles and systems involved in resolution
: ML incidents usually involve a broader range of staff during troubleshooting and resolution, including business/product and management/leadership. ML systems have a broad impact on multiple systems, and are generally built on and fed by multiple complex systems. This leads to a likely diverse set of stakeholders for any incident. ML outages often impact multiple systems because of their role in integrating with and modifying other parts of your infrastructure.

Unclear timeline/resolution
: Many ML incidents involve impact to quality metrics that themselves already vary over time. This makes the timeline of the incident and the resolution more difficult to specify precisely.

To develop a more intuitive and concrete understanding of why these differences show up in this context, let's consider a few example outages of ML systems.

Anatomy of an ML-Centric Outage

These examples are drawn from real experiences by the authors but do *not* correspond to individual, specific examples that we have participated in. Nonetheless, our hope is that many people with experience running ML systems will see familiar characteristics in at least one of these examples.

As you read through them, play close attention to some of the following characteristics that may differ substantially from other kinds of outages:

Architecture and underlying conditions
: What decisions did we make about the system before this point that could have played a role in the incident?

Impact start
: How do we determine the start of the incident?

Detection
: How easy is it to detect the incident? How do we do it?

Troubleshooting and investigation
: Who is involved? What roles do they play in our organization?

Impact
: What is the "cost" of the outage to our users? How do we measure that?

Resolution
: How confident are we in the resolution?

Follow-up
: Can we distinguish between fixing and improving? How do we know when the follow-up from the incident is done and prospective engineering is taking place?

Keep these questions in mind while you consider the stories presented later in this chapter.

Terminology Reminder: Model

In Chapter 3, we introduced distinctions among the following:

Model architecture
: The general approach to learning

Model (or configured model)
: The specific configuration of an individual model plus the learning environment, and the structure of the data we will train on

Trained model
: A specific instance of one configured model trained on one set of data at a point in time

This distinction matters particularly because we often care about which of these has changed, to possibly be implicated in an incident. We will try to be clear in the following sections which we're referring to.

Story Time

We tell the following stories within the framework of our invented firm, YarnIt, in order to help them resonate with you. But they are all based on, or at least inspired by, real events that we've observed in production. In some cases, they are based on a single outage at a single time, and in others they are composites.

Story 1: Searching but Not Finding

One of the main ML models that YarnIt uses is a search ranking model. Like most web stores, customers come to the site and click links offered to them on the front page, but they also search directly for the products they're looking for. To generate those search results, we first filter our product database for all of the products that roughly match the words that the customer is looking for, and then rank them with an ML model that tries to predict how to order those results, given everything we know about the search at the time it's performed.

Ariel, a production engineer who works on search system reliability, is working on the backlog of monitoring ideas. One of the things the search team has been wishing it monitored and trended over time is the rate that a user clicks one of the first five links in a search result. The team members hypothesize that that might be a good way to see whether the ranking system is working optimally.

Ariel looks through the available logs and determines an approach for exposing the resulting metric. After doing a week-on-week report for the past 12 weeks to make sure that the numbers look reasonable, Ariel finds some initially promising results. From 12 weeks ago to 3 weeks ago, Ariel sees that the top five links are clicked by customers around 62% of the time. Of course, that could be better, but a substantial majority of the time we're finding *something* that the users are curious about within the first few results.

Three weeks ago, however, the click rate on the first five links started going down. In fact, this week it's only 54%, and Ariel notes that it appears to still be dropping. That's a huge drop in a very short period of time. Ariel suspects that the new dashboard is flawed and asks the search reliability team to take a look. Instead, the team confirms: the data looks correct, and those numbers are really concerning!

Detection has occurred.

Ariel declares an incident and notifies the search model team since it might be a problem with the model. Ariel also notifies the retail team, just to check that we're not suddenly making less money from customers who are searching for products (as opposed to browsing for them) and also asks the team to check for recent changes to the website that would change the way results are rendered. Ariel then digs into the infrastructure for the search reliability team itself: what has changed on its end? Ariel finds—and the search model team confirms—that no changes have been made to the model configuration in the past two months. There have also been no big changes in the data or accompanying metadata used by the model—just the normal addition of customer activity to the logs.

Instead, one of the search model team members notes something interesting: they use a *golden set* of queries to test new models daily, and they've noticed that in the past three weeks the golden set is producing incredibly consistent results—consistent enough to be suspicious. The search model is normally updated daily by retraining the same model on searches and resulting clicks from the previous day. This helps keep the model updated with new preferences and new products. It also tends to produce a little instability in the results from the golden set of queries, though that instability is normally within reasonable bounds. But starting three weeks ago, *those* results became remarkably stable.

Ariel goes to look at the trained model deployed in production. *It's three weeks old, and has not been updated since that point.* This explains the stability of the golden queries. It also explains the drop-off in user click behavior: we're probably showing fewer good results on new preferences and new products. Indefinitely, of course—if we keep the same, stale model, we'll eventually be unable to correctly recommend anything new. So Ariel looks at the search model training system, which schedules the search model training every night. It has not completed a training run in over three weeks, which would definitely explain why there isn't a new trained model in production.

We have a proximal cause for the outage, but at this point we don't know the underlying cause, and there's no obvious simple mitigation: without a new trained model in production, we cannot improve the situation. This is also a very rough proximal start of impact.

The training system is distributed. A scheduler loads a set of processes to store the state of the model, and another set of processes to read the last day's search logs and update the model with the new expressed preferences of the users. Ariel notes that all of the processes trying to read logs from the search system are spending most of their time waiting for those logs to be returned from the logging system.

The logging system accesses raw customer logs via a set of processes called *log feeders* that have permission to read the relevant parts of the logs. Looking at those log-feeder processes, Ariel notices a group of 10 of them and sees that each is crashing and exiting every few minutes. Diving into process crash logs, Ariel sees that the log feeders are running out of memory, and when they can't allocate more memory, they crash. When they crash, a new log-feeder process is started on a new machine, and the training process retries its connection, reads a few bytes, and then that process runs out of memory and crashes again. This has been going on for three weeks.

Ariel proposes that they try increasing the number of log-feeder processes from 10 to 20. Spreading the load from the training jobs around might prevent the jobs from crashing. They can also look at allocating more memory to the jobs if needed. The team agrees, Ariel makes the change, the log-feeder jobs stop crashing, and the search training run completes a few hours later.

The outage is mitigated as soon as the training run completes and the new trained model is put into production.

Ariel works with the team to double-check that the new trained model loads automatically into the serving system. The query golden set performs differently than the one from three weeks ago but performs acceptably well. Then they all wait a few hours to accumulate enough logs to generate the data they need to make sure that the updated trained model is really performing well for customers. Later, they analyze the logs and see that the click-through rate in the first five results is now back to where it should be.

At this point, the outage is resolved. Sometimes there is no obvious mitigation stage, and mitigation and resolution take place at the same time.

Ariel and the team work on a review of the incident, accumulating some post-outage work they'd like to perform, including the following:

- Monitor the age of the model in serving and alert if it's over a certain threshold of hours old. "Age" here might be wall-clock age (literally, the timestamp on the file) or data age (how old the data is that the model is based on). Both of these are mechanically measurable.
- Determine our requirements for having a fresh model and then distribute the available time to the subcomponents of the training process. For example, if we need to get a model updated in production every 48 hours at the most, we might give ourselves 12 hours or so to troubleshoot problems and train a new model, so then we can allocate the remainder of the 36 hours to the log processing, log-feeding, training, evaluation, and copying to serving portions of the pipeline.
- Monitor the golden query test and alert if it is unchanged as well as if it's changed too much.
- Monitor the training system *training rate* and alert if it falls below a reasonable threshold such that we predict we will miss our deadline for training completion based on the allocated amount of time. Selecting what to monitor is difficult, and setting thresholds for those variables is even harder. This is covered briefly in "ML Incident Management Principles" on page 274, and covered previously in Chapter 9.
- Finally, and most importantly: monitor the top-five-results click-through rate and alert if it falls below a certain threshold. This should catch any problem

that affects the quality as perceived by users, but not caught by any of the other causes. Ideally, the metric for this should be available at least hourly so that we can use it while troubleshooting future problems, even if it's stable on only a day-by-day basis.

With those follow-up items scheduled, Ariel is ready for a break.

Stages of ML incident response for story 1

This outage, although quite simple in cause, can help us start to see the way that ML incidents manifest somewhat differently for some phases of the incident response lifecycle:

Pre-incident
: The training and serving system was a somewhat typical structure, with one system producing a trained model and periodically updating it, and another using that model to answer queries. This architecture can be very resilient, since the live customer-facing system is insulated from the learning system. When it fails, it is often because the model in serving is not updated. The underlying log data is also abstracted away in a clean fashion that protects the logs but still lets the training system learn on them. But this interface to the logs is precisely where the weakness in our system occurred.

Trigger
: Distributed systems often fail when they pass a particular threshold of scaling that sharply reduces performance, sometimes referred to as a *bottleneck*. In this case, we passed the threshold of performance of our log-feeder deployment and did not notice. The trigger was the simple growth of the data, corresponding growth of the training system requirements, and the business need to consume that data.

Outage begins
: The outage begins three weeks before we notice it. This is unfortunate, and why good monitoring is so important.

Detection
: ML systems that are not well instrumented often manifest systems problems as only quality problems—they simply start performing less well and get gradually worse over time. Model quality changes are often the only end-to-end signal that something is wrong with the system infrastructure.

Troubleshooting
: ML troubleshooting often involves a broader set of teams than some other kinds of outages, precisely because they often manifest as publicly visible quality problems. Until the problem can be narrowed down, it's prudent not to make

assumptions about the kind of outage we're experiencing—it could be a systems problem, a model problem, or just a drift in our ability to correctly predict the world.[3] Sometimes it's the world changing faster than we can keep up with—more on this in story 3. Not all quality degradations are even outages. Future stories will show a yet broader cast of characters involved in troubleshooting.

Mitigation/resolution
: The fastest and least risky steps to mitigate the problem, in this case, involved training a new model and successfully deploying it to our production search serving system. For ML systems, especially those that train on lots of data or that produce large models, no such quick resolution might be available.

Follow-up
: We can add a rich set of monitoring here, much of which is not easy to implement but will benefit us during future incidents.

This first story shows a fairly simple ML-associated outage. We can see that outages can present as quality problems of models not quite doing what we expect or need them to do. We can also start to see the pattern of broad organizational coordination that is required in many cases for resolution. Finally, we can see that it is tempting to be ambitious in specifying the follow-up work. Keeping these three themes in mind, let's consider another outage.

Story 2: Suddenly Useless Partners

At YarnIt, we have two types of business. The first part of our business is a first-party store selling knitting and crocheting products. But we also have a marketplace for recommending products from other partners who sell them through our store. This is a way that we can make a wider variety of products available to our customers without having to invest more in inventory or marketing.

When and how to recommend these marketplace products is a little tricky. We need to incorporate them into our search results and discovery tools on the website as a baseline, but how should we make recommendations? The simplest approach would be to list every product in our product database, include all actions that touch them in our logs, and add them to our main prediction model. A notable constraint is that each of these partners requires that we separate their data from every other partner; otherwise, they won't let us list their products.[4] As a result, we'll have to

3 This drift, as well as the contents of the golden set, are also key places where unfair bias can become a factor for our system. See Chapter 6 for a discussion of these topics at more length.

4 This kind of restriction on data commingling is somewhat common. Companies are sensitive about their commercially valuable data (for example, who bought *X* after searching for *Y*) being used to benefit their competitors. In this case, these companies may even regard YarnIt as a competitor (although one who sends them significant business that they value).

train a separate model per partner, and extract partner-specific data into isolated repositories, though we can still have a common feature store for shared data.

YarnIt is ambitious enough to plan for a potentially *very* large number of partners—somewhere between five thousand or five million—and so instead of a setup optimized for a few large models, we need a setup optimized for thousands of tiny models. As a result, we built a system that extracts historical data from each partner and puts it into a separate directory or small feature store. Then at the end of every day, we separate out the previous day's deltas and add them to our stores just before starting training. Now our main models train quickly, and our smaller partner models train quickly as well. Best of all, we're compliant with the access protection demanded by our partners.

The pre-incident is complete at this point. The stage is set, and the conditions for the outage have been set. It may be obvious by this point that several opportunities exist for things to go wrong.

Sam, a production engineer at YarnIt, works on the partner training system. Sam is asked to produce a report for CrochetStuff, a partner, in advance of a business meeting. While preparing the report, Sam notices that the partner in question has zero recent conversions (sales) recorded in the ML training data but that the accounting system reports that it's selling products every day. Sam produces a report and forwards it to colleagues who work on the data extraction and joining jobs for some advice. In the meantime, Sam leaves this fact off the report on data to the partner team and simply includes the sales data.

The detection happens here. No computer system detected the outage, which means that it may have been going on for an indefinite amount of time.

Data discrepancies in counts like this happen all the time, so the data extraction team does not treat Sam's report as a high priority. Sam is reporting a single discrepancy for a single partner, and the team files a bug and plans to get to it in the coming week or so.

The incident is unmanaged and continuing chaotically. It might be small, or it might not. No one has determined the extent of the impact of the data problem yet, and no one is responsible for coordinating a quick and focused response to it.

At the business meeting, CrochetStuff notes that its sales are down 40% week-on-week and continuing to drop daily. Their reports on page views, recommendations, and user inquiries are all down, even though when users *do* find the products, the rate at which they purchase continues to be high. CrochetStuff demands to know why YarnIt suddenly stopped recommending all of its products!

By this point, we have had internal detection, an internal partner advocate, customer reports, and a possible lead of what is happening. This is a lot of noise, but sometimes we don't declare an incident until many people independently notice it.

Sam declares an incident and starts working on the problem. The logs of the partner model training system clearly report that the partner models are successfully training every day, and there are no recent changes to either the binaries that carry out the training or the structure and features of the models themselves. Looking at the metrics coming from the models, Sam can see that the predicted value of every product in the CrochetStuff catalog has declined significantly every day for the past two weeks. Sam looks at other partners' results and sees exactly the same drop.

Sam brings in the ML engineers who built the model to troubleshoot what is happening. They double-check that nothing has changed and then do some aggregate checks on the underlying data. One of the things they notice is what Sam noticed originally: there are no sales for any partners in the last two weeks in the ML training data. The data all comes from our main logs system and is extracted every day to be joined with the historical data we have for each partner. The data extraction team resurrects Sam's bug from a few days before and starts looking at it.

Sam, who needs to find a fast mitigation for the problem, notes that the team stores older copies of trained models for as long as several months. Sam asks the ML engineers about the consequences of just loading an old model into serving for now. The team confirms that while the old trained model versions won't have any information about new products or big changes in consumer behavior, they will have the expected recommendation behavior for all existing products. Since the scope of the outage is so significant, the partner team decides it is worth the risk to roll back the models. In consultation with the partner team, Sam rolls back all of the partner trained models to versions that were created two weeks earlier, since that seems to be before the impact of the outage began. The ML engineers do a quick check of aggregate metrics on the old models and confirm that recommendations should be back to where they were two weeks ago.[5]

5 This is a risky way to test this hypothesis. It would have been better to roll out a single model first to validate that the old models performed better and didn't have another catastrophic problem. But this is a defensible choice that people make during high-stakes outages.

At this point, the outage is mitigated but not really resolved. Things are in a pretty unstable state—notably, we cannot build a new model with our accustomed process and have it work well—and we still need to figure out the best full resolution as well as how to avoid getting ourselves into this situation again.

While Sam has been mitigating, the data extraction team has been investigating. It finds that while the extractions are working well, the process that merges extracted data into the existing data is consistently finding no merges possible for any partners. This appears to have started about two weeks ago. Further investigation reveals that two weeks ago, in order to facilitate other data analysis projects, the data management team changed the unique partner key, used to identify each partner in its log entries. This new unique key was included in the extracted data, and because it differed from previous partner identifiers, the newly extracted logs could not be merged with any data extracted prior to the key being added.

This is now a reasonable root cause for the outage.

Sam requests that a single partner's data be re-extracted and that a model be trained on the new data in order to quickly verify that the system will work correctly end to end. Once this is done, Sam and the team are able to verify that the newly extracted data contains the expected number of conversions and that the models are now, again, predicting that these products are good recommendations for many customers. Sam and the data extraction engineers do some quick estimations on how long it will take to re-extract all of the data, and Sam then consults with the ML engineers on how long it will take to retrain all of the models. They arrive at a collective estimate of 72 hours, during which they will continue to serve recommendations from the stale model versions that they restored from two weeks prior. After consulting with the retail product and business team, they all decide to carry out this approach. The partner team drafts some mail to partners to let them know about the problem and a timeline for resolution.

Sam requests that all partner data be re-extracted and that all partner models be retrained, preferably from scratch, starting with the beginning of data we have. They monitor the process for three days, and once it is done, verify that the new models are recommending not only the older products but also newer products that didn't exist two weeks prior. After careful checking, the new models are deemed to be good by the ML engineers and are put into production. Serving results are carefully checked, along with many folks doing live searches and browsing to verify that partner listings are actually showing up as expected. Finally, the outage is declared closed, and the partner team drafts an update to partners letting them know.

At this point, the outage is resolved.

Sam brings the team together to go over the outage and file some follow-up bugs so that they can avoid this kind of outage in the future and detect it more quickly than they did this time. The team considers rearchitecting the whole system so that they can eliminate the problem of having two copies of all of the data, with slightly different uses and constraints, but decides that they still don't have a good idea about how to meet their performance goals for both systems if they are unified.

They do file a set of bugs related to monitoring the success of data extraction, data copying, and data merging. The biggest problem is that they don't have a good source of truth for the question, how many lines of data should be merged? This failure happened for an entire class of logs, and the team was quickly able to add an alert for "log lines merged must be greater than zero." But during the investigation, a series of less catastrophic failures were also found, and to catch those, we would need to know the expected number of logs per partner to be merged and then the actual number that were merged.

The data extraction team settles on a strategy of storing the count of merged log lines by partner by day and comparing today's successes to the trailing average of the last *n* days. This will work relatively well when partners are stable but will be noisy when they experience big changes in popularity.

Two years later, this alerting strategy is still unimplemented as a result of challenges in implementing it without unnecessary noise. The strategy may be a good idea but, given the dynamic retail environment, has proven unworkable, and the team still lacks good end-to-end rapid detection of this kind of log extraction and merging failure, except in the catastrophic case. However, a heuristic they did implement a few months in—a hook that triggers on any relevant change to the partner configurations and notifies an engineer to potentially expect breakage—has at least increased ongoing awareness of such a change as a potential trigger for outages.

Stages of ML incident response for story 2

Many of the characteristic stages that this incident went through are similar to those of any distributed systems incident. It has prominent differences though, and the best way to see those with some context and nuance is to walk through the partner training outage and look at the ML-salient features that occur during each section:

Pre-incident

Most of what went wrong was already latent in the structure of our system. We have a system with two authoritative sources for the data, one of which is an extracted version of the other, with incremental extracts being applied periodically. ML systems most commonly fail because of problems with data and metadata. We will dig into the tactics for observing and diagnosing outages across systems with coupled data and ML in "ML Incident Management Principles" on page 274.

Trigger

The data schema was changed. It was changed very far away from where we observed the problem, which obviously made it difficult to identify. It is important to think about this outage as a way of identifying the assumptions we have made about our data throughout the processing stack. If we can identify those assumptions and where they are implemented, we can avoid creating data processing systems that can be damaged by changes to those assumptions. In this case, it should have been impossible to change the schema of our main feature store without also modifying or at least notifying all downstream users of that feature store. Explicit data schema versioning is one way to achieve this result.

Outage begins

The outage begins when one internal system processing data uses another internal system that processes data in a way that is no longer consistent with its structure. This is a common hazard for any large distributed pipeline system.

Detection

ML systems quite commonly fail in ways that are detected first by end users. One challenge with this is that ML systems are often accused of failure, or at least not working as well as we might hope, even under normal operations, and so it may seem reasonable to disregard the complaints of users and customers. The primary method of noticing this particular outage is a common one: the recommendations system wasn't making recommendations of the same quality as it used to. With ML system monitoring, keeping the high-level, end-to-end, coarse-grained picture in mind is particularly useful—with the central question being, have we substantially changed what the model is predicting over the past short while? These kinds of end-to-end quality metrics are completely independent of the implementation and will detect any kind of outage that substantially damages models. The challenge will be to filter that signal so that not too many false positives occur.

Troubleshooting
: Sam needs to work with multiple teams to understand the scope and potential causes of the outage. We have commercial and product staff (the partner team), ML engineers who build the model, data extraction engineers who get the data out of the feature store and logs store and ship it to our partner model training environment, and production engineers like Sam coordinating the whole effort. Troubleshooting ML outages really has to start not with the data but with the outside world: what is our model saying, and why is that wrong? There is *so* much data that starting by "just looking through the data" or even "doing aggregate analysis of the data" is likely to be a long and fruitless search. Start with the model's changed or problematic behavior, and it will be much easier to work backward to why the model is now doing what it is doing.

Mitigation
: With some services, it is possible to simply restore an older version of the software while a fix is prepared. While this may inconvenience any users depending on new features, everyone else can continue unaffected.

ML outages can only sometimes be mitigated by restoring an older version of the model, because their job is to help computer systems adapt to the world, and there's no way to restore a snapshot of the world as it used to be.

Additionally, quickly training up new models often requires more computing capacity than we have available. As was the case with our partner model outage, no cost-free quick mitigation exists. To determine which mitigation was the best option, the decision ultimately needed to be made by the product and business staff most familiar with our partners, users, and business. This level of escalation to business leaders happens sometimes for non-ML services but much more frequently for ML services. Most organizations that rely on ML to run important parts of their business will need to cultivate technical leaders who understand the business, and business leaders who understand the technology.

Resolution
: Sam makes sure that the data in the partner training system is correct (at least in aggregate, and spot checks seem to confirm that it looks good). New models are trained. When we are ready to deploy them, there's actually no simple way to determine whether the new models "fix" the problem. The world continues to change while we are working on resolving this problem. So some previously popular products may be less in vogue now. Some neglected products may have been discovered by our users. We can look at the aggregate metrics to see whether we are recommending partner products at closer to the rate that we did previously, but it won't be identical. Sometimes people use a golden set of queries here to see if they can produce a "correct" set of recommendations for pre-canned results. This can increase our confidence somewhat but adds the new problem that we

will want to continuously curate this golden set of queries to be representative of what our users search for. Once we do that, we will not necessarily have stable results over very long periods of time.[6]

Follow-up

After-incident work is always difficult. For a start, the people with direct knowledge are tired, and may have been neglecting their other work for some time by this point. We have already paid the price of the outage, so we might as well get the value for it. While monitoring bugs are typically included in post-incident follow-up, it is incredibly common for them to languish (in some cases for years) for ML-based systems. The reason is relatively simple: it is extremely difficult to monitor real data and real models in a high-signal, low-noise way. Anything that is overly sensitive will alert all the time—the data is different! But anything that is overly broad will miss complete outages of subsets of our services. These problems exist for most distributed systems but are characteristic for ML systems.

While this outage was technically complex and somewhat subtle in its manifestation, many ML outages have very simple causes but still show up in difficult-to-correlate ways.

Story 3: Recommend You Find New Suppliers

We have models for several aspects of our business at YarnIt. The recommendations model in particular has an important signal: purchases. Simply put, we recommend a product in every browsing context where users tend to purchase that product when it is offered to them. This is good for our users, who more quickly find products that they want to buy, and for YarnIt, which will presumably sell more products more quickly.

Gabi is a production engineer who works on the discovery modeling system. One unusually pleasant summer day, Gabi is working through configuration cleanups that have been lingering and addressing requests from other departments. Customer support staff sent a note that they have been tracking a theme in feedback on the website for the past couple of weeks, saying that the recommendations are "weird." Subjective impressions like this by some customers are generally pretty hard to take

6 Incident response has a pair of useful concepts. *Recovery point objective* (*RPO*) is the point in time that we are able to restore the system to full functionality after recovering from the outage, ideally immediately prior to the outage. *Recovery time objective* (*RTO*) is how long it will take to restore the system to functionality from an outage. ML systems certainly have an RTO; retraining a model, copying an old version—these take time. But the problem is that most ML systems have no meaningful notion of an RPO. Occasionally, a system runs entirely in the "past" on preexisting inputs, but most of the time ML systems exist to adapt our responses to current changes in the world. So the only RPO that matters is "now" for ever-changing values of "now." A model trained "a few minutes ago" might be good enough for "now" but might not. This significantly complicates thinking about resolution.

any concrete action on, but Gabi files the request into a "pending follow-up" section for later follow-up.

No spoilers! We definitely cannot say whether incident detection has happened yet at this point.

Further in the incoming requests, Gabi spots an unusual problem report. The website payments team tells Gabi that Finance is reporting a big drop in revenue. Revenue is down 3% for the past month on the site. That might not seem like a big drop, but after further digging, the team finds that last week versus four weeks ago is down closer to 15%! The payments team has checked the payment-processing infrastructure, and found that customers are paying for carts successfully at the same rate they historically have. They note, though, that the carts have fewer average products than they used to, and in particular fewer people are purchasing products from recommendations than expected. This is why the payments team has contacted Gabi. Seeing numbers this big, Gabi declares an incident.

Incident detected and declared.

Gabi asks the financial team to double-check the week-versus-four-weeks-ago comparison for the past several weeks, and asks for a more detailed timeline of revenue for the past several weeks. Finally, Gabi asks for any product, category, or partner breakdowns available. Gabi then asks the payments team to verify its numbers about recommendations added to carts as well as to provide any breakdowns it can. In particular: does the team see a particular type of cart that has fewer recommendations than others or that has changed more recently?

Meanwhile, Gabi starts looking at aggregate metrics for the application, just trying to figure out some basic questions. Are we showing recommendations at all? Are we showing recommendations as often as we have in the past, and for all the queries and users and products that we did in the past, and in the same proportions across user subpopulations? Are we generating sales from recommendations at the same rate as we typically have? Is there anything else salient about the recommendations that is obviously different?

Gabi also starts doing the normal production investigation, focusing particular attention on what changed in the recommendations stack recently. The results are not promising for finding an obvious culprit: the recommendations models and binaries to train the models are unchanged in the last six weeks. The data for the model is updated daily, of course, so that's something to look at. But the data schema in the feature store hasn't changed in several months.

Gabi needs to continue troubleshooting but takes time to compose a quick message to the finance and payments teams that asked for help with this issue. Gabi confirms what is known so far: the recommendations system is running and producing results, and there are no recent changes to be found, but the quality of the results has not been verified. Gabi reminds them to inform their department heads if they have not already, which seems wise given the amount of money the company appears to be losing.

No obvious software, modeling, or data updates correlate with the outage, so Gabi decides that it's time to dig into the recommendations model itself. Gabi sends a quick message to Imani, who built the model, asking for help. As Gabi is explaining to Imani what they know so far (fewer products purchased, fewer recommendations purchased per checkout, no system changes to speak of), the note from customer support comes to mind. Customers complaining about "weird" recommendations, if the timeline matches up, certainly seems relevant.

Customer support staff confirm that they started getting the first sporadic complaints just over three weeks ago but that they have been intensifying and became especially pointed in the last week. Imani thinks this may be worth investigating and asks Gabi to grab enough data to trend some basic metrics on the recommendations system: number of recommendations per browse page, average hourly delta between expected "value" of all recommendations (probability that a customer will purchase a recommended product times the purchase price), and the observed value (total value of recommended products ultimately purchased). Imani grabs a copy of recent customer queries and product results in order to use them as a repeatable test of the recommendation system. The recommendation system uses the query that a user made, the page that they are on, and their purchase history (if we know it) to make recommendations, so this is the information that Imani will need to query the recommendation model directly.

Without more information, we have to worry that by doing this, Imani may have violated the privacy of YarnIt's customers. Search queries may contain protected information like user IP addresses, and any collection of search queries contains the additional problem that when correlated with each other for a given user, they reveal even more private information.[7] Imani definitely should have consulted with privacy and data protection professionals at YarnIt, or better yet, not even had direct, unmonitored access to the queries to make this kind of a mistake.

Imani extracts out about 100,000 queries and page views and sets up a test environment where they can be played against the recommendation model. After a test run through the system, Imani has recommendations for all of the results and has stored a copy of the whole run so that it can be compared to future runs if they need to modify or fix the model itself.

Gabi comes back and reports something interesting. Just over three weeks ago, the number of recommendations per page dropped slowly. For one week, the difference between expected value and observed value of each recommendation declined only a bit. By two weeks ago, the recommendation count plateaued at just under 50% lower than it has been. But then the value of the recommendations began to drop significantly compared to the expected value. That decline continued through last week. Two weeks ago, the observed value of the recommendations hit a low value of 40% of their expected value. Even more strangely, though, the gap between the expected value and observed value of the recommendations started narrowing a week ago, but at the same time the number of recommendations shown began falling again, so that now we seem to be showing very few recommendations at all, but those that we do show seem to be relatively accurately valued. Something is definitely wrong, and it is starting to look like it's the model, but no clear diagnosis is flowing from this set of facts. Figure 11-1 shows a graphical representation of this set of changes over time.

7 IP addresses are probably personal information or PII in many jurisdictions, so caution must be exercised. This is not always widely understood by systems engineers or operators, especially those who work in countries with looser legal governance frameworks. Additionally, search queries that can be correlated to the same user demonstrably reveal private information by means of the combination of the queries. Famously, see Wikipedia's "AOL search log release" page (*https://oreil.ly/U2bCP*) for context.

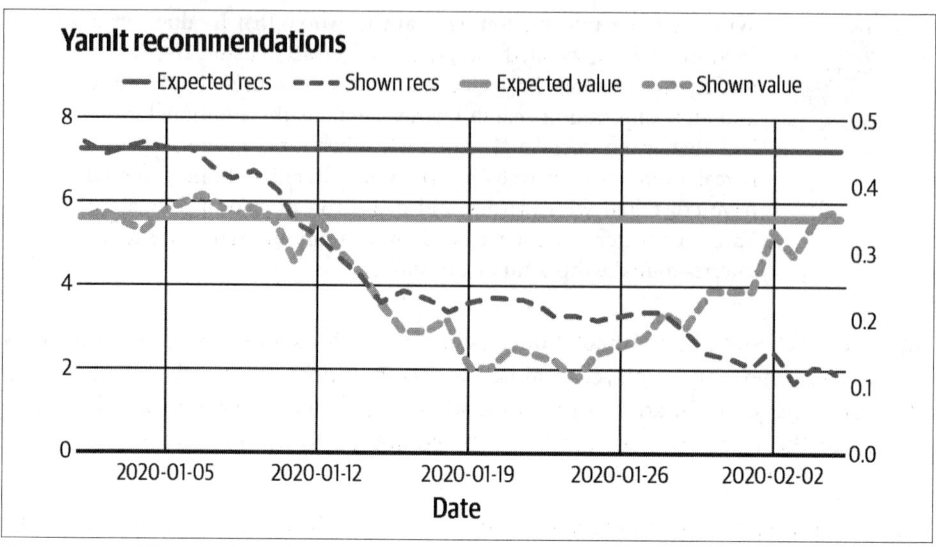

Figure 11-1. The number of expected and shown recommendations as well as their average expected value as they change over the period of the incident

Imani continues to build the QA environment to test hypotheses. On a hunch, Gabi and Imani grab another 100,000 queries and page views from a month ago (before there was any evidence of a problem) as well as a snapshot of a model from every week in the last six weeks. Since the model retrains daily, even though the configuration of the model is exactly the same day after day, each day the model has learned from the things the users did the preceding day. Imani plans to run the old and new queries against each of the models and see what can be learned.

Gabi pushes for a quick test first: today's queries versus a month-old model. Here's the thinking: if that works, there's a quick mitigation (restore the old model to serving) while troubleshooting continues. Gabi is focused on solving the lost revenue problem as quickly as possible. Imani runs the tests, and the results are not that promising and difficult to evaluate. The old model makes different recommendations than the new model and does seem to make slightly more of them. But the old model still makes many fewer recommendations against today's queries than it did against the queries a month ago.

Without something quite a bit more concrete, Gabi isn't comfortable that changing the model to an older one will help. It might even do more damage to our revenue than the current model. Gabi decides to leave the recommendation system in its current state. It's time to send another note to the folks in finance and payments about the current status of the troubleshooting. The payments and finance contacts both report that their bosses want a lot more information about what's going on. Gabi's colleague Yao, who has been shadowing the investigation and is familiar with the recommendations system, is drafted to handle communications. Yao promptly sets up a shared document with the state as it is known so far and links to specific dashboards and reports for more information. Yao also sends out a broad notice to senior folks in the company, notifying them of the outage and the current status of the investigation.

Imani and Gabi finish running the full sweep of old and new queries against older and newer models. The results are different for each pair, but nothing broadly systematic stands out that might explain the differences, and the general metrics match the weird pattern described previously. Imani decides to forget the model for a second and focus instead on the queries and page views themselves. Imani wants to figure out how they have changed in the last month, thinking maybe the problem is with the model's ability to handle a shift in user behavior rather than something being wrong with the model itself.

Imani spot-checks the queries, but there are 100,000 of them in each of two batches, and whatever might be substantially different about them is not exactly obvious. Gabi, meanwhile, produces two reports. The first looks exclusively at the search queries that customers used to get to the product pages they ended up on. Gabi tokenizes the search queries and just counts the appearances of each word. While that's running, Gabi takes the product pages the customers ended up on and assigns each to a large category (yarn, pattern, needles, accessories, gear), and then to subcategories within those, according to the product ontology (built by another team). Gabi lines up the two pairs of reports and looks for the biggest differences between the user behavior four weeks ago and today.

The results are shockingly obvious: compared to four weeks ago, users have increasingly been looking for very different products. In particular, they are now looking for lightweight yarns, patterns for vests and smaller items, and smaller-gauge needles. Imani and Gabi stare at the results, and the problem suddenly seems so obvious. What happened four weeks ago? It got very hot in the northern hemisphere, where the majority of YarnIt's customers are based. The heat came earlier than usual and significantly decreased the interest most customers had in knitting with chunky, warm wool.

Imani points out, however, that that doesn't explain the decrease in recommendations, only the change in what the recommendations should be. This still leaves the question of why aren't we just recommending good hemp and silk yarn instead of wool? Gabi walks through a few queries to the recommendation engine by hand, using a command-line tool built for troubleshooting like this, and notices something. The recommendation engine test instance is set to log many more details than the production instance. One of the things it's logging at a pretty high rate is that many candidate recommendations are disqualified from being shown to users because they're out of stock.

Yao gets an update from Imani and Gabi, updates some of the shared doc, and publishes some information to the increasingly large group of people waiting to find out how the company is going to fix this problem. Someone from the retail team sees the note about many recommendations being out of stock and mentions to Yao that YarnIt did lose several important suppliers recently. One of the biggest, KnitPicking, is a popular supplier of fashionable yarns, many of which happen to be lightweight. In fact, KnitPicking was one of the largest suppliers of those weights of yarn at those price points. Yao gets more details on the timing of the supply problems, adds it to the doc, and reports back to Gabi.

This is an interesting state for the incident to be in. We have a likely root cause but no obvious way to mitigate it or resolve it.

Imani and Gabi have a solid hypothesis on the weird recommendations. The recommendations system is configured with a minimum threshold of expected value for each recommendation it shows so that it won't show terrible recommendations when it doesn't have any good ones. But it takes a while for a recommendation's expected value to adjust, especially when it hasn't been shown often recently. Imani concludes that the system quickly learned that few people wanted heavyweight wool yarns. But once those were understood to be poor recommendations, it took a while for the system to cycle through many other products until it finally concluded that we really don't have very much stock in the products that our customers currently want to buy.

Gabi, Imani, and Yao schedule a meeting with the heads of retail and finance to discuss what they have learned and ask for guidance on how to proceed. Oddly, the current state seems to be that the recommendations system is now moderately good for current circumstances. It recommends few products for most customers on most page views since we don't have much of what most of our customers want right now. The loss of revenue was as much due to supply problems as it ever was due to the recommendations system. Presented with the facts as they are known, the head of retail asks the team to verify its findings to be certain but agrees that fixing the supply problem is the highest priority. The finance lead nods and goes off to sharply reduce projections for how much money we will make this quarter. There is no obvious change to the recommendations model that can improve the situation, given our supply shortfalls and the way that the weather has impacted our customers' preferences.

At this point, the outage is probably over, since we've decided not to change the system or model.

The team gathers the next day to review what happened and what they can learn from it. Here are some of the proposed follow-up actions:

- Monitor and graph the number of recommendations per page view, revenue per recommendation, and percentage gap between expected value of all recommendations per hour and the actual sales value.
- Monitor and alert on high rates of candidate product recommendations unavailable (for whatever reason: out of stock, legal restrictions, etc.). We can also consider monitoring stock levels directly if we can find a high-signal way to do so, although ideally this would be the responsibility of a supply-chain or inventory management team. We should be careful here not to be overbroad in our monitoring of other teams' work to avoid burdening our future selves with excessive alerting. We should think about monitoring user query behavior, in aggregate, directly as well so that we might be able to detect significant shifts in query topics and distribution. This kind of monitoring is generally good for graphing but not for alerting—it's just too hard to get right. Finally, we can work more closely with the customer support team to get them tools to investigate user reports like these. If the support team had a query replicator/analyzer/logger, it may have been able to generate a considerably more detailed report than "customers say they get weird recommendations." This kind of "empower another team to be more effective" effort often pays off much more than pure automation.
- Review ways to get the model to adjust more quickly. Taking so many days to converge on the right recommendation behavior isn't reasonable. The overall stability of the model has been perceived to be of value, but in this case it ended up showing bad recommendations to users for many days and making it harder for the production team to troubleshoot problems with it. Imani wants to find a way to improve the responsiveness to new situations without making the model overly unstable.
- We should treat this as an opportunity to think about what the model should do when it doesn't have any good recommendations. This is fundamentally a product and business problem rather than an ML engineering problem: we need to figure out the behavior we want the model to exhibit and the kinds of recommendations we think we should surface to users under these circumstances. At a high level, we would like to keep making money at a reasonable rate with good margins, even when we do not have the products that our customers want the

most. Figuring out whether there's a way to identify a product recommendation strategy to do that is a hard problem.

- Finally, it's clear that some exogenous data to the ML system should be always available to make troubleshooting situations like these easier. In particular, the production engineers should have revenue results in aggregate and broken down by product category in the product catalog, by geography, and by the original source of the user viewing the product (search result or recommendation or home page).

Many of these follow-ups are quite ambitious and unlikely to be completed in any reasonable amount of time. Some of them, though, can be done fairly quickly and should make our system more resilient to problems like this in the future. As always, figuring out the right balance and understanding the trade-offs in implementing those is precisely the art of good follow-up, though we should favor the ones that make problems faster to troubleshoot.

Stages of ML incident response for story 3

Although this incident had a somewhat different trajectory from that of the previous two stories, we can see many of the same themes appear. Rather than repeat them, let's try to focus on any additional lessons we can learn from this outage:

Pre-incident
: No obvious significant failure in the architecture or implementation of our system led to this outage, which is interesting. We definitely could have made choices that would have made the outage progress differently, and more smoothly for our users, but in the end we cannot recommend products we don't have, and sales were going to go down. There may be a model that could produce better recommendations under these circumstances (rapid change in demand combined with an inventory problem), but that falls more under the heading of continuous model improvement rather than incident avoidance.

Trigger
: The weather changed, and we lost a supplier. This is a tough combination of events to directly detect, but we can certainly try with some of the monitoring efforts that were picked.

Outage begins
: In some ways, there is no outage. That is what is most interesting about this incident. An outage can be understood to be a failing of the system such that it yields an incorrect result. It's appropriate to describe the "weird recommendations" period as an outage, but one with only minimal costs since the main impact was probably to annoy our users a bit. But the loss in revenue wasn't caused by the

recommendations model nor was it preventable by it. Likewise, the outage won't end until the weather changes or we source a new supply of lightweight yarns.

Detection
: The earliest sign of the outage was the customer complaints about weird recommendations. That's the kind of noisy signal that probably cannot be relied on, but as noted, we can get the support team better tools so that it can report problems in more detail. Other, less obvious, signals may have had a higher accuracy that we could use for detection, but even figuring them out is a data science problem.

Troubleshooting
: The process of investigating this outage includes some of the hallmarks of many ML-centric outage investigations: detailed probing at a particular model (or set of models or modeling infrastructure) coupled with broad investigation of changes in the world around us. The investigation might havc proceeded more quickly if Sam had followed up on the detailed timeline of revenue from the finance team. With the breakdown of revenue changes by product, category, or partner, we should have been able to see a sharp shift in consumer behavior combined with a sharp rise and then drop in sales from KnitPicking (as our stock in its products ran low). It is sometimes difficult to remember that clarity about an outage might come from looking more broadly at the whole situation rather than more carefully at a single part of it.

Mitigation/resolution
: Some outages have no obvious mitigation. This is tremendously disappointing, but occasionally there's no quick way to restore the system to whatever properties it previously had. Moreover, the only way to actually resolve the core outage, and get our revenue back on track, is to change what our users want or to fix the products that we have available to sell. One thing the team didn't think about, probably in part because it was focused on troubleshooting the model and resolving the ML portion of the outage, was that there may have been other, non-ML ways of mitigating the outage: what if our system showed out-of-stock recommendations and invited customers to be notified when we had those (or similar) products available? In that case, we might have avoided some of the lost revenue by shifting it forward in time and also reduced the weird recommendations served to customers. Sometimes mitigations can be found outside of our system.

Follow-up
: In many cases, follow-up from an ML-centric incident evolves into a phase that doesn't resemble "fix the problem" so much as "improve the performance on the model." Post-incident follow-up often devolves into longer-term projects, even for non-ML-related systems and outages. But the boundary between a "fix" and "ongoing model improvement" is particularly fuzzy for ML systems. One

recommendation: first define your model improvement process clearly. Track efforts that are underway and define the metrics you plan to use to guide model quality improvement. Once an incident occurs, take input from the incident to add, update, or reprioritize existing model improvement work. For more on this, see Chapter 5.

These three stories, however different in detail, demonstrate common patterns for ML incidents in their detection, troubleshooting, mitigation, resolution, and ultimately post-incident follow-up actions. Keeping these in mind, it is useful to take a broader view of what is happening to make these incidents somewhat different from other outages in distributed computing systems.

ML Incident Management Principles

While each of these stories is specific, many of their lessons remain useful across various events. In this section, we will back away from the immediacy of the stories and distill what they, and the rest of our experience with ML systems outages, can teach us in the long term. We also offer a specific list of recommendations for you to follow to get ready for and respond to incidents.

Guiding Principles

Three overarching themes that appear across ML incidents are so common that we list them here as guiding principles:

Public
: ML outages are often detected first by end users, or at least at the very end of the pipeline, all the way out in serving or integrated into an application. This is partly true because ML model performance (quality) monitoring is difficult. Some kinds of quality outages are obvious to end users but not obvious to developers, decision makers, or SREs. Typical examples include anything that affects a small sample of users 100% of the time. Those users get terrible performance from our systems all the time, but unless we happen to look at a slice of just those users, aggregate metrics probably won't show anything wrong.

Fuzzy
: ML outages are less sharply defined in two dimensions: in impact and in time. With respect to time, determining the precise start and end of an ML incident is often difficult. Although there may well be a traceable originating event, establishing a definitive causal chain can be impractical. ML outages are also unclear in impact: it can be hard to see whether a particular condition of an ML system is a significant outage or just a model that is not yet as sophisticated or effective as we would like it to be. One way to think about this is that every model starts out very basic, doing only some portion of what we hope it can do one day. If our

work is effective, the model gets better over time as we refine our understanding of how to model the world and improve the data the model uses to do so. But there may be no sharp transition between "bad" and "good" for models. There is often only "better" and "not quite as good." The line between "broken" and "could be better" is not always easy to see.

Unbounded

ML outage troubleshooting and resolution involve a broad range of systems and portions of the organization. This is a consequence of the way that ML systems span more technical, product, and business arms within organizations than non-ML systems. This isn't to say that ML outages are necessarily more costly or more important than other outages—only that understanding and fixing them usually involves broader organizational scope.

With the three big principles in mind, the rest of this section is organized by role. As we have stated, many people working on ML systems play multiple roles. It is worth reading the principles for each role, whether you expect to do that work or not. By structuring the lessons by role, we can bring out the particular perspective and organizational placement particular to that role.

Model Developer or Data Scientist

People working at the beginning of the ML system pipeline sometimes don't like to think about incidents. To some, that seems like the difficult "operations" work that they would rather avoid. If ML ends up mattering in an application or organization, however, the data and modeling staff will absolutely be involved in incident management in the end. They can do certain things to get ready for that.

Preparation

A few concrete steps, taken in advance, can significantly improve the ability of an organization to respond to an incident. Among these are the following:

Organize and version all models and data

This is the most important step that data and modeling staff can take to get ready for forthcoming incidents. If you can, put all training data in a versioned feature store, with clear metadata spelling out where all the data came from and which code or teams are responsible for its creation and curation. That last part is often skipped: we will end up performing transformations on the data we put into the feature store, and it is critical that we track and version the code that performs those transformations. Additionally, if we can, we should store intermediate artifacts from all training runs in the metadata system as well. Finally, it is useful to store historical versions of the models in ready-to-serve format. As you saw, these can be terribly useful for quick mitigation in the case of a rapid and unexplained decline in model quality.

Specify an acceptable fallback
When we first start, the acceptable fallback might be "whatever we're doing now" if we already have a heuristic that works well enough. In a recommendations case, this might be "just recommend the most popular products" with little or no personalization. The challenge is that as our model gets better, the gap between that and what we used to do may get so large that the old heuristic no longer counts as a fallback. For example, if our personalized recommendations are good enough, we may start attracting multiple (potentially very different) groups of users to our applications and sites.[8] If our fallback recommendation is "whatever is popular," that might produce truly awful recommendations for every different subgroup using the site. If we become dependent on our modeling system, the next step is to save multiple copies of our model and periodically test falling back to them. This can be integrated into our experimentation process by having several versions of the model in use at any one time, with (for example) a primary, a new, and an old model.

Decide on useful metrics
The final bit of preparation that is most useful is to think carefully about model quality and performance metrics. We need to know whether the model is working, and model developers will have a set of objective functions that they use to determine this. Ultimately, we want a set of metrics that detect when the model stops working well that are independent of how it is implemented. This turns out to be a more challenging task than it might seem, but the closer we can approximate this, the better. Chapter 9 addresses the topic of selecting these metrics in a little more detail.

Incident handling

Model developers and data scientists play an important role during incidents: they explain the models as they currently are built. They also generate and validate hypotheses about what might be causing the problems we are seeing.

To play that role, model and data folks need to be reachable; they should be available off-hours on an organized schedule such as an on-call rotation or equivalent. They should not expect to be woken up frequently, but they might well be indispensable if they are.

Finally, during incident handling and triage, model and data staff may be called upon to do custom data analysis and even to generate variants of the current model to test hypotheses. They should be ready to do so, but also prepared to push back on any

8 Of course, these different recommendations might mean the model is picking up on proxies for unfair bias, and model designers and operators should use fairness evaluation tools to regularly look for this bias.

requests that require violating user privacy or other ethics principles. See "The Ethical On-Call Engineer Manifesto" on page 284 for more detail on this idea.

Continuous improvement

Model and data staff should work to shorten the model quality evaluation loop as a valuable but not dominant priority. Chapter 5 provides more details, but the idea here is similar to any troubleshooting: the shorter the delay between a change and an evaluation of that change, the faster we can resolve a problem. This approach will also pay notable benefits to the ongoing development of models, even when we're not having an outage. To do this, we'll have to justify the staffing and machine resources to get the training iterations, tools, and metrics required. It won't be cheap, but if we're investing in ML to create value, this is one of the best ways for this part of our team to deliver that value, with the least risk of multiday outages.

Software Engineer

Some, but not all, organizations have software engineers who implement the systems software to make ML work, glue the parts together, and move the data around. Whoever is playing this role can significantly improve the odds that incidents go better.

Preparation

Data handling should be clean, with clear provenance and as few versions of the same data as possible. In particular, when there are multiple "current" copies of the same data, this can result in subtle errors detected only in drops in model quality or unexpected errors. Data versioning should be explicit, and data provenance should be clearly labeled and discoverable.

It is helpful if model and binary rollouts are separate and separable. The binaries that do the inference in serving, for example, and the model that they are reading from, should be pushed to production independently, with quality evaluations conducted each time. This is because binaries can affect quality subtly, as can models. If the rollouts are coupled, troubleshooting can be much more difficult.

Feature handling and use in serving and training should be as consistent as possible. Some of the most common and most basic errors are differences in feature use between training and serving (called *training-serving skew*). These include simple differences in quantization of a feature, or even a change in certain features' contents altogether (a feature that used to be income becomes zip code, and chaos ensues immediately, for example).

Implement or develop tooling as much as possible (sometimes test development is done by specialist test engineers, but this is organizationally specific). We will want tooling for model rollout and model rollback and for binary rollout and rollback.

We should have tools to show the versions of the data (reading from the metadata) in every environment, and tools for customer support staff or production engineers (SREs) to read data directly for troubleshooting purposes (with appropriate logging and audit trails to respect privacy and data integrity guarantees). Where possible, find tooling that exists for your framework and environment already, but plan to implement at least some. The more tooling that exists that works, the lower the burden on software engineers during incidents.

Incident handling

Software engineers should be a point of escalation during incidents, but if they have done their jobs well, they should be alerted only rarely. Software failures will occur in the model servers, data synchronizers, data versioners, model learners, model training orchestration software, and feature store. But as our system gets more mature, we will be able to treat this as a few large systems that can be well managed: a data system (feature store), a data pipeline (training), an analytics system (model quality), and a serving system (serving). Each of these is only slightly harder for ML than for non-ML problems, and so software engineers who do this well may have very low production responsibilities.

Continuous improvement

Software engineers should work regularly with model developers, with SREs/production engineers, and with customer support in order to understand what is missing and how the software should be improved. Most common improvements will involve resilience to big shifts in data and thoughtful exporting of software state for more effective monitoring.

ML SRE or Production Engineer

ML systems are run by someone. Larger organizations may have dedicated teams of production engineers or SREs who take responsibility for managing these systems in production.

Preparation

Production teams should be staffed with sufficient spare time to handle incidents when they come up. Many production teams fill their plate with projects, ranging from automation to instrumentation. Project work like this is enjoyable and often results in lasting improvements in the system, but if it is high priority and deadline driven, it will always suffer during and after an incident. If we want to execute project work effectively, we need to have spare capacity.

We will also need training and practice. Once the system is mature, large incidents may happen infrequently. The only way that our on-call staff will gain fluency with the incident management process itself, but also with our troubleshooting tools and techniques, is to practice. Good documentation and tooling helps, but not if on-call staff can't understand the docs or find the dashboards. Some systems are sufficiently dynamic as to provide their own, regular, opportunities for this practice (this is a polite way of noting that some of our systems are broken fairly often). In this case, teams should be sure to regularly share the incident management lead role when the opportunity arises. For systems where this is not the case, scheduling regular, intentional, small breakages during business hours is one good approach.[9]

Production teams should conduct regular architectural reviews of the system to think through the biggest likely weak spots and address them. These might be unnecessary data copies, manual procedures, single points of failure, or stateful systems that cannot easily be rolled back.

Setting up monitoring and dashboards is a topic unto itself and is covered more extensively in Chapter 9. For now, we should note that monitoring distributed throughput pipelines is extremely difficult. Since progress is not reducible to a single value (the oldest data we're still reading, newest data we have read, how fast we're training, how much data is left to read), we need to make decisions based on changes in data distribution in the pipeline.

We will need to set up SLOs and defend them. As noted, our systems will be behaving in complex ways with multiple dimensions of variable performance along the "somewhat better" and "somewhat worse" axis. To pick thresholds, the first thing we'll need to do is define SLIs that we want to track. In ML, these are generally slices (subsets) of the data or model. Then we'll pick a metric for how those are performing. Since these metrics will change over time, if our data is normally distributed, we can pick thresholds by their distance from the median.[10] If we update that periodically but not too often, we will continue to be sensitive to large shifts while ignoring longer-term trends. This may miss outages that happen slowly over weeks or months, but it will not be overly sensitive.

Production engineering teams should educate themselves about the business that they are in. This seems ancillary but isn't. ML systems that work make a difference for the organizations that deploy them. To successfully navigate incidents, SREs or production engineers should understand what matters to the business and how ML interacts with that. Does the ML system make predictions, avoid fraud, connect clients, recommend books, or reduce costs? How and why does it do that, and why does that matter to our organization? What are the specific objectives that our

9 For much more on this topic, see *Chaos Engineering* (*https://oreil.ly/R38WM*) by Casey Rosenthal and Nora Jones (O'Reilly, 2020).

10 The statistics covering this and techniques for setting thresholds are covered well in Mike Julian's *Practical Monitoring* (*https://oreil.ly/xsRZZ*) (O'Reilly, 2017), especially Chapter 4.

organization is trying to accomplish, and how are they measured? Or even more basically, how is our organization put together? Where is an organizational chart (for a sufficiently large organization)? Answering those questions ahead of time prepares a production engineer for the necessary work of prioritizing, troubleshooting, and mitigating ML outages.

Finally, we need as many objective criteria to trigger an incident as possible. The hardest stage of an incident is before it is declared. Often many people are concerned. There is pervasive and disconnected evidence that things are not going well. But until someone declares an incident and engages the formal machinery of incident management, we cannot manage the incident directly. The clearer the guidelines we determine in advance, the shorter that period of confusion.

Incident handling

Step back and look at the whole system. ML outages are seldom caused by the system or metric where they manifest. Poor revenue can be caused by missing data (on the other side of the whole system!). Crashes in serving can be caused by changes in model configuration in training or errors in the synchronization system connecting training to serving. And, as we've seen, changes in the world around us can themselves be a source of impact. This is a fairly different practice than production engineers normally employ, but it is required for ML systems outages.

Be prepared to deal with product leaders and business decision makers. ML outages rarely stop at the technical team's edge. If things are going wrong, they usually impact sales or customer satisfaction—the business. Extensive experience interacting with customers or business leaders is not a typical requirement for production engineers. ML production engineers tend to get over that preference quickly.

The rest of incident handling is normal SRE/production incident handling, and most production engineers are good at it.

Continuous improvement

ML production engineers will collect many ideas about how the incident could have gone better. These ideas range from monitoring to rapidly detect the problem that we're not yet doing to system rearchitectures that would avoid the whole outage in the first place. The role of the production engineer is to prioritize these ideas.

Post-incident follow-up items have two dimensions of prioritization: value and cost/feasibility of implementation. We should prioritize work on items that are both valuable and easy to implement. Many follow-up items will fall into the category of "likely valuable but extremely difficult to implement." These are a separate category that should be reviewed regularly with senior leads but not prioritized alongside other tactical work, since working on them in that setting will never make sense.

Product Manager or Business Leader

Business and product leaders often think that following and tracking incidents is not their problem, but rather one for the technical teams. Once you add ML to your environment in all but the most narrow ways, your awareness of it likely becomes critical. Business and product leaders can report on the real-world impact of ML problems, and can also suggest which causes are most likely and which mitigations are least costly. If ML systems matter, business and product leaders should and will care about them.

Preparation

To the extent possible, business and product leaders should educate themselves about the ML technologies that are being deployed in their organization and products, including, and especially, the need to responsibly use these technologies. Just as production engineers should educate themselves about the business, business leaders should educate themselves about the technology.

We should know two critical things: first, how does our system work (what data does it use to make what predictions or classifications), and second, what are its limitations? There are many things that ML systems can do but also many they cannot (yet, perhaps). Knowing what we cannot do is as critical as knowing what we're trying to do.

Business and product leaders who take a basic interest in the way ML works will be astoundingly more useful during a serious incident than those who do not. They will also be able to directly participate in the process of picking ML projects worth investing in.

Finally, business leaders should ensure that their organization has the capacity to handle incidents. This largely means that the organization is staffed to a level capable of managing these incidents, trained in incident management, and invested in the kind of spare time necessary to make space for incidents. If it has not, it is the job of the business leader to make space for these investments. Anything else creates longer, larger outages.

Incident handling

Business leaders rarely have an on-call rotation or other systematized way of reaching them urgently, but the alternative is "everyone is mostly on call most of the time." Culturally, business leaders should consider formalizing these on-call rotations, if only to enable themselves to take a vacation with freedom. The alternative is to empower another rotation of on-call staff to make independent decisions that can have significant revenue consequences.

During the actual incident, the most common problem that business leaders will face is the desire to lead. For once, they are not the most valuable or knowledgeable person. They have the right to two things: first, to be informed, and second, to offer context of the incident's impact on the business. They do not generally usefully participate directly in the handling of the incident; they're simply too far removed from the technical systems to do so. Many business leaders should consider proxying their questions through someone else and stay off direct incident communications (chat, phone, Slack) as a way of avoiding their natural desire to take over.

Continuous improvement

Business leaders should determine the prioritization of work after outages and should set standards for what completion of those items means. They can do that without having particular opinions about how, exactly, we improve. But, rather, they can advocate for general standards and approaches. For example, if we rank follow-up work items in priority order (Highest Priority through Nice to Have), we can prioritize work on the Highest Priority bugs ahead of the High Priority ones, and so on. And we can set guidelines that if all of the Highest Priority items are not done after a particular period of time, we have a review to figure out whether anything is blocking them and what we can do, if anything, to speed up implementation.

Similarly, product teams have a huge role in specifying, maintaining, and developing SLOs. SLOs should represent the conditions that will meet a customer's needs and make them happy. If they do not, we should change them until they do. The people to own the definition and evolution of those values are principally the product management team.

Special Topics

We haven't yet addressed two important topics that show up during the handling of ML incidents. This section delves into those subjects.

Production Engineers and ML Engineering Versus Modeling

Given that many ML systems problems present as model quality problems, a minimum level of ML modeling skill and experience seems required by ML production engineers. Without knowing something about the structure and functioning of the model, it may be difficult for those engineers to effectively and independently troubleshoot problems and evaluate potential solutions. The converse problem also appears: if there is no robust production engineering group, we might well end up with modelers responsible for the production serving system indefinitely. While both of these outcomes may be unavoidable, they are not ideal.

This is not completely wrong, but it's also entirely situationally dependent. Specifically, in smaller organizations, it will be common to have the model developer, system developer, and production engineer be a single person or the same small team. This is somewhat analogous to the model in which the developer of a service is also responsible for the production deployment, reliability, and incident response for that service. In these cases, obviously expertise with the model is a required part of the job.

As the organization and services get larger, though, the requirement that production engineers be model developers vanishes entirely. In fact, most SREs doing production engineering on ML systems at large employers never or rarely train models on their own. That is simply not their expertise and is not a required, or even useful, expertise to do their jobs well.

ML SREs or ML production engineers do need certain ML-related skills and knowledge to be effective. They need basic familiarity with what ML models are, how they are constructed, and above all, the flavor and structure of the interconnected systems that build them. The relationship of components and the flow of data through the system is more important than the details of the learning algorithm.

Let's say, for example, that we have a supervised learning system that uses TensorFlow jobs scheduled at a particular time of day to read all the data from a particular feature store or storage bucket and to produce a saved model. This is one completely reasonable way to build an ML training system. In this case, the ML production engineer needs to know something about what TensorFlow is and how it works, how the data is updated in the feature store, how the model training processes are scheduled, how they read the data, what a saved model file looks like, how big it is, and how to validate it. That engineer does not need to know how many layers the model has or how they are updated, although there's nothing wrong with knowing that. They do not need to know how the original labels were generated (unless we plan to generate them again).

On the other side of the same coin, suppose we have settled on a delivery pipeline in which an ML modeling engineer packages their model into a Docker container, annotates a few configuration details in an appropriate config system, and submits the model for deployment as a microservice running in Kubernetes. The ML modeling engineer may need to understand the implications of how the Docker container is built and how large the container is, how the configuration choices will affect the container (particularly if there are config errors), and how to follow the container to its deployment location and do some cursory log checking or system inspection to verify basic health checks. The ML modeling engineer probably does not, however, need to know about low-level Kubernetes choices like pod-disruption budget settings, DNS resolution of the container's pod, or the network connectivity details between the Docker container registry and Kubernetes. While those details are important, especially in the case where infrastructure components are part of a failure, the ML modeling engineer won't be well suited to address them and may need to rely on handing off those types of errors to an SRE specialist familiar with that part of the infrastructure.

Detailed knowledge of model building can certainly be extremely helpful. But the biggest reliability problem that most organizations run into is not a lack of knowledge about ML. It is rather a lack of knowledge and experience building and productionizing distributed systems. The ML knowledge is a nice addition rather than the most important skill set.

The Ethical On-Call Engineer Manifesto

We've written a lot in this chapter about how performing incident response is different and more difficult when ML is involved. Another way in which ML incident response is hard is how to handle customer data when you're on call and actively resolving a problem, a constraint we call *privacy-preserving incident management*. This is a difficult change for some to make, since today (and decades previous), on-call engineers are accustomed to having prompt and unmediated access to systems, configuration, and data in order to resolve problems. Sadly, for most organizations and in most circumstances, this access is absolutely required. We cannot easily remove it and still allow for fixing problems promptly.

On-call engineers, in the course of their response, troubleshooting, mitigation, and resolution of service outages, need to take *extra* care to ensure that their actions are ethical. In particular, they must respect the privacy rights of users, watch for and identify unfair systems, and prevent unethical uses of ML. This means carefully considering the implications of their actions—not something easy to do during a stressful shift—and consulting with a large and diverse group of skilled colleagues to help make thoughtful decisions.

To help us understand why this should be the case, let's consider the four incident dimensions in which ethical considerations for ML can arise: the impact (severity and type), the cause (or contributing factors), the troubleshooting process itself, and the call to action.

Impact

Model incidents with effects on fairness can wreak truly massive and immediate harm on our users, and of course reputational harm to our organization. It doesn't matter whether the effect is obvious to production dashboards tracking high-level KPIs or not. Imagine a bank loan approval program that is accidentally biased. Although the data supplied in applications might omit details on the applicants' race, there are many ways the model could learn race categories from the data that is supplied and from other label data.[11] If the model then systematically discriminates against some races in approving loans, we might well issue just as many loans—and show about the same revenue numbers on a high-level dashboard—but the result is deeply unfair. Such a model in a user-facing production system could be bad for both our customers and our organization.

In ideal circumstances, no organization would employ ML without undergoing at least a cursory Responsible AI evaluation as part of the design of the system and the model.[12] This evaluation would provide clear guidelines for metrics and tools to be used in identifying and mitigating bias that might appear in the model.

Cause

For any incident, the cause or contributing factors to the outage can have consequences for the ethically minded on-call engineer. What if the cause turns out to be a deliberate design decision that is actually hard to reverse? Or the model was developed without enough (or any) attention being paid to ethical and fairness concerns? What if the system will continue to fail in this unfair way without expensive refactoring? Insider threat is real,[13] don't forget, but we don't need to imagine that malice aforethought took place for these kinds of things to happen: a homogeneous team, strongly focused on shipping product to the exclusion of all else, can enable it purely by accident. Of course, all of this is enhanced by the current lack of explainability of most ML systems.

Troubleshooting

Ethics concerns (generally, privacy) often arise during the troubleshooting phase for incidents. As you saw in story 3, it is tempting—maybe sometimes even required—to look at raw user data while troubleshooting modeling problems. But doing so directly exposes private customer data. Depending on the nature of the data, it might have other ethical implications as well—consider, for example, a financial system that includes customer investment decisions in the raw data. If a staff member has access to that private information and uses it to direct their own personal investments, this is obviously unethical and in multiple jurisdictions would be seriously illegal.

11 Models can learn race from geographic factors in areas that have segregated housing, from family or first names in places where those are racially correlated, from educational history in places that have segregated education, and even from job titles or job industry. In a racially biased society, many signals correlate with race in a way that models can easily learn even without a race label in the model.

12 This topic is covered much more extensively in Chapter 6.

13 See, for example, "Famous Twitter Accounts Hacked: Insider Threat or Social Engineering Attack?" (*https://oreil.ly/lc6kp*) by Clare O'Gara, though the incidents that make the papers are almost by definition a small subset of the ones that actually happen.

Solutions and a call to action

The good news is that a lot of these problems have solutions, and getting started can be reasonably cheap. On the one hand, we've already spoken about the generally underweighted role that diverse teams can play in ensuring an organization against bad outcomes. Fixing those generally involves fixing the process that produced them rather than mitigating a specific one-time harm.

But a diversity of team members is not, by itself, a solution. Teams need to adopt the use of Responsible AI practices during the model and system design phase in order to create consistent monitoring of fairness metrics and to provide incident responders a framework to evaluate against. For deliberate or inadvertent access to customer data during incident management, restricting that access by default with justification, logging, and multiple people in charge over the data (to act as ethical checks on each other) is a reasonable balance of risk versus reward. Other mechanisms that are useful to avoid the construction of flawed models are outlined in Chapter 6.

Finally, though it is not within our authority to declare it unilaterally—nor would we wish to—we strongly believe there is an argument for formalizing such a manifesto, and promoting it industry-wide. The time will come—if it is not already here—when an on-call engineer will discover something vital and worthy of public disclosure, and may be conflicted about what to do. Without a commonly understood definition of what a whistleblower is in the ML world, society will suffer.

Conclusion

An ML model is an interface between the world as it is and as it changes on the one hand, and a computer system on the other. The model is designed to represent the state of the world to the computer system and, through its use, allow the computer system to predict and ultimately modify the world. This is true in some sense of all computer systems but is true at a semantically higher and broader sense for ML systems.

Think of tasks such as prediction or classification, whereby an ML model attempts to learn about a set of elements in the world in order to correctly predict or categorize future instances of those elements. The purpose of the prediction or classification always is, and has to be, changing the behavior of a computing system or an organization in response to the prediction or classification. For example, if an ML model determines that an attempted payment is fraud, based on characteristics of the transaction and previous transactions that our model has learned from, this fact is not merely silently noted in a ledger somewhere—instead, the model will usually reject the transaction after such a categorization is made.

ML failures specifically occur when a mismatch arises among three elements: the world itself and the salient facts about it, the ML system's ability to represent the world, and the ability of the system as a whole to change the world appropriately. Failures can occur at each of those elements or, most commonly, in combination or at the intersections between them.

ML incidents are just like incidents for other distributed systems, except for all of the ways that they are not. The stories here share several common themes that will help ML production engineers prepare to identify, troubleshoot, mitigate, and resolve issues in ML systems as they arise.

Of all of the observations about ML systems made in this chapter, the most significant is that ML models, when they work, matter for the whole organization. Model and data quality have to therefore be the mission of everyone in the organization. When an ML model goes bad, fixing it will sometimes require the whole organization as well. ML production engineers who hope to get their organizations ready to manage these kinds of outages would do well to make sure that the engineers understand the business, and the business and product leaders understand the technology.

CHAPTER 12

How Product and ML Interact

As companies rush to use ML capabilities to meet customer needs, they are eager to leverage cutting-edge research to tackle a wide variety of business applications. Many of the product teams and business managers, still anchored in traditional software product development methodologies, find themselves in a new and unfamiliar territory: building ML products.

Building your first ML product can be overwhelming. It's not just a question of getting ML right, difficult enough in itself; rather, the integration of the ML into the rest of the product (and the rest of the business) requires many things that need to work together. Among these, data collection practices and governance, the quality of data, definition of product behavior, UI/UX, and business goals all contribute to the success of an ML-based product or feature.

Different Types of Products

One of the important and useful features of ML is that it can be applied to many types of products. It can be used in analytics applications to derive insights about business trends and metrics. It can be rolled into an appliance or device to be shipped to consumers. Sophisticated ML systems are built into self-driving cars to detect other objects and to make decisions about driving. The breadth of ML applications is huge and growing. As a result of this huge diversity of use cases, organizations focused on integrating ML into their existing or new products face an extremely steep learning curve and numerous choices about their implementation.

We don't cover most of these different kinds of ML in this chapter. Specifically, it is not feasible for this chapter to seriously consider each of the many common types of ML-product integrations that exist. Instead, we will focus on the use case we have been discussing throughout the book: *yarnit.ai* (our ecommerce web store) wanting to integrate ML.

Considering this use case specifically has some nice aspects, and not just because it extends an example that is touched on in every chapter of this book. The *yarnit.ai* set of use cases includes some from the backend (e.g., consuming browsing and purchase logs to predict interest in a product) and many frontend integrations as well (e.g., looking up what products this customer might be interested in right now). It has some complexity without being overwhelmingly difficult to understand for most readers. Of course, it clearly can't be an appropriate analogue for every other possible system. Nonetheless, we hope there is something of use to you here.

Agile ML?

Outside certain well-defined environments, many contemporary software engineering teams have moved to developing in iterative, focused, short loops called *sprints*—as per the Agile Manifesto (*https://agilemanifesto.org*). However, applying Agile to ML systems is far from straightforward. Agile, as a methodology, has several foundational characteristics: short feedback loops, customer-oriented stories or story points, and estimation geared toward a small team working on those story points.

Integration of ML into the product violates most of these assumptions. The feedback loops are long (sometimes months or years) and come indirectly from the data rather than directly from the customer. Small team execution is less useful because the integration often involves people across the company (see Chapter 13 for more perspective on this). Model development can have arbitrarily long delays while we integrate the model into the product and wait for the results to show up in the data. Additionally, we cannot build first and validate later since even the building phase needs to be based on the data. There's one final, and perhaps most important point: ML models are neither perfectly reproducible nor stable over time. As has been pointed out, we can train the same model multiple times with different results, or we can leave the same model in place but have the world change sufficiently that our value is no longer the same. ML models are never "done," so ML product integration has limited determinism.

Still, even though Agile approaches may not be super well suited for ML, adopting a standard ML development lifecycle, as discussed in this chapter, can facilitate your organization taking advantage of ML systems to power great customer experiences and increase your revenues.

ML Product Development Phases

To manage uncertainty during the product development lifecycle, ML projects need to be highly iterative from the beginning. The most important phases of an ML product development project are discovery and definition, business goal setting, MVP construction and validation, model and product development, deployment, and support and maintenance (Figure 12-1).

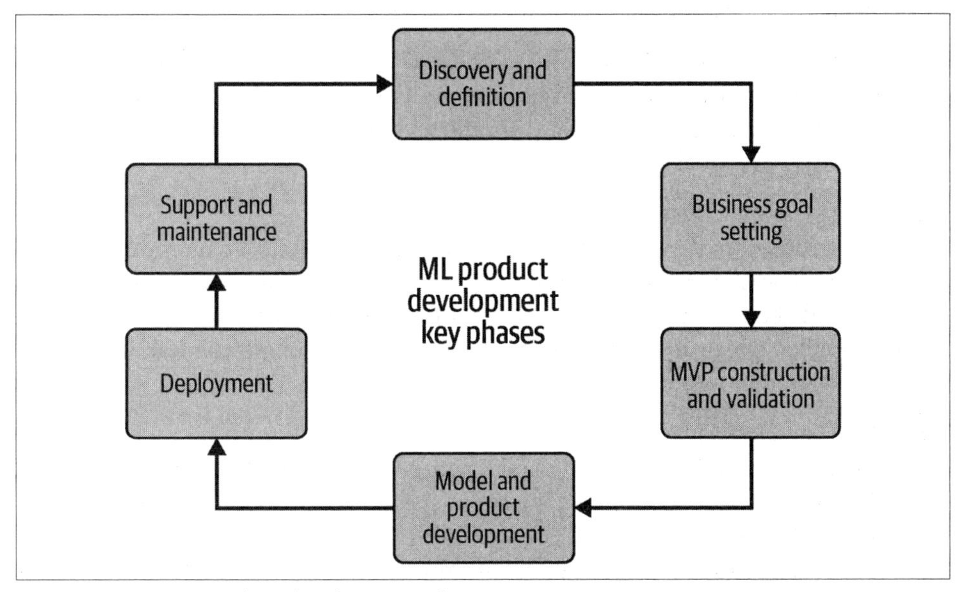

Figure 12-1. ML product development phases

Discovery and Definition

The development of an ML product should always start with *discovery and definition*. Skipping this step is tempting, particularly if you come from a build-first, validate-later background, but in an ML context, this step is critical and helps product teams handle the uncertainty early. This begins by defining the problem space, and understanding and framing the business problem and the desired outcome. It finishes by mapping the problems to a solution space, which is not always an easy task. Technical teams are looking for a quantitative baseline to measure feasibility. On the other hand, business is struggling to achieve clarity and looking for clear cost-benefit analysis at a time when the outcomes are unclear. Balancing stakeholder needs, business-problem framing, defining goals and outcomes, and building relationships with the business will be key.

Given that our product is designed to display results to humans in the end, there are no shortcuts to good old-fashioned user research. Conduct thorough user research to identify the pain points of the user and prioritize them according to their needs. This helps build a user journey map, identifying critical workflows and potential roadblocks. Further, the roadmap is useful for defining the processes that need to be modified for the ML solution to work in the first place. You can do a market-sizing exercise to estimate the business potential. Then the next question is, how do we know if ML can help address our user problems? Numerous ML applications exist but, at its core, ML is best suited for making decisions or predictions. In general, our style of human-facing ML system will have one or more of the following characteristics as deployed in production:

Complex logic that's impractical to solve with human-defined rules
: For example, search engines often have multiple phases of ranking that happen in series, such as initial retrieval from the text index, primary ranking using text similarity, contextual ranking, and personalized ranking.

Large-scale personalization
: If the problem is expected to scale to thousands of users or more, it could be a good use case for ML. In the *yarnit.ai* web store example, we expect to have thousands of customers take advantage of a new offer/discount within three to six months, which implies being able to support personalization across a large user population.

Rules that change quickly over time
: If the rules generally remain static year after year, a heuristic solution is preferred. However, if the business's success depends on quick adaptation and rule changes, ML is a good route. For example, if users of a web store are writing product reviews constantly, the algorithm for recommending relevant products needs to adapt in real time and is amenable to an ML solution.

A clear evaluation metric
: For example, we want the model to provide recommendations that result in a sale. In search, typing "kids hat" should return lists of patterns and yarn normally associated with kids colors or knitting gauge needles that are associated with hats.

No requirement for 100% accuracy
: If business success can be achieved with a high probability of accuracy rather than with perfection, ML is a good option. For example, recommendation systems will not be considered faulty if users don't always want what is served. Users can still have a great experience, and the ML model can learn from the lack of sales to deliver improved recommendations in the future.

Business Goal Setting

The selection of ML requirements takes a careful combination of design and business goals. Product managers (PMs) need to understand the end user and the long-term business goals to drive value. For example, we want to create improvements to the web store that encourage customers to find and buy more products they want, but really we also want that to happen repeatedly, every time they return. Ultimately, we do this by focusing on the end user and developing a deep understanding of a specific problem the product can solve. Otherwise, there will be a risk of developing an extremely powerful system only to address a small (or possibly nonexistent) problem.

ML is rarely error free, and developing without guardrails can have serious consequences. We need to acknowledge that getting it wrong could carry huge costs and that understanding the cost of getting it wrong is a significant part of building an ML product. For example, say we are trying to predict and cancel orders on behalf of customers when they reach out to the support team about order errors and/or billing issues. The cost here is that errors in the model might wrongly interpret the user's intent as a cancellation, thereby having financial consequences for the user and the company. On the other hand, in the case of a recommender system to suggest similar products on the product details page, for a user with no purchase history, the impact of bad recommendations might be just low conversion rates, and perhaps a vague sense that our web store is not very trustworthy or useful. We must then think about how to improve the model, but the consequences of getting it wrong are not catastrophic in our use case (although they clearly are in many other ML products).

ML relies to a large extent on probabilities. Therefore, there will always be a chance that the model gives the wrong output. PMs are responsible for anticipating and recognizing the consequences of a wrong prediction. One great way to anticipate consequences like the ones noted previously is to test, test, and test some more. Understand what makes up the probabilities computed by the model. We need to use the business purpose of ML in order to think about the desired precision (and recall) of the model.

Once all of the consequences of getting a prediction wrong are identified, the PM needs to make sure relevant *safety nets* are defined and built into the product to mitigate the risks. Additionally, PMs may think of ways to change the underlying product (change interaction flow, modify the input and output data paths, etc.) in order to make incorrect predictions much less likely. As such, safety nets can be intrinsic and extrinsic, and they are ingrained in all products. ML is not an exception. As described in the following list, defining the safety nets and business performance metrics that are relevant to the product we are building is a critical step before introducing ML into the product:

Intrinsic safety nets
: These recognize the impossibilities that are fundamental to the nature of the product. For example, at our web store, a user cannot cancel an order if no order has been made in the first place. Emails received from such users, with no order numbers and with a subject of "canceling orders" can be ignored by the model that is trying to learn the reasons for order cancellations. However, it's a good idea to have a customer support agent look into that case. A useful activity is to map out the user journey for the product and identify the states that the user can

go through. This helps weed out impossible predictions. Intrinsic safety nets are invisible to the user.

Extrinsic safety nets
Extrinsic safety nets are visible to the user. They can take the form of confirming user intent or double-checking the potential outcome. Some message systems have a model that tries to detect the intent of a message and suggest replies to its users. In most cases, however, these systems do not automatically assume the reply is correct and send it without a human confirming the choice. More commonly, these systems ask the user to pick from a list of potential replies.

Business performance goals
Along with the general ML model performance metrics that we've discussed in Chapter 8, it is extremely important for PMs to clearly define the business performance metrics to measure the success of the ML systems in production. To evaluate business performance, it is necessary to start with a product or feature goal. For example, increasing revenue could be a great business goal for our *yarnit.ai* web store. Once this is defined, a product metric should be assigned to evaluate success. The best metrics are specific and measurable. Specific metrics reduce ambiguity and increase focus. Also, the easier it is to measure success, the more certain we can be that we're achieving the results we set out for in the first place.

These metrics can vary from product to product. For example, the following are a few important business metrics to track for ML-based recommendations on an ecommerce store like *yarnit.ai*:

Click-through rate (CTR)
The number of product page views from a recommendations list divided by the total number of products displayed on the recommendations list. For example, on the shopping cart page, this metric would help determine the success of the ML model powering the "Frequently bought together" list.

Conversion rate
The number of add-to-cart events from a recommendation list divided by the total number of products displayed on the recommendations list. For instance, on the product details page, this metric would help determine the success of the ML model powering the "Compare with similar items" list.

Average order value (AOV)
The average value of orders from all purchase events. AOV is equal to the total revenue divided by the number of orders.

Recommender-engaged AOV
: The average value of orders that include at least one item selected from a recommendations list. This is calculated from the recommender-engaged revenue divided by the number of orders with at least one item that was selected from a recommendations list.

Total revenue
: The total revenue from all recorded purchase events. This value includes shipping and taxes.

Recommender-engaged revenue
: The revenue for purchase events that include at least one catalog item selected from a recommendations list. This value includes shipping and taxes and any discount applied.

Also, ecommerce businesses usually track a lot more business metrics,[1] including customer acquisition cost (CAC), customer lifetime value (CLV), customer retention rate (CRR), refund and return rates (RrR), and cart abandonment rate, as well as vanity metrics such as social media engagement, website traffic, and page views.

MVP Construction and Validation

Investing in ML models in order to integrate with our product is likely to be expensive. To figure out whether the ML integration into our product will work, we need to answer two questions: (1) can we make a model that works (or works well enough), and (2) can we integrate that model into our product in a compelling and useful way? Building a dataset with the right features and labels, training the model, and putting it in production can range from a few weeks to a few months. We will want to get a signal of usefulness early on to validate whether the model will work. This can be accomplished through careful offline evaluation of the model we have built, analyzing the responses to various common use cases we expect from the application.

To evaluate the utility in the user interaction, it is a good idea to fake that interaction first, for a small set of users. Sometimes this is referred to as launching a *minimal viable product* (*MVP*) with a fixed set of rules or heuristics (without real ML models in place) to prove the point that the feature will really solve the customer needs.

For example, considering the use case of personalization, have a list of items ready for a user to select based on what they have selected last. Simple rule-based engines are often the first steps to evolution into a more complex ML model. Here are some examples of rule-based recommenders:

1 Learn more about the general ecommerce business metrics in "7 E-Commerce Metrics to Help You Measure Business Success" (*https://learn.g2.com/e-commerce-metrics*) by Anastasia Stefanuk.

- If the user bought a knitting pattern, they probably need the yarn for that pattern as well as needles and other knitting-related supplies.
- If a user buys new yarn every fall as the weather gets colder, we should recommend it to them every fall in their part of the world.
- If a user always pays by credit card, we surface that as the default payment option next time.

It will obviously be impossible to write rules to cover every case, and that is when ML can be used best, but a few simple rule-based proxies go a long way toward validating the outcome of the ML approach. The idea is to test whether users respond positively to ML. While these techniques might not give the best results, they are important for getting a signal. Getting a signal early on can save time and effort, and help correct the vision and direction of the product. This is your best shot at guaranteeing returns on the investment put into building an ML system.

Model and Product Development

With a clear set of goals and targets developed through previous stages, the next step is to build the models and integrate them with customer-facing features. For a general introduction to this, see Chapters 3 and 7. But in general, this is a collaborative endeavor with design, engineering, ML researchers, business owners, and PMs working together, and PMs can drive a lot of value by continuing deliberate stakeholder management. The build will go from simple to complex—PMs should prepare to drive the product through train, test, and validate cycles; manage standups with scientists who may or may not be making progress; and navigate the iterative dependence between design, science, and engineering teams. We discuss the roles and responsibilities of various teams in Chapter 13.

Deployment

In the production deployment stage, the ML system is introduced to the infrastructure, where it will serve the live customer traffic, and gather feedback data that is fed into the ML training pipelines to improve the models. Feedback loops help measure the impact of a model and can add to the general understanding of usability. In the context of an ML system, feedback is also important for a model to learn and become better. We discuss various model serving architectures and best practices in Chapter 8.

Feedback loops are an important data collection mechanism, which yield labeled datasets that can be directly plugged into the learning mechanism. In our *yarnit.ai* web store example, the feedback loop could be quite simple—does the guest click on the recommendations, and is this recommendation bought?

Rolling out to production should be heavily tied to observing the business metrics (discussed earlier in this chapter) in a controlled manner, to understand the impact and to inform decisions about increasing the rollout or turning it off and investigating unexpected results. In particular, one challenge is that the specific initiatives or projects within a large company that ultimately contain the day-to-day work of building an ML model may have many layers of separation from the higher-level business goals. Personalization is a good example: at a high level, personalization may be about increasing a metric like average revenue per user or increasing the growth of new user acquisitions. But when a model is developed, it is likely to be optimized around improving various low-level user experience metrics like scores of engagement.

PMs have a crucial job to be a translator between these two domains and to constantly work to improve the methods of user observability that can help connect improvements in the lower level with hypotheses around impact in the higher levels. So it's extremely important for product teams to have a plan for how an ML system will be deployed incrementally. We discuss various methods of model evaluations in Chapter 5.

Support and Maintenance

ML systems designed to integrate into products are not complete on their first shipping version. Arguably, they are never complete because they are attempting to model the state of the world and deliver useful value about the state of the world in the product. While developing and deploying ML systems can be relatively affordable, maintaining them over time can be more difficult and expensive than commonly assumed.[2]

Organizations that are serious about integrating ML into their products need to be serious about continuing the maintenance of those ML models and the infrastructure that produces them. This is at least part of why the Agile approach doesn't fit ML product integrations well: they never really ship, or rather they never finish. As our product changes, as the needs of our customers change, as our understanding of the business changes, and as the world changes, we will need to keep developing and shipping models.

2 This paragraph may be a slight overstatement for emphasis. While the work of a functional ML product requires ongoing maintenance, and significantly more maintenance than a comparable non-ML project, it is still true that most of the work occurs toward the first half of the project. As the model matures (and perhaps as the behavior or situation it seeks to model also stabilizes), some organizations will reach a steady state, as further investments in model improvements have a diminishing return.

It is a feature of ML systems that it's not usually clear when the work a team is doing is "maintenance" or "development," so conceptions of development work that rely on a clean separation between them are tricky to apply.

Build Versus Buy

It is important in this section that we carefully scope the question of whether to build or buy. Models are local to the organization that owns the data and the infrastructure. In general, models must come from your learning on your own data.[3] But the tools and infrastructure we use to create our models are a different story. The world of ML has no shortage of tools available, both open source and commercial. The decision of build versus buy is usually a trade-off balancing cost, risk, customization, long-term resource availability, intellectual property (IP), and vendor lock-in. However, in the age of ML, we have more dimensions to this traditional question to consider, including make-or-buy models, data processing infrastructure/tools, and the end-to-end platform that holds it all together. Here are some considerations to account for when deciding what's best for the business.

Models

As we've stated, models are generally local and not as easily acquired outside our own organization. That said, many industry-specific ML models and applications are available. Vendor-provided solutions may present a significant time- and effort-saving potential, but the key aspects to assess build versus buy for models are discussed next.

Generic use cases

Buying a prebuilt solution requires the data and process currently used in our organization to be highly compatible with the expected input/output and behavior of that solution. If a use case's details, data, or processes are fairly specific to the organization, the effort spent adjusting internal systems and processes to match the vendor-provided solution's specifications may wipe away the benefit from buying something ready-made. If a packaged solution is truly generic, it can provide great benefit. Simple reading of this paragraph might sound as though this would be a comparatively rare happening: how likely is the existence of a generic model that does what we need without any need of retraining or extension on our data?

As ML service providers multiply, the odds here are getting better all the time. Consider some incredibly useful but really generic use cases:

---

3 One exception here is for transfer learning, whereby we get a general model (say, one that does image recognition) from an external provider and then train it on our own data. But even then, the final model is ours.

Object recognition in images
: Tell me what is in a picture.

Text-to-speech and speech-to-text in multiple languages
: Carry out spoken interaction with users.

Sentiment analysis
: Tell me how positive or negative a set of text is about its subject.

In all of these examples, pretrained models can work well across a large number of use cases, including ours at YarnIt. In these cases, we might be able to skip training a model entirely.

Company's data initiatives

The benefit of work done on deploying a specific ML use case goes way beyond that use case itself. Often, first use cases are the start of a full data initiative. As a result, ML models and/or application build-or-buy decisions should also take into account the company's data strategy: sometimes the acquisition of expertise and technology are actual ends in and of themselves.

Data Processing Infrastructure

Data processing and infrastructure tools and technologies have evolved and are still constantly evolving at a great pace. Traditional vendors, open source, and cloud-based solutions compete with each other on every segment of the data infrastructure, processing, and storage stack. Build or buy has taken a slightly different meaning and is usually more centered around delegating responsibility as follows:

Support and/or maintenance
: Typically full open source versus an independent third-party software vendor

Operations
: Cloud versus on-premises

Accessibility
: Packaged cloud tools versus native cloud tools

When it comes to infrastructure for data processing, the main decision drivers are no longer about use cases but instead about internally available skills and vendor lock-in risk factors. Commercial and open source platforms both have their pros and cons.

In the ML space, many commercial platforms are new, and are from new providers and lack any long track record or history. As such, they may present significant opportunities but also significant risks.

On the other hand, open source solutions, whether for ML or general-purpose software, may be widely adopted, and in many cases have a longer track record than some of the commercial solutions. But they also have downsides. In particular, they may have slower release cycles, but more importantly, they rarely offer a full solution to a problem. Open source ML solutions are generally great point solutions within a broader context but still require that we do the work necessary to integrate them into our environment and assemble a complete solution.

End-to-End Platforms

At the present state of maturity, essentially no ML platforms are available that provide relatively well-integrated solutions starting with data and ending with a model that is available in a serving system. This isn't necessarily a huge problem right now. Models, data processing infrastructure, human resources, and business value all evolve at different paces in various ways. Orchestrating everything together to ensure sustainable success in ML requires significant effort. But again, here, the decision drivers are different, now based on the fundamentals of the organization's data approach:

Long-term or one-off data initiatives
: ML tools and technologies are evolving, with no sign of reducing the pace of innovation. This presents a serious problem, similar to the one that we have seen in other contexts: technology investments must have time to pay off. And if the utility of a platform is gone before we have finished getting the implementation cost out of it, we may realize that we've made a mistake. While this has sometimes been a significant concern with distributed computing or data storage platforms, it is particularly difficult in the ML space, where tools and technologies may become obsolete in the time it takes to properly implement them.[4] If the initiative is meant to be long-term, acquiring a full platform able to accommodate the fast-evolving data technologies may be relevant, while for short-term initiatives, it may be possible to assemble based on today's components.

Expected scale of the data initiative
: On small-scale ML projects, little overarching cohesion or coherence is required, which leads to a more natural "build" scenario. If the broader vision is to expand data to augment a significant part of the organization's activity, it will require the inclusion of many skills and company processes to successfully execute on that vision. In this case, a buy approach is likely more relevant.

4 A bit of nuance exists here. It is true that algorithms change quickly (certainly yearly), as new approaches are tried and found to be more successful at some types of problems, Some ML platforms, such as TensorFlow and PyTorch most notably, are maintained over longer periods of time (those two since 2015 and 2016, respectively). However, even in that time, both have changed significantly, and other ML platforms have arisen that may be more effective ways to solve the same problems. So even the most stable ML platforms and tools are stable only over modest periods of time.

For example, commercial ML platforms not only allow teams to complete one data project from start to finish one time, but also introduce efficiencies everywhere to scale. That includes features for the following:

- Spending less time cleaning data and doing other parts of the data processing flow that don't provide direct business value
- Smoothing production issues and avoiding reinventing the wheel when deploying models on a daily basis
- Improving documentation and reproducibility to ease compliance with some regulatory requirements

Scoring Approach for Making the Decision

Once we are clear on the problem we're trying to solve and that we need ML specifically to solve the problem, at a minimum we need to consider each of the following factors to evaluate whether building or buying ML is the right choice for the business:

Alignment
: How well does the product or technology meet our needs? Are our objectives met? Are customizations required? Are data privacy and security requirements met?

Investment
: Is there a specific net present value (NPV) or ROI that must be achieved? What is the total cost of ownership including human capital, software, and hardware?

Time
: How fast can we develop and deploy the solution in production?

Competitive advantage
: Does the technology provide an asset or IP that is proprietary? Does this IP offer increased value to customers, employees, and investors?

Maintenance and support
: What is the effort and cost of support for human resources, hardware, and software?

We can make a decision about which parts of our platform to build or to buy based on our weight of each of these factors.

Making the Decision

Whether building or buying, incorporating ML technologies as a key business tool is a strategic decision that should not be made quickly or without all the components in place to support either endeavor. It's important to carefully evaluate how a decision will affect the company's long-term goals and roadmaps.

Additionally, don't forget to convene with key stakeholders to review decision criteria and determine the pros and cons. We discuss more details on the roles and responsibilities of various teams and key stakeholders in Chapters 13 and 14. It's essential to include your leadership, product, and sales teams to gather consensus and feedback for effective change management. Finally, talk to various vendors to weigh in on the options. Be transparent with them and get their honest opinion about your choices and situation. We could achieve the results through either path but need to leverage the results of the previous discovery steps to help make the right solution for the organization's unique needs.

Sample YarnIt Store Features Powered by ML

ML can be used in web stores like *yarnit.ai* to meet various business goals. ML can increase revenue by improving conversion rates and average order value, can increase profit margins, and can even improve customer loyalty. Today's consumers don't want to be treated as one of many customers. They prefer a highly personalized experience. Product recommendations is an area in which a lot of features can be powered by ML models. Just as humans get to know someone and are then able to choose what the best birthday gift for them would be, ML models can leverage data including the product catalog, search queries, viewing history, past purchases, items placed in the shopping cart, products recommended on social media, location, customer segments/buyer personas, and so on. The following are a few example recommendation use cases.

Showcasing Popular Yarns by Total Sales

This nonpersonalized technique is not based on a user's individual choices but rather on collective preference. Recommendations could be displayed on the home page based on criteria such as the following:

- The number of yarns or patterns purchased (e.g., when a new pattern is released during holiday season and everyone rushes to get it).
- How much time shoppers spent viewing a specific type of yarn or pattern.
- Number of views and purchases in the user's country or area. Often we need to consider cultural or seasonal aspects when operating the web store in multiple countries or regions.

In this example, although we may not be really personalizing the customer's experience, showcasing popular items by each category (yarn types, patterns, brands, etc.) allows us to target first-time users who have no account history yet (a.k.a. the *cold-start problem*). This has the obvious benefit of relying only on time-tested database technologies, which are much more common and reliable than ML.

Recommendations Based on Browsing History

Just as its name states, users will be getting new product suggestions based on the products they've already viewed. For example, if the user is searching and browsing for "wool yarns," we can display products by popular brands that specialize in making wool-based yarns and patterns.

Cross-selling and Upselling

Most modern web stores have some infrastructure designed to *cross-sell* and *upsell*. Both of these are aimed at helping users choose the best items possible while also increasing revenue even if we're selling to a new customer. For example, when the customer is looking for "baby yarns" on the product page, showing cross-selling recommendations like "Patterns featuring this yarn" and/or "Popular baby clothes featuring this yarn" can not only help increase the average order value but also save a lot of time for customers. Similarly, showing special discounts on bigger sizes or bundles might help customers save money as well as increase revenues for our business.

Content-Based Filtering

We could leverage the metadata of the products to power recommendations. Accuracy of the metadata will have a greater impact on the quality of such recommendations.

For example, as shown in Figure 12-2, many web stores have a "Similar products" feature either on the product detail page or before/after checkout. In our *yarnit.ai* web store, if our user is always buying "red" and "blue" yarns that are of type "cotton" from the brand "xyz," we could consider recommending the same color and type of yarns from a different brand, "abc," when the recommended items also have similar quality ratings and "abc" is running a special sale event.

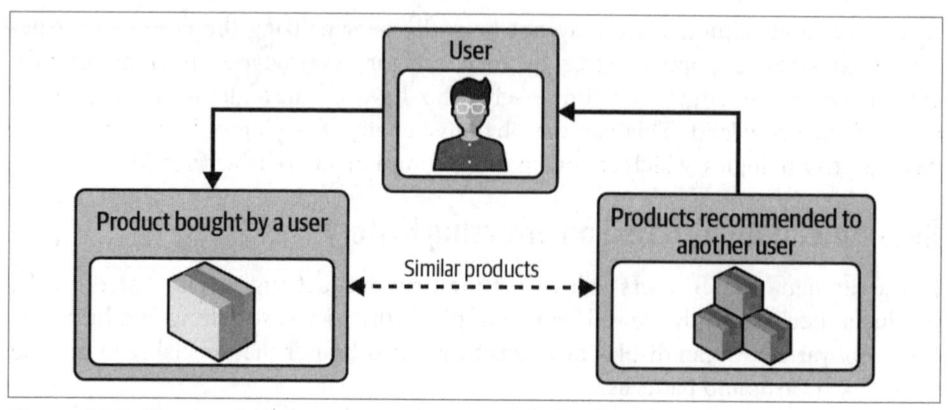

Figure 12-2. Content-based recommendations with the "Similar products" feature

Collaborative Filtering

Collaborative filtering might be considered the most popular of all product recommender methods. This technique relies solely on how other users have positively interacted with a product (either a view or a purchase or a positive rating on a purchase). The method stems from the idea that people with similar past preferences will probably like the same things in the future too. On top of this, it trusts real choices people have made as opposed to simple ratings, which can be just an estimated guess.

For example, as shown in Figure 12-3, user A and user B have similar personas in terms of their browsing and purchase histories as well as product feedback on the web store. Using the persona similarities, when user B is on a "blanket yarn" product details page, we could show the tools and accessories like "markers" and "needles" that were bought by user A along with the same "blanket yarn" product. Because both users have similar tastes/personas, showing them contextual and relevant recommendations would improve not only the user experience but also the chances to upsell, which directly impacts the overall revenue for the business.

Most major web stores use both the collaborative and content-based filtering techniques to improve the accuracy of the personalized recommendations. To evaluate whether a recommender model will work, we can do a simple A/B test (*https://oreil.ly/13N37*) by starting with a couple of predictions and a hypothesis like "This algorithm will improve engagement and/or conversion by *x*%"(multivariate testing is also an option). Each alternative will be based on a separate recommender technique.

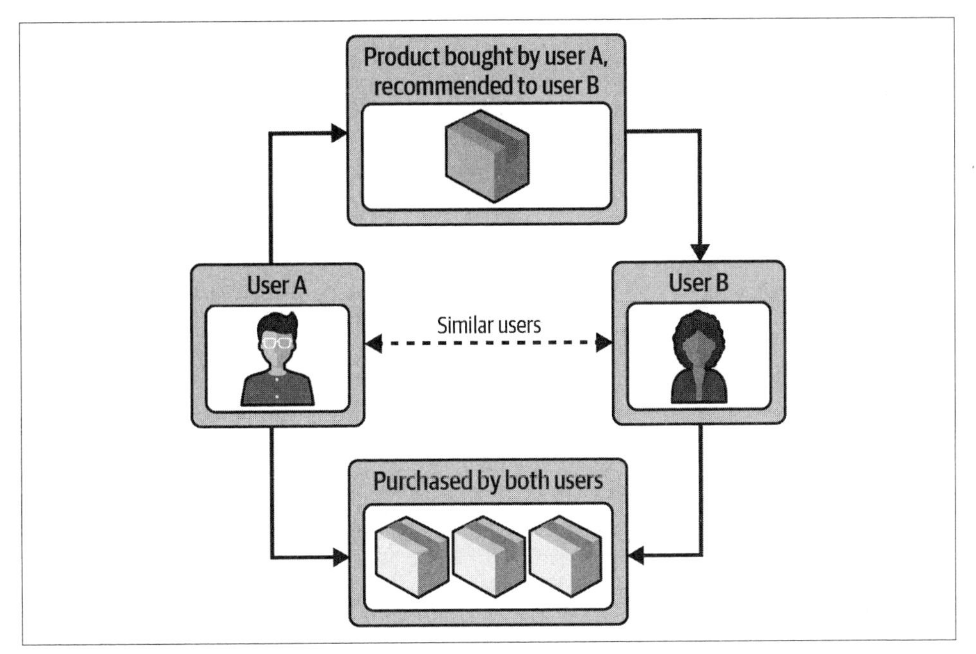

Figure 12-3. Product recommendations using collaborative filtering

Conclusion

For ML projects to have a massive business impact, PMs and business owners need to be relentlessly focused on asking the right questions at every stage. Rather than dive directly into technical details and ML implementations, teams must ensure they understand the business problems and goals as specifically as possible. This requires speaking with stakeholders and translating the needs into technical requirements. Once the business goals are understood, we must first assess whether ML is really needed to achieve those goals. If ML is the solution, the very next question that needs to be answered is whether we should integrate with an existing solution provided by an external service provider (buy) or invest in building the ML systems ourselves (build). In either case, a lot of planning and coordination is needed to integrate the ML solutions into customer-facing products.

Clearly defining business goals and measurable product metrics, choosing appropriate measures of success, and deploying solutions iteratively are each important steps toward building great ML products. But ML has a host of other requirements, many of which are different from the requirements of traditional software application development. Success here looks like having a high degree of awareness of the new situation that ML presents you with, while keeping a clear eye on what is actually of business benefit, and driving toward that—potentially over bumpy roads.

CHAPTER 13
Integrating ML into Your Organization

Integrating any significant new discipline into an organization often looks more like an exercise in irregular gardening than anything else: you spread the seeds around, regardless of whether the ground is fertile or not, and every so often come back to see what has managed to flourish. You might be lucky and see a riot of color in the spring, but without more structure and discipline, you'll more likely be greeted by something barren.

Getting organizational change right is so hard for plenty of *general* reasons. For a start, an effectively infinite amount of material is available on how to change organizations and cultures. Even choosing from this plethora of options is daunting, never mind figuring out how best to implement whatever you settle on.

In the case of ML, though, we have a few *domain-specific* reasons this is true, and arguably these are more relevant. As is rapidly becoming a cliché, the thing that is fundamentally different about ML is its tight coupling with the nature and expression of *data*. As a result, anywhere there is data in your organization, there is something potentially relevant to ML. Even trying to enumerate all the areas of the business that have or process data in some way helps to make this point—data is everywhere, and ML follows too. Thus ML is not just a mysterious, separate thing that can be isolated from other development activities. For ML to be successful, *leaders need a holistic view of what's going on, and a way to influence what's being done with it—at every level.*

What is particularly frustrating and counterintuitive about this situation is that almost every change management methodology recommends starting out small, in order to manage the risk of trying to do too much at one time. Though this is greatly sensible in most cases, and your first ad hoc experiments can generally be done without too much overhead, success in a small pilot guarantees nothing about how ML implementation on a larger scale might work well. Avoiding siloization is hard enough in most organizations, but is particularly crucial for ML.

However, there is good news: it is absolutely practical to start small, to grow a successful pilot, and to make sure the opportunities and risks of ML are handled correctly as you grow. But you have to be deliberate about the way you do it, which is why we should talk about our frameworks and assumptions first.

Chapter Assumptions

We have written this chapter with assumptions that we would like to make clear before we begin. Each is detailed in this section.

Leader-Based Viewpoint

Our first assumption—which might already be clear—is that this chapter and Chapter 14 are unapologetically written for the organizational leader. Though there are points of relevance to data scientists, ML engineers, SREs, and so on, this chapter is most urgently addressed to those responsible for the health, structure, and outcomes of their organization, on a scale from a team (two or more people) to a business unit or company (hundreds or thousands of people).

Detail Matters

Generally speaking, organizational leaders don't engage in detail and directly with implementation and management of *any* change project, except in cases of strong need. ML needs to be a little different. As per the preceding assumption, since doing ML well involves understanding the principles of how it works, what use it makes of data, what counts as data, and so on, leaders need to know this before the decisions they make are going to be sensible.

Our main observation here is that by default, leaders are not going to pick up ML-relevant knowledge as part of their regular management activity, and so there needs to be an explicit mechanism for doing so. This is in opposition to the bulk of conventional managerial theory, which asserts that most teams can be managed effectively with a handful of representative KPIs and a good understanding of team dynamics. We currently believe that ML is sufficiently complex, new, and potentially impactful that being aware of the details matters—though we expect this will change as time goes on.[1] For the moment, though, leaders need to understand ML basics and

1 For what it's worth, we don't venerate leaders who are continually involved in the details—sometimes it's better and sometimes it's worse—but we do believe that it's necessary for leaders to understand the trade-offs at this point in ML's evolution. At the very least, organizational leaders need to know the business metric being optimized and need to have a means of measuring whether the ML system is optimizing that metric effectively. Understanding some of the details of the implementation as well as the process for measuring the effectiveness will allow leaders to do so with confidence.

be able to access practitioners, which in turn will help them assess the likelihood of success, and make outcomes better.

Outside of training programs, our main recommendation to achieve this level of understanding is that ML should not be siloed, either as a standalone technology-driven effort, or in any other way. Given it can touch everything, a strong siloization would invert the flow of information and control, and could introduce big risk management problems. ML is naturally a horizontal activity.

ML Needs to Know About the Business

Our third assumption is that the complexity of the *business* is a direct input to how ML is conceived and implemented. It's not just about including ML folks in department circulars about goals and performance—it is much more pervasive and holistic than that. ML practitioners need to be more aware of broad business-level concerns and state than the average product developer.

A couple of examples will make this clearer:

Example 1

An ML developer at YarnIt wants to make a business impact, specifically on the web sales part of the business. They work to build a model identifying products that are underperforming in terms of sales. The model recommends these products in particular contexts in order to increase their sales. A model like this might be successful in a variety of ways. It might identify new purchasers for these products or simply remind people who used to purchase these products that they might want to purchase them again.

But now there's a problem in our case: it manages to find cashmere wool yarn, a prestige item that YarnIt has in very low supply. Since this model does not have features for (can't effectively represent or understand) margin or inventory, it manages to sell all of it. Cashmere yarn is in short supply in the industry, and replacing it will take weeks, or even months. Although very few customers purchase this yarn in large quantities, many of them may buy a little bit from time to time as part of larger orders. Another effect that you sometimes see in ecommerce is that when a web shop doesn't have a particular thing in stock, customers sometimes take their whole order elsewhere—and YarnIt experiences this too. So now YarnIt is losing sales on *other* products because of their lack of stock on *this* product.

Example 2

Another case exposes a cross-organizational privacy and compliance problem. The recommendations and discovery teams train models to help customers get the most out of the *yarnit.ai* site and help YarnIt make the most money from customers. Among the features that the models use is information about the browser the customers are using to access the website, so the team uses a limited set of information from the `User-Agent` string provided by the browser—just enough to determine the browser and platform.

Meanwhile, the web design team has been working on new interactions and wants a better sense of what browser configurations customers are using. To get all of the information required, the team decides, *without* talking to the ML folks, to just log and track the full browser information. As a result, the modeling teams start using the full browser `User-Agent` in their models without knowing it.

The problem is that the full contents of the `User-Agent` plus location information (that the model also has) often uniquely identify a single person. So now we have a model that has the ability to target individual people. This violates some privacy governance policies of YarnIt and compliance requirements in the countries where YarnIt operates as well, posing a serious risk for the whole organization.

In both of these cases, the individual teams did nothing wrong *in their context*. But acting on incomplete information about the effect of their choices on other teams led to bad outcomes. As a result, leaders need to be aware of how ML functions in their organization, so they can provide the vital coordination and broad oversight that would otherwise be missing. The big question is really how best to provide this.

Here is one structural way to think about it: you need to be able to centralize the portions of the ML work where oversight and control are most important, to liberate those portions of the work where domain-specific concerns are most important, and to provide an *integration point* where these workstreams can meet. It is at that integration point that oversight, steering, and communication should take place.

The good news is that most organizations already have some venues where cross-cutting conversations take place. If you are lucky, it will be natural for such conversations to take place within (for example) product management, where customer lifecycle management is a normal matter of concern. If it will not fit into an existing meeting, workstream, or venue of some kind, you will have to create a new one. But, however it is implemented, you will need to have these kinds of conversations going on.

The Most Important Assumption You Make

We base our material on all of the preceding assumptions. However, your organizational change effort also has assumptions—even, or especially, if you think it doesn't—and those assumptions are close to you and therefore may ultimately determine the success of your ML effort.

The most important one, the one that's hugely valuable to examine before you embark on your ML journey, is strongly related to the question of what you're trying to achieve with your ML project. It is what you assume ML can do for you.

The Value of ML

ML can do more for your business than just make you more money.[2] Implementing ML could mean that you could improve civic engagement, raise more funds for disaster relief, or figure out which bridges most urgently need maintenance.[3] But for business leaders, it usually means making more money and making customers happier (and happy customers generally lead to more money for the business), or sometimes reducing costs via automation.

As an example, let's go back to YarnIt. The CEO started hearing about ML some years ago. Some of the early ideas for how to put ML to use included the following:

- Power the search results on the website by including a model of which products a given customer (or web session owner if they weren't logged in) was likely to purchase, given their previous behavior and interest.
- Help customers discover new products, including devoting a substantial portion of the front page to those selected by a model purpose-built to do so.
- Manage inventory. ML can model the supply chain constraints and inventory levels and propose optimal reordering for products to ensure that YarnIt has stock of the appropriate mix of products, given financial, sales, storage, and supply constraints.
- Improve profitability by adding a product and order margin as a feature to many of the other models.

Notice that these ideas have different timescales and different levels of organizational intrusiveness. Some of them can be easily tacked on to what is already being done and how it's done. Changing the ranking of the search results will probably have a measurable impact on customer satisfaction and sales, but can be done relatively quickly (weeks or months, not quarters or years) and will not require particular participation from the broader organization to implement. Changing the way inventory and supply chain work will take longer and will require much broader participation from other departments. If done well, it has the possibility of really transforming the overall efficiency of YarnIt as a company, perhaps akin to just-in-time supply chains. But it is not a change that can be led by an ML engineer.

2 This observation isn't limited to for-profit businesses. Leaders looking to implement ML often hope for it to improve the thing they already do: make more money, give out more food, pay for more housing. ML can do these things, but it can also transform the way you think about running the organization as a whole.

3 During the writing of this book, a prominent bridge collapsed in Pittsburgh, Pennsylvania, where some of us live. Although the main problem with physical infrastructure in the US is simply that the country doesn't spend enough money, it is also true that prioritizing where to spend limited resources might be amenable to an ML application.

Even facing these implementation challenges, organizational leaders are usually broadly sold on the value of ML. Indeed, ML has the potential to realize additional value based on data already collected by the organization. It is the kind of once-in-a-generation technological change that really can transform the way organizations function—hence the necessity to examine carefully what you think ML can do for you, figuring out what subset of that you want to achieve, and writing all that down before you start.

Significant Organizational Risks

Suppose you have assessed what ML can do for you, decided on the specific form it will take, written down your assumptions, and are eagerly rubbing your hands together, anticipating the wonderful changes that are going to bring you fame and fortune right now. "What next?" is the obvious question. Unfortunately, before we get started, it's as important to understand the risks as it is the value. Otherwise, you won't be in a position to make a well-founded decision that involves prioritizing one over the other.

ML Is Not Magic

While most business leaders have some appreciation of the value and potential of ML, they do not necessarily understand the risks equally well. As a result, you can sometimes see a leadership perspective emerging that treats ML practitioners as almost magical in what they can achieve. Yet no one—at the leadership level, anyway—understands how or why. By misunderstanding the scope and mechanisms of ML, leadership also overlooks the scope of the impact of those projects across the organization. The greatest danger in this case is that the risks become invisible, or effectively become externalities—in other words, someone else's problem. That is a recipe for inevitable, though perhaps arbitrarily deferred disaster.

Mental (Way of Thinking) Model Inertia

Transforming the way an organization works is never a simple proposition, and thousands of pages have been written on that topic. Here, we confine ourselves to saying that implementing ML is just like any other change, in that it requires stakeholder management and obtaining buy-in from those affected, but also unlike other changes in that the total set of stakeholders is likely to be much larger.

As a result, that component of the problem that is a function of the number of stakeholders (for example, the pure logistics of figuring out who needs to be involved) obviously gets larger. More importantly, though, any component that involves persuasion, communication, and, in particular, understanding the way people *model* the change is also dramatically increased in importance.

When driving a significant change, just showing up to all your meetings with the talking points of the senior leader involved is rarely going to work. You are not going to persuade people to change their behavior based on a characterization of the situation that is only from a senior leader's point of view. In particular, the key issue is of *mindset*, and what *mental models* are used by both leaders and practitioners throughout the organization to represent what's going on and how to react to it. If the plan assumes that everyone will just move to understanding things the way you (or a certain individual) do, the plan will probably be short-lived.

But sometimes the new way of doing things is genuinely the correct way of responding to a particular situation. If it *is* the correct way to proceed, and yet the mental models are not changing, you, the changer, have to undergo the burden of persuasion. To be most effective, that persuasion has to be accompanied by a motivation or set of motivations, and those need to speak to the mental models of the audience. For example, perhaps the audience believes there's nothing real to be gained by spending the effort to make the data of their teams readable by others; or perhaps they're scared by the prospect of being shown to be worse than other teams; or perhaps they profoundly believe that the only thing that matters is getting product features out to the public as fast as possible, and spending effort on literally anything else doesn't matter. Either way, your proposal for change will require the mental models of your audience to be solicited, understood, and addressed.

Ultimately, for most practical concerns, implementing ML requires serious stakeholder management and a large concerted effort to shift mental (way-of-thinking) models.

Surfacing Risk Correctly in Different Cultures

Obviously, if benefits are clear at the leadership level, but risks are invisible, this leads to risk management taking place in an ad hoc, underfunded fashion, or potentially even deliberately not taking place. Those risks could even be misrepresented, particularly in negatively oriented cultures. It may be useful to review the organization typology suggested by Dr. Ron Westrum in the context of software engineering organizations to understand the implications of this situation in more detail.[4]

If we could simplify somewhat, Westrum suggests that organizations can be broadly characterized as power-oriented, rule-oriented, or performance-oriented. Of these, the only organizational culture that experiences *structural* problems when implementing ML is the rule-oriented, bureaucratic culture. Why? On one hand, *power-oriented* organizations tend to crush novelty, and as a result are much less likely

4 Westrum's original paper is "A Typology of Organizational Cultures" (*https://oreil.ly/T9Jje*). The Google Cloud Architecture Center (*https://oreil.ly/POd9Y*) reviews this in the context of DevOps, but most of the points are relevant to ML production engineering as well.

to implement ML on their own in any serious way. On the other hand, *performance-oriented* cultures have an openness to novelty, cooperation, communication, and risk sharing. These environments are likely to tolerate the kinds of open coordination that successful ML implementations require.

On the contrary, *rule-oriented* organizations tolerate novelty but punish people when it goes wrong. Failure leads to negative consequences for those seen to have failed, organizations have narrow (and thoroughly defended) responsibilities, and coordination is minimal. In these organizations, we expect that ML will be able to gain a foothold, but when anything goes wrong or becomes difficult, those "at fault" will be punished and the innovation will promptly cease. Unfortunately, such behavior makes it very hard to adequately model and respond to the cross-cutting risks that go with ML; significant losses may well result.

Siloed Teams Don't Solve All Problems

Another common risk is for ML teams and projects to be treated equivalently to the way other new kinds of work are treated, and a common instinct is to start a new siloed team to do that work, leading to its separation in the organization. This is a common way to reduce startup friction in order to build something and demonstrate results. However, it does present a problem for ML, since implementing it at all usually requires help from multiple departments or divisions. But more importantly, because of the broad scope of impact that ML projects can have, successfully deploying ML requires organizational change to support structure, processes, and the people needed to keep it reliable. It is definitely possible to keep the scope too narrow for success.

Implementation Models

Having discussed some risks involved in introducing ML to an organization, let's focus on the nuts and bolts—how to actually get it done.

A small implementation project probably starts with applying ML to something that is integral to your organization's success. This involves creating and curating data sources, assembling teams with the right expertise in the problem space *and* in ML, and making the horizontal regulation mechanisms you'll need to track progress and steer the ship. Throughout this process, it will probably be advantageous to proceed with the blissful optimism that's required when you know trouble will come find you, eventually—though you don't know quite when or what sort.

Start out by picking a metric of some kind: ideally, it will be something *useful* but not *critical* in your system. That way, you can get valuable experience from the implementation, and the work will involve assembling the cross-connections between teams that you'll need for future expansion, but if things get messed up, the likelihood of significant disaster is lowered.

Let's consider an example at YarnIt. The implementation team will probably consider a few options. Using ML to help with search ranking is one appealing place to start. But the team notices that sales coming directly from the search results page represent significant revenue. This makes this an appealing place to apply ML eventually, but a risky place to get started. After looking for other parts of the site and seeing that all of them are revenue sensitive or revenue critical, the team takes a different approach: what if we *add* ML-produced results where there is nothing right now? The team members look around the *yarnit.ai* site and notice that several pages do not show recommendations to end users but could. They decide to add recommended products to the cart-add confirmation page that users see when they add an item to their shopping cart—or to put it another way, we take that moment where the users have demonstrated their interest in one product to recommend other products.

This is a good place to start: purely additive, with low risk and at least a chance of a reasonable return. Purchase intent is already present, and the change is not too intrusive to existing customer workflows. So the team pursues this "People who bought *X* also bought *Y*" model and decide to measure it by collecting click-through rates on those recommendations and comparing them to click-through rates on search results. Of course, once the team knows more about how to do this, another possible option is to adopt a more traditional approach that combines looking at what is achievable/feasible and what the expected yield would be, rather than focusing on minimizing intervention risk.

Remembering the Goal

Though it is important to preserve flexibility, particularly in the conduct of the implementation, we also have to remember the goal—to experiment with ML in order to build capacity in the organization. Hopefully, you'll achieve the business metric improvement you selected, but even if you don't, the project can still be successful overall: you can still learn a lot and try another approach if things don't work out.

But it is important to walk that delicate line between not becoming distracted by what you encounter along the way, and not being too rigid about what you'd decided previously.

Writing down your strategic goals, and the context that led to those goals, can be useful to refer back to when you are amidst troubleshooting tactical issues or handling an incident.

Greenfield Versus Brownfield

A fundamental question that often arises when you're starting a new set of activities in your organization is whether you're doing it from nothing (also known as *greenfield*) or you have an existing system, team, or business process to handle (also known as *brownfield*). In practice, almost all implementations are brownfield, because most organizations get most value out of improving a system they already have. In general, though simplifying considerably, transformation projects go easier the more of a greenfield situation it is or can be made to be.

A common intuition is that it is easier to build on something that already exists and is (somewhat) functional. But in fact, a crucial measure of success for new initiatives is how much opposition it attracts. Opposition is more commonly encountered in brownfield situations, where someone else's career success can often depend on nothing changing.

For just those reasons, most implementation projects that expect to meet significant opposition usually try to start a new team or function that covers previously uncovered responsibilities. Our view is that because of the strong interconnected nature of ML, it is not realistic to expect that relative isolation to continue for long. Eventually—and probably sooner than you think—you'll talk to someone else you need to ask permission from.

Our best guidance here, as previously, is to start with a metric that makes sense, since that's the easiest story to tell—successful transformations almost always require good stories. Then use that to determine how greenfield or brownfield your project needs to be, while acknowledging that most things are brownfield.

ML Roles and Responsibilities

Doing ML work well involves a dizzying array of skills, focus areas, and business concerns. We find one useful way to structure this knowledge is by thinking, as always, of the flow of data within your organization. For example:

Business analysts or business managers
: These roles are responsible for the operations of a particular line of business as well as the financial results from that line of business. They have the data and desire needed to make ML successful, but if it goes badly, their ability to do their job will suffer as a result of bad information.

Product managers
: These roles set the direction for the product and determine how ML will be incorporated into existing products. They help us decide what, if anything, we will do with the data. There may also be ML-specific product managers who guide what we implement as well.

Data engineers or data scientists
: These people understand how to extract, curate, manage, and track data as well as how to extract value from it.

ML engineers
: They build and manage models and the systems that produce them.

Product engineers
: They develop the products that we are trying to improve with ML. They help us understand how to add ML to the product.

SREs for ML or MLOps staff
: They lead the overall reliability and safety for the deployment of ML models. They improve existing processes for building and deploying models, propose and manage the metrics to track our performance over time, and develop new software infrastructure to enforce model reliability. These roles wrap around the entire process and are some of the only engineers looking at the process from end to end.

Each of these roles may be combined with others in a smaller organization. They are functions to think about filling.

How to Hire ML Folks

Hiring talented ML staff is difficult right now and is likely to stay difficult for the foreseeable future. The growth of demand for ML skills has far outstripped the supply of educated, experienced staff. This affects all employers, but the most prestigious of ML companies (generally large tech organizations) continue to hire most of the new graduates and experienced staff. This leaves other organizations in difficult circumstances.

The usual recommendations for how to proceed in this case include options like attempting to reach potentially qualified candidates earlier in the cycle, making sure that the operation of the recruitment process is generally strong, communicating with the candidate regularly, selling the candidate on the advantages that the company in question has, and so on. While all of those are true, useful, require effort, and might well work for you, they are standard approaches. If the market is particularly hot, doing all of those well *still* might not work.

We recommend another approach. Reframe the problem as dividing staffing between those who really require ML knowledge and experience immediately, and those who can learn it on the job as they go. Most situations, and indeed startup programs, require only one or two experienced ML researchers or practitioners. These experienced employees can help design models to meet the organization's goals and can also specify the systems needed to build and deploy those models. But the staff needed to manage the data, integrate ML into the product, and maintain the models in production can all be folks who are talented in other ways, but are learning ML deeply on the job. (You can even buy books like this to help bring those employees up to speed more quickly!)

So, having partitioned the problem, we still have the question of how most organizations can hire those first few experienced ML researchers and engineers to seed this process. The standard playbook involves a mix of hiring contractors or consultants from experienced firms, paying for a single superstar with real experience and credentials who is willing to teach, and betting on junior rising stars, while understanding that the path will be bumpy. These are practical options when your organization has a desire to produce ML but does not have prestige or money to compete against bigger firms.

Now that we've considered some of the concrete challenges that organizations face adapting ML specifically, let's take a step back and consider the problem from the perspective of traditional organizational design.

Organizational Design and Incentives

Making an organization function well, given what it is supposed to do—often called *organizational design*—is a difficult art that involves a mixture of strategy, structure, and process. The key point for leaders is that reporting structures are often the least important part of successful organizational designs. Other much more powerful aspects and determinants affect behavior.

Before we dive in even deeper, it is worth acknowledging that organizational design is a technical and often jargon-filled topic. For some leaders, especially those at smaller organizations, it may be difficult to see the forest for the trees in the proceeding sections. Talk of strategy and process and structure can be difficult to map onto the main actual tasks: hiring the right people and getting ML added to your application. Ultimately, though, the main lesson is that thinking about the way your organization currently works, and how that will change, hugely improves your chances of doing ML successfully.

We can choose from numerous models to understand how to change an organization in order to achieve a certain goal. This section is not designed to provide a complete review of all of them. Rather, we will select one common approach to thinking about organizations, the Star Model (*https://oreil.ly/y1xts*) by Jay R. Galbraith, and apply it specifically to the challenge of implementing ML in an organization (Figure 13-1).

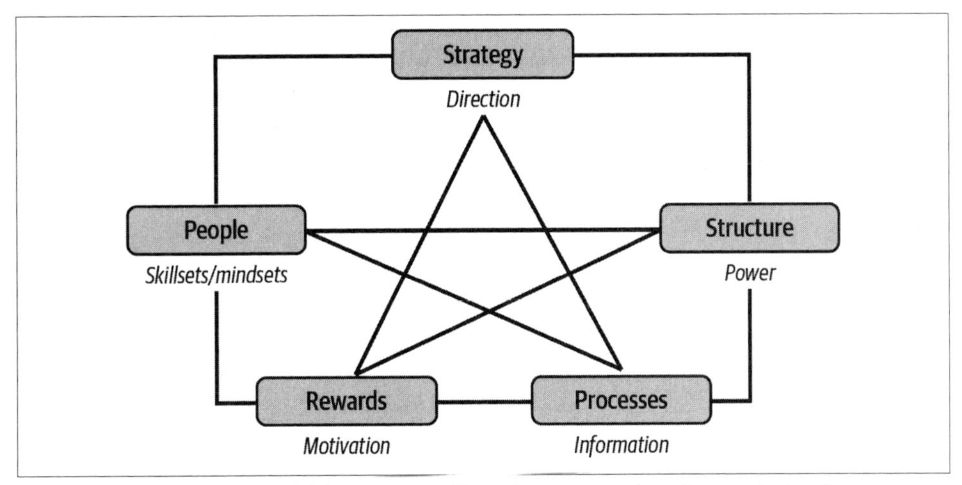

Figure 13-1. The Star Model (© Jay R. Galbraith. Reprinted with permission.)

In this model, strategy, structure, processes, rewards, and people are all design policies or choices that can be set by management and that influence the behavior of the employees in the organization.

This model is useful because it goes beyond the reporting structure or organization chart, where most leaders tend to start and end their change efforts. Galbraith points out that "most design efforts invest far too much time drawing the organization chart and far too little on processes and rewards." This model allows you to take that observation and then think about whether all of the interconnected aspects are affected or can be changed to support the requirements better. Policies, processes, people, and rewards policies can then be adjusted to support your structure and strategy.

Let's review each of these in the context of an organization trying to implement ML.

Strategy

The *strategy* is the direction that your organization is trying to go. It drives your business or organizational success model. It affects which parts of the organization are given attention or funded, and how the organization is measured or considered successful.

"Best-in-class machine learning for the yarn distribution industry" could be a strategy that identifies ML as a primary focus for YarnIt, but also might limit where ML is deployed if we insist on only "best-in-class" ML.[5] Another strategy of "machine

5 "Kind of reasonable most of the time" ML can actually be an improvement on existing algorithmic deployments in a variety of situations.

learning in all aspects of the product" might mean the organization funding new and innovative ways of using ML everywhere, with more tolerance for lower-quality results to start with. On the other hand, if we set a strategy to "increase sales by diversifying approaches including the use of ML," we might consider it as more experimental or less important than other more traditional ways of increasing sales.

Structure

Structure describes who has power in an organization. You may also think of it as the organization chart or reporting structure because it identifies formal oversight authority. (It can be very different, of course; in other places, the authority may lie within the team, where certain technical leaders must support a decision before it is implemented.)

One way to think about choices for organizational structure, and the one that Galbraith identifies, is that it includes functional, product, market, geographic, and process structures:

Functional
: This structure organizes the company around a specific function or specialty (for example, centralizing ML implementation in a single team).

Product
: This structure divides staff into separate product lines. In this case, the ML teams would be distributed into the individual product teams.

Market
: The company is organized by the customer market segment or industry they sell to. For YarnIt, this might be by type of crafter (knitter, weaver, or crocheter).

Geographical
: This structure organizes by territory: the product has a dependency on region, location, or even distribution economics (such as where the food comes from). The only obvious reason to consider this structural approach for ML would be governance and compliance with local laws. This is probably not how we would structure an ML implementation.

Process
: Also sometimes known as a *horizontal organization*, this structure aggregates power in all of the people who develop and deploy processes in an organization. This may be a good model for ML teams that work across various product lines but need to create standards and processes for the organization.

Leaders will generally have a mental model for the way the organization works and the approach they should use to effectuate change. For example, think of a mental model that, to start a new function, the senior leader must be hired first. With this mental model, a leader will tend to centralize ML functions around a specific senior leader—and this has obvious drawbacks if the right leader is not to be found, or centralization doesn't fit well with (say) the existing engineering culture. Similarly, a siloed ML function might work better for senior leaders to maintain control of, but would inhibit progress of ML on other engineering teams. Ultimately, leaders will probably need to shift their mental model of the way things work, depending on the chosen ML strategy. (No one-size-fits-all structure exists, though for those desiring a this-size-fits-some approach, we cover structure implementation choices in detail in Chapter 14.)

Processes

Processes constrain the flow of information and decisions through an organization, and hence are critical to the way ML will work. They can be used to address issues in the structure as necessary. The Galbraith framework defines two types of processes: *vertical processes* allocate scarce resources (e.g., budget), and *horizontal processes* manage workflows (e.g., customer order entry and fulfillment end to end).

One potential way to begin adding ML to your organization is to treat the introduction as a vertical process, with decisions made centrally but implemented throughout the organization. That works well if the leaders have mastered their dependencies and connections. If that's not the case, you can get disconnected decisions. For example, if we fund an ML training and serving team to add a new ML feature to our application, do we also fund teams to curate all of the data, or to handle model quality measurement over time or fairness? If we do, we might end up duplicating centralized functions in our local scope, which is inefficient and potentially friction-increasing.

Once the organization has several ML projects implemented, centralizing the infrastructure from those projects to fulfill specific workflows may add robustness and reliability. For example, many model teams will start by providing their own infrastructure end to end, but eventually we might have many modeling teams providing models that integrate into our application. At that point, we could centralize serving for some of those models, think about building a central feature store, and so start establishing common aspects of the ML organizational infrastructure regardless of the model team.

Rewards

Rewards are both financial and nonmonetary. While most organizations will find it difficult to compete for ML talent on a financial basis alone, it might make more sense for an organization to compete on mission, culture, or growth. Most employees value recognition, status, or career opportunity. They also value applying their skills autonomously to create something of value. The autonomous part is tricky because ML staff need to be independent, but it is also critical that they be subject to the kind of governance that the organization needs to ensure that the ML it deploys is fair, ethical, effective, and compliant with relevant laws. Aligning rewards to not just raw execution of business goals, but also reliability, fairness, and robustness will help create the right incentives to not overlook these areas.

One other surprising point should be noted about rewards for ML skill and knowledge. Recall that ML is likely to impact most parts of our organization. One thing that should be considered is rewarding staff throughout the organization for learning more about ML. If the sales staff, accounting staff, buyers, and product managers all have a basic education in ML, the organization may well be much more effective in the long run.

We expect that ML expertise will continue to be scarce indefinitely into the future, with consequent effects on compensation, difficulty of hiring, etc.—see "How to Hire ML Folks" on page 317 for suggestions for active approaches here.

People

Finally, we need to consider the collection of factors that influence the human beings in our organization. This includes the mindsets and skills that those people need. It also includes human resource policies of recruiting, selection, rotation, training, and development of people. For example, flexible organizations need flexible people. Cross-functional teams require people who can cooperate with each other and are "generalists" that understand multiple aspects of the organization.

Given how rare ML education and skills are at present, most organizations should consider hiring staff who can learn on the job rather than only those already qualified. While this is true across the map of ML staffing, it is especially true in the territory of SRE. An ML production engineer benefits much more from solid reliability and distributed systems skills than they do from ML skills. In these roles, ML is the *context* in which the work happens, but not always the *content* of that work.

Finally, the organization will need people who can work through the ambiguity of problems caused by ML without stopping at a root cause of "the ML model said so." That's a fine place to start, but people will need to be able to think creatively about the way ML models are built, how changes in the world and in the data impact them, as well as how those models impact the rest of their organization. Some of that perspective and approach will come from ML education and skills, but some comes from a curious and persistent approach to problem-solving that not everyone starts with.

A Note on Sequencing

The preceding topics were separated out for clarity of explanation and ease of illustration. Though separation of concerns is a powerful technique much used in computer science, in real-world organizational work, everything is entangled and intertwined in a way that can often make it practically impossible to change things by just exerting control over a single dimension. The good news is, it often turns out that's what you want.

Changing one single dimension of the preceding Star Model elements is unlikely to result in success by itself. A strategy change divorced from process change will almost certainly result in just effectively the same output. Swapping in a new set of people who will learn the old culture has a good chance of developing a new set of workers who will behave as the old ones did. Financially rewarding new behavior while still allowing old behavior to be easy to accomplish (because all the processes are optimized for that) won't change anything in and of itself. And so on; the grim reality is, successful change often relies on pushing across many fronts in parallel.

However, you don't have to move forward at the same pace across all of these fronts at the same time, or with the same intensity. That's the second piece of good news—you can *sequence* this. Announce you're changing strategy, then processes, then rewards. Deal with them one at a time, but touch them all—at least, the ones that matter for your organization. Tell everyone the timescale you're following and the criteria you're using to evaluate success. Communicate your intentions, but acknowledge that not everything is going to change at once—it won't—but loudly and publicly commit to the overall goal. That increases credibility for the change and gets you supporters inside the organization.

Conclusion

We can't provide context-free recommendations for which precise dimensions of change to push on, since so much depends on your local situation. However, we can recommend at least thinking about the following:

- What does the organization care about? Driving change by trying to accomplish something the organization doesn't care about is more likely to be ignored, but also much less likely to get resources. Your work is overall more likely to be successful if you're aligned with those concerns.
- What are people doing today, and how does it need to change? Many plans for change originate from high-level staff disconnected from the day-to-day experiences of other staff members. Your change plan will have a much better chance of success if you take a step back and look at what they do and why they do it.
- How easy will it be to do the new thing rather than the old thing? If the new thing is harder to do than the old thing, everyone may well agree it's vitally important to do it, but the change will be slower and more difficult, if it happens at all. Make it easier to do the right thing and harder to do the wrong thing.

- Finally, acknowledge that change will take time. As we've said, displaying organizational vulnerability not only gets you more support from people who appreciate realism, but also allows people to manage their own reaction to the change better. Just don't forget to keep a regular communications cadence up: a big-bang announcement followed by nothing for ages often causes people to wonder whether the momentum has stalled.

If you want to see worked examples of the preceding points, see Chapter 14.

CHAPTER 14

Practical ML Org Implementation Examples

Organizations are complex entities, and all of their different aspects are connected. Organizational leaders will face new challenges and changes in their organization as a result of adopting ML. To consider these in practice, let's look at three common organizational adoption structures and how they apply to the organizational design questions we have been considering.

For each of these scenarios, we will describe how the organizational leader has chosen to integrate ML into the organization and the impact of that choice. Overall, we will consider the advantages and likely pitfalls each choice has, but in particular, we'll consider the way that each choice affects the process, rewards, and people aspects (from the Star Model introduced in Chapter 13). Organizational leaders should be able to see enough details in these implementation scenarios to recognize aspects of their own organizations, and be able to map these into their own organizational circumstances and strategies.

Scenario 1: A New Centralized ML Team

Let's say that YarnIt decides to incorporate ML into its stack by hiring a single ML expert who develops a model to produce shopping recommendations. The pilot is successful, and sales increase as a result of the launch. The company now needs to make some decisions about how to expand on this success and how (and how much!) to invest in ML. The YarnIt CEO decides to hire a new VP to build and run the ML Center of Excellence team as a new, centralized capability for the organization.

Background and Organizational Description

This organizational choice has significant advantages. The team can specialize, ample opportunities for collaboration and working together exist, and the leadership has clear scope to prioritize the work on ML across the company. Reliability experts can be in the same organization if their scope is limited to ML systems. The centralization also creates a significant nexus of influence: the leaders of the ML organization have more standing to advocate for their priorities across all of YarnIt.

As the group grows and the projects diversify, more of YarnIt will need to interact with the ML organization. This is where the centralization becomes a disadvantage. The ML team cannot be too distant from the rest of the business, as it will take the team longer to see opportunity, to deeply understand the raw data, and to build good models. An ML team siloed away from individual product teams is unlikely to be successful if it doesn't have the support of those product teams. Even worse, placing these two functions (ML and product development) completely separately in the organizational chart might encourage the teams to be competitive instead of cooperative.

Finally, a centralized organization may not be usefully responsive to the needs of the business units requesting help to add ML to their products. When it comes to productionizing ML, the business units likely will not understand the reliability needs of the ML teams and not understand why reliability processes are being followed (thus slowing delivery).

While these pitfalls exist for a solely centralized ML team, the organization can always evolve. This scenario goes through the Star Model as if it uses only a centralized team, but we will also illustrate an evolution of a centralized team doing infrastructure in scenario 3. Another possible evolution is that the centralized team educates and enables others to increase ML literacy within the rest of the organization.

Process

As we have mentioned, the impact of introducing ML in an organization tends to be pervasive. To address some of the disadvantages of the centralized organization, introducing processes can help distribute (or decentralize) decisions or knowledge. These processes include the following:

Reviews by key stakeholders
: These reviews, presented by the ML team on a regular basis, should ensure approval of the current modeling results as well as an understanding by business leaders of the way the system is adapting to the business. A separate science of exactly which metrics to include in these reviews is necessary, but the metrics should be complete and need to include the improvements as well as the costs of any given model implementation. Key business stakeholders may also review the priority of the various ML team efforts as well as the ROI and use cases for those efforts.

Independent evaluation of changes

One issue that can crop up in a centralized ML team is that all the changes become dependent on one another and may be held up by other changes. If the team instead evaluates the changes independently, ensuring accuracy of each model as it changes production by itself, then changes can be available more quickly. Often a model may improve performance on average but may hurt performance on specific subgroups. Judging whether these trade-offs are worth it can often require significant analysis and digging, and may require judgment based on business goals that are not easily reflected by simple metrics like predictive accuracy.

De-risked testing of combinations of changes

In large model development teams, several improvements often are developed in parallel and these then might be launched together. The question is whether the updates play well together. This creates a need to test candidate changes in combination, in addition to individually.[1] The caveat is that it might not be possible to test all combinations (because of resource/opportunity cost). It is important to create a process to vet candidate changes, and to determine which are useful to test in combination. This may be in the form of a go/no-go meeting to discuss model launches and how the teams are testing in combination with other changes. The meeting may then facilitate further testing or a decision to launch.

While this is easier in a centralized ML model team, introduce processes so that people from the various business units can also evaluate the changes. For the centralized ML team, this may be the product/business team that requested a change or feature, or the support team that may be affected by the changes.

Rewards

To ensure a successful implementation of the ML program, we need to *reward interaction between business and model builders*. Staff members need to be evaluated on the effectiveness of their cross-organizational collaboration and alignment to business outcomes rather than simply the effectiveness of accomplishing the narrow mission of their own department or division or team. Seeking input and formal reviews by other organizations, providing feedback, and communicating plans should all be rewarded behaviors. The way these are rewarded is culturally dependent per organization, but they should be recognized.

1 In most of the literature, this is presented as an algorithmic or technical problem. Of course, it is those things but it is also very much an organizational problem. If we don't have the decision and management framework to evaluate changes separately and strategize about how to deploy them, we will not be able to correctly prioritize that work.

Rewards may be monetary (bonuses, time off) but can also include promotions and career mobility. At YarnIt, we might add evaluative characteristics like "effective influencer" to the characteristics we look for in successful employees.

People

An *experimentation mindset* in product leaders is required, both to appreciate the value created by ML as well as to tolerate some of the risks. Product leaders need to understand that tuning the ML model for the organizational objectives may take some experimentation and that negative impacts will almost certainly occur along the way.

Bigger Isn't Always Better

The YarnIt ML team begins work on a project whose success definition is to create *bigger carts* (shopping carts containing more products worth more money). The idea is to increase the number of large purchases, since customers who make large purchases are more profitable to the organization. The team measures its success by the size of the shopping cart created per user impacted by the new model, since that's the metric the team has easily available.

A problem arises: the new larger carts are abandoned at a much higher rate than expected. Users seem to be creating large carts but then never checking out and purchasing the products. This cart abandonment rate is escalated to a senior leader in sales who starts troubleshooting the problem immediately, first with the web and payments teams (might be a UI problem or payments processing problem) and then ultimately with the ML team.

Together they develop a theory of what might be happening: the ML model is successful at convincing users to put more in their carts, but some users balk when they see the total size of the purchase and decide to abandon the cart instead of purchasing some of the products they can afford. At this point, everyone can collaborate on a solution.

Sales can generate an acceptable target for cart abandonment for the ML team to include in its model optimization, the web UI team can think of ways to make it easy to check out parts of a cart rather than a whole cart, and the product team can think about remarketing to users who abandoned carts, asking whether they might want to purchase just some of the products. And overall the ML team can evaluate whether the "large carts'" effort did, in fact, optimize revenue.

The end result will be happier users and a more profitable YarnIt, but only by working through this set of challenges rather than simply rejecting the poorly performing ML model right away.

YarnIt, and all organizations implementing ML, need to hire for a mindset of *nuance* in order to be successful. Leaders need to be tolerant of complexity and comfortable working across organizational boundaries even outside their own scope of authority. Notably, leaders also must have the confidence to represent the complexity of these impacts, without overly simplifying or sanitizing, upward to their leadership. YarnIt's CEO does not need to hear that ML is magic and will solve all problems, but instead how business objectives are being achieved with ML as a tool. The goal is not to do ML but to move the needle for the business. While ML is powerful, the nuance of ML is to derive value by minimizing the negative impacts.

The people in this centralized team need to be trained about quality, fairness, ethics, and privacy issues if they do not already have expertise in these areas.

Default Implementation

An oversimplified default implementation of the centralized model is as follows:

- Hire a new leader with ML modeling and production skills.
  - Hire ML engineering staff to build models.
  - Hire software engineering staff to build ML infrastructure.
  - Hire ML production engineering staff to run the infrastructure.
- Establish implementation plans with product teams (source of data and integration point for new models).
- Establish regular executive reviews of the whole program.
- Plan compensation by successful implementation and compensate both ML staff and product area staff.
- Start a privacy, quality, fairness, and ethics program to establish standards and compliance monitoring for those standards.

Scenario 2: Decentralized ML Infrastructure and Expertise

YarnIt might decide to invest in several experts across the organization, rather than a single senior leader. Each department will have to hire its own data scientists, including the shopping recommendations and inventory management teams. Essentially YarnIt will allow data science and simple implementations of ML to appear wherever there is sufficient demand and a department is willing to pay for it.

Background and Organizational Description

This approach is much faster, or at least it is faster to get started. Every team can hire and staff projects according to its own priorities. The ML experts as they are hired will be close to the business and products, and will thereby have a great understanding of the requirements, goals, and even politics of each group.

There are risks. Without a central place for ML expertise, especially in management, developing a deeper understanding of what YarnIt needs to do to be successful at ML will be harder. Management will not understand what specialized tools and infrastructure will be needed. There is likely to be a bias for trying to solve ML problems with existing tools. It will be hard to understand when the ML team is advocating for something it really needs (model-specific quality-tracking tools like TensorBoard), as opposed to something that might be nice to have but may not be required (GPUs for some model types and sizes or cloud training services that offer huge scale but also large costs). Additionally, each team will repeat some of the same work: creating a robust and easy-to-use serving system that can share resources across multiple models, and monitoring systems to keep track of training progress for models and to ensure they complete training. All of this duplication can be expensive if it is avoidable.

If some of these teams are doing work that spans the other products, and they probably will be, troubleshooting and debugging become much harder. When product or production problems arise, YarnIt will need to figure out which team's model is responsible or, worse yet, get multiple teams together to debug an interaction among their models. A proliferation of dashboards and monitoring will make this exponentially more difficult. Uncertainty about the impact of any given model's change will go up.

Finally, YarnIt will struggle to ensure that it has a consistent approach. In terms of ML fairness, ethics, and privacy, just a single bad model can harm their users and damage our reputation in public. YarnIt also may be duplicating authentication, integration into IT, logging, and other DevOps tasks with this organizational structure.

While real trade-offs exist, this decentralized approach is exactly the right one for many organizations. It reduces startup costs while ensuring that the organization gets targeted value out of ML immediately.

Process

To make this structure effective, organizations should focus on processes that can introduce consistency without introducing too much overhead. These processes include the following:

Reviews by senior stakeholders
: Model developers should still participate in reviews by senior stakeholders. It is a really useful practice for model developers to create write-ups of each proposed model development objective and finding that they come up with. These internal reports or whitepapers, while short, can record in some detail the ideas that have

been tried and what the organization has learned from them. Over time, these create organizational memory and enforce rigor in evaluation similar to the ML equivalent of the rigor in a code review for software engineers. YarnIt should create a template for these reports, possibly generated by collaboration among some of the first groups to start using ML, and a standard schedule for reviews by a small group with representatives beyond just the organization implementing ML.

Triage or ML production meetings
ML model developers should meet weekly with production engineering staff and stakeholders from the product development group to review any changes or unexpected effects of the ML deployments. Like everything, this can be done well or poorly. A bad version of this meeting may not have all the relevant points of view, might be based on incidental problems rather than well-understood systematic ones, might delve too much into problem-solving, or simply might last too long. Good production meetings are short, focus on triage and prioritization, assign ownership of problems, and review past assignments for updates and progress.

Minimum standards for technical infrastructure
YarnIt should establish these minimum standards to ensure that models all pass certain tests before launching into production. These tests should include baseline tests, such as "Can the model serve a single query?" as well as more sophisticated ones involving model quality. Even simple changes such as standardized URLs could help drive consistency internally (and anything that helps to make things easier to remember and behave the same is useful in the complex, quickly changing world of ML).

Rewards

To balance the decentralizing effects of this approach, YarnIt senior management will need to *reward consistency and published quality standards*. For example, leaders should reward employees for timely write-ups and careful reviews that are published in a widely available corpus. Each team will have local priorities that it will tend to prioritize, so it is necessary to reward behaviors that balance increased velocity with consistency, technical rigor, and communication.

One specific factor to note is that in this scenario, YarnIt is less likely to have staff with significant ML experience. One useful reward is to encourage staff to attend (and present at) conferences related to their work.

People

In this deployment, YarnIt should look for people who can *think both locally and globally*, balancing the local benefits against possible disadvantages to the company (or vice versa). Skills such as influencing without authority and collaborating across organizational lines may be useful to explicitly call out and reward.

The organization will still need people who *care about quality, fairness, ethics, and privacy issues* and can influence the organization—this is true in every deployment scenario. The difference here is that in this case, these staff members will have to develop local implementations to achieve quality, fairness, ethics, and privacy while also developing broad standards and advocating for their implementation across the company.

Default Implementation

Here's an oversimplified default implementation of the decentralized structure scenario:

- Each team hires experts in their own business units:
  - Hire ML engineering staff to build models directly with the product teams.
  - Hire or shift software engineering staff to build ML infrastructure.
  - Hire ML staff or shift production engineering staff to run the infrastructure.
- Develop a practice of internal reports of findings for review by senior stakeholders.
- Establish company-wide technical infrastructure standards.
- Run weekly triage or ML production meetings to review changes.
- Start a privacy, quality, fairness, and ethics program to establish standards and compliance monitoring for those standards.

Scenario 3: Hybrid with Centralized Infrastructure/ Decentralized Modeling

YarnIt started its implementation via the centralized models, but as the organization matures and ML adoption spreads throughout the company, the company decides to revisit that model and consider a hybrid structure. In this case, the organization will maintain some centralized infrastructure teams and some ML model consulting teams in the central organization, but individual business units are free to hire and develop their own ML modeling experts as well.

These distributed ML staff members might start by relying heavily on the central modeling consultants but over time will grow more independent. All of the teams will be expected to use and contribute to the central ML production implementation, however.

By centralizing the investment and use of infrastructure, YarnIt will continue to benefit from efficiency and consistency. But decentralizing at least some of the ML expertise will increase the speed of adoption and improve alignment between the ML models and the business needs.

Note that many organizations evolve into this hybrid model. As a leader, it might be wise to plan for this evolution.

Background and Organizational Description

The disadvantages of this hybrid implementation draw from each of the centralized and decentralized ML organizational structures. Inefficiencies might exist in decentralizing staffing and implementing ML throughout the organization. The business units might not understand ML well and might have particularly bad implementations. This can be especially problematic if the failures relate to privacy or ethics. Meanwhile, the centralized infrastructure might create friction for the decentralized modeling teams. And, the centralized infrastructure may feel more complex and be costlier. However, the longer the company is around, the more that the centralized infrastructure model will pay off.

Process

One way to think about the impact of this hybrid implementation is to reconsider the cart abandonment example in "Bigger Isn't Always Better" on page 328. In this case, provided the modeling team lives in the web store product team, these team members are more likely to notice the problem quickly and to realign the metrics of the model with sales instead of simply maximizing cart size. They will also think of possible *new features* like "purchase half this yarn now, and set aside half of it for later in one month." In all of these cases, ML has triggered the conversation about how to be more customer friendly.

But say the problem occurs across organizational divisions: a web store model causes a problem for purchasing. In those cases, resolution is likely to be much slower, as the purchasing team tries to convince the web team that the model is causing problems. In these cases, organizational culture will have to support cross-team model troubleshooting and even development.

Consider the processes recommended for scenario 1 and 2 to see if they may help alleviate possible disadvantages to this implementation:

- De-risked testing of combinations of changes
- Independent evaluation of changes

- Reviews:
  - ML team(s) findings documentation and review
  - Business reviews of model outcomes
  - Triage or ML production meetings
- Minimum standards for technical infrastructure
- Training or teams for quality, fairness, ethics, and privacy issues

Rewards

In the hybrid scenario, YarnIt senior management should reward business units for utilizing the centralized infrastructure, to prevent them from developing their own, duplicative infrastructure. Centralized infrastructure teams should be rewarded for meeting the needs of the other business units. Measuring and rewarding adoption, while also encouraging use of central infrastructure in almost all cases, makes sense.

Central infrastructure teams should have a plan to identify key technology developed in the business units and extend its use to the rest of the company. And from a career development perspective, ML modelers from the business units should be able to rotate onto the central infrastructure team for a period of time to understand the services available and their constraints, as well as to provide an end-user perspective to those teams.

People

To function well across YarnIt, all of these teams will need to have a company-wide perspective of their work. The infrastructure teams need to build infrastructure that works and that is genuinely useful and desirable for the rest of the company. The ML teams embedded with the business need to have a mindset that cooperation is best, so they should be looking for opportunities to collaborate across divisions.

Default Implementation

Here is an oversimplified default implementation of the centralized infrastructure / decentralized modeling model:

- Hire a centralized team (leader) with ML infrastructure and production skills:
  - Hire software engineering staff to build ML infrastructure.
  - Hire ML production engineering staff to run the infrastructure.
- Each product team hires experts in their own business units:
  - Hire ML engineering staff to build models directly with the product teams.

- Develop a practice of internal reports of findings for review by senior stakeholders.
- Establish company-wide technical infrastructure standards.
- Plan compensation by successful implementation and compensate not only for meeting business goals but also for efficiency of utilizing the central infrastructure.
- Select processes that will aid in cross-organizational collaboration such as cross-team ML findings reviews.
- Start a privacy, quality, fairness, and ethics program to establish standards and compliance monitoring for those standards.

Conclusion

Introducing ML technologies into an organization for the first time is difficult, and the best path will necessarily be different from organization to organization. Success will require thinking, in advance, about the organizational changes necessary to ensure success. This includes being honest about the missing skills and roles, process changes, and even entire missing suborganizations. Sometimes this can be solved by hiring or promoting the right new senior leader and charging them with the implementation. But often the necessary organizational changes will span the whole company.

Table 14-1 summarizes various organizational structures, and their impacts and requirements around people, process, and rewards.

Some teams, such as the production engineering or software engineering teams, will not require significant ML skills to start being effective. But they will benefit from study groups, conference attendance, and other professional development activities. We also need to build ML skills among business leaders. Identify a few key leaders who can understand the benefits and complexities of adding ML to your infrastructure.

The biggest barrier to success, though, is often the tolerance of the organization's leadership for risk, change, and details, as well as for sticking with ML because it might take a while to manifest returns. To make progress with ML, we have to take risks and change what we do. And ensuring that the teams are focused on business results will help adoption. This involves altering processes that work well and changing the behavior of successful teams and leaders. It often means risking successful lines of business as well. To understand the risks being taken, leaders have to care about some of the details of the implementation. Leaders need to be tolerant of risk but plainly care about the ultimate reliability of their ML implementation.

Table 14-1. Summary of structures and requirements

	Centralized ML infrastructure and expertise	Decentralized ML infrastructure and expertise	Hybrid with centralized infrastructure and decentralized modeling
People	Specialized teams with clear focus, nexus of influence on ML priorities and investments. Experimentation mindset is required. Leaders/senior members need to be effective collaborators and influencers outside of their own organization. Teams need to be the champions of ML quality, fairness, ethics, and privacy across the company.	ML expertise is spread across various teams and is often both duplicated and sparse. Leaders/senior members need to encourage and enforce internal communities. Siloed decisions will cause bad/inconsistent customer experiences and thereby significant business impacts. Teams across all product areas need to gain expertise on ML quality, fairness, ethics, and privacy.	Centralize ML infrastructure and modeling for common/core business use cases but encourage individual model development for specific needs. Avoids duplication and improves team efficiencies and consistency, especially at scale. ML quality, fairness, ethics, and privacy need to be in the DNA across all departments.
Process	Needs a lot of cross-functional collaboration for making decisions and sharing knowledge. Key stakeholders across business units need to review proposals and results, and launch plans collectively. Decentralized/independent model evaluation is needed to ensure and measure business goals and impacts. Validate changes in combinations to avoid unintentional regressions and establish go/no-go review meetings.	Needs a lot of documentation around best practices, knowledge, evaluation, and launch criteria to maintain consistency, *or* a deliberate decision not to maintain any outside of local team scope (which is problematic for ML). Key stakeholders across business units need to review proposals and results, and launch plans collectively. Well-structured and moderated go/no-go meetings are needed to avoid delays. Establish standards for technical infrastructure for ML pipelines.	Needs cross-functional collaboration and decent documentation between infrastructure and individual product teams on a project/program basis. Establish cross-functional teams with clear accountability. Regular cross-functional syncs should occur at project/program level. Key stakeholders across business units need to review proposals and results, and launch plans collectively.
Rewards	On top of overall quality and meeting business goals, individual/team performance needs to be measured based on the effectiveness of cross-functional collaboration. Establish mechanisms to compensate both ML and product teams together for successful AI feature launches.	On top of overall quality and meeting business goals, individual/team performance needs to be measured based on consistency, published quality standards, and operating internal ML communities.	On top of overall quality and meeting business goals, individual/team performance needs to be measured based on reusability, evolution of common infrastructure, and speed of execution.

CHAPTER 15
Case Studies: MLOps in Practice

This book has laid out principles and best practices for MLOps, and we've done our best to provide examples throughout. But there is nothing like hearing stories from folks working in the field to help see how these principles play out in the real world.

This chapter provides a set of case studies from different groups of practitioners, each detailing a specific issue, challenge, or crisis that they have lived through from an MLOps perspective. Each story was written by the practitioners themselves, so we can hear in their own words what they went through. We can see what they faced, how they dealt with it, what they learned, and what they might do differently next time. Indeed, it is striking to see how things as deceptively simple as load testing, or as seemingly unrelated as a launched update to an entirely different mobile app, can cause headaches for those in charge of daily care and feeding of ML models and systems. (Note that some of the details may have been glossed over or omitted to protect trade secrets.)

1. Accommodating Privacy and Data Retention Policies in ML Pipelines

By Riqiang Wang, Dialpad

Background

The automatic speech recognition (ASR) team at Dialpad is responsible for the end-to-end speech transcription system that generates live transcripts for various AI features (collectively known as *Dialpad AI*) for our customers across the world. Various subcomponents of our AI system heavily rely on the ASR outputs to make further predictions, so any error in the transcripts gets propagated to other downstream natural language processing (NLP) tasks, such as real-time assists or named entity recognition (NER). Therefore, we continually aspire to improve the ASR models in our ML pipelines.

Problem and Resolution

In 2020, our system was achieving great accuracy for typical North American dialects, but our benchmarking as well as anecdotal evidence showed that other dialects were often mistranscribed. As we expanded our business to other major English-speaking countries like the United Kingdom, Australia, and New Zealand, we needed to at least reach the same bar set for North American dialects. Consequently, we started looking into how to improve ASR accuracy for specific dialects, or even the plethora of dialects within North America. This included transfer learning experiments and using specialized lexicons, but on their own they were not enough. Privacy is at the center of everything we do at Dialpad, which is also a major challenge in most of the modern ML ecosystems. In this case study, we discuss a couple of challenges we've come across and the solutions we've implemented as we worked toward deploying a model for multiple dialects while respecting user privacy.

In a departure from the relevant literature, we mainly use the term *dialects* instead of *accents* because we recognize variations beyond the accent (i.e., the sound of the speech). For example, New York and New Zealand dialects differ also in vocabulary, phrasal expressions, and even grammar. We want to ideally address all these aspects in making ASR more inclusive.

Challenge 1: Which dialects?

At Dialpad, we value user privacy, and to power various AI features, we need massive amounts of data for model training. Therefore, accommodations have to be made within our Dialects pipeline. Specifically, we keep as little metadata related to calls as possible, and remove calls when needed for privacy reasons.

But for training a good Dialects model, we wanted to know which of our users speak with a given dialect, be it British, Australian, or others. Thus, we needed as much metadata as feasible so we could sample accordingly for each dialect.

We first considered ideas such as letting human transcribers annotate the accent being spoken in each call, but then realized that crucially, it is extremely difficult to pinpoint accents without having experience with them, especially with non-native speakers, who are more likely to have idiolects than dialects (i.e., each speaker has their own way of speaking). Then we thought about letting users self-report dialects, but such classification self-evidently raises data privacy concerns that cut against our motivating goal of inclusiveness.

Solution: Get rid of the concept of dialects!

Ultimately, we realized that regardless of a given user's dialect, we just wanted our model to do better with the speech that it was then struggling with. The ASR model did well with North American dialects because we had been feeding it North American speech, so we could also improve this existing model by adding undersampled data, building a model agnostic of dialects. We ended up simply getting more data that our model was doing poorly on, filtered by the model's own confidence measure. We manually transcribed this underrepresented data and trained a new model with this dataset plus the original training dataset.

Within a few rounds of model tuning and evaluations, the ASR models started performing better on the underrepresented dialects test set that we manually curated, without any changes to the training techniques or model architecture. More importantly, this extra dialect dataset was only a tiny fraction of the larger original training data, but made a significant difference in performance. This shows the importance of intentionality and diversity in data collection. It also suggests that we can rely on confidence/uncertainty measures as a pseudo-diversity measure for data collection, when true diversity is difficult to measure.

Challenge 2: Racing the clock

Making Dialpad's no-cost, customizable data-retention policy available as standard to all customers means that a customer can request their data to be deleted anytime, or schedule new data to be available for only a specific period of time. These substantial privacy wins, however, require equally substantial cleverness across the entire ASR system in terms of model testing and experimentation, and especially so for the Dialects ML pipeline that consists of multiple steps: collection of audio, transcription, data preparation, experimentation, and final productionization. These steps can together span over multiple quarters, longer than the lifespan of some of the collected data. That means training data and test sets are not constant, making it difficult to reproduce experimental results, sometimes leading to delays in training the models and launching the desired improvements for customers.

Late in the process of rolling out the new Dialects model, we saw that it performed well on multiple test sets, but performed significantly worse with one single test set across multiple internal trials (compared to the model in production, released six months earlier). This halted deployment of the model while we investigated why. We used multiple methods, including training the new Dialects model from scratch and checking data partitioning (after a previous misadventure inadvertently mispartitioning between training data and test data).

We also wanted to reproduce the results from the production model by using the same process to train a model, but 11 months later, the data subject to retention policies had begun expiring, and we didn't have the exact training dataset anymore.

This made it difficult to reproduce the results of past model builds, and we gained only inconclusive results. Ultimately, the key insight to resolving the discrepancy was that the previous model that had performed well on the test set was actually *in use* during the time the data from production was taken to make the test set. Since our human transcribers create test-set transcripts by editing production model transcripts, this means that the reference transcript of this test set is biased toward the output of the old model. We will never know for sure, however, because the transient nature of data subject to arbitrary data retention regimes compounds the problem of building and maintaining a sufficient corpus for underrepresented data.

Solutions (and new challenges!)

We see this experience as a good step toward integrating our respect for data privacy goals with the rigor of reproducible R&D. By the end of this project, we had created a separate, specialized data team that handles ASR or the NLP team's data tasks, redefining our whole data collection and annotation preparation process. The data team's task is to standardize our test-set creation process, creating dynamic test sets that provide high reproducibility even if some test data needs to be removed because of data retention policies. For example, data with time-based retention policies are no longer considered when creating a test set, and the data team also handles manual data deletions by backfilling, while monitoring how performance metrics on our test sets change over time.

The team has also standardized training data collection: instead of each ASR engineer writing their own query to get data from our database, we can now submit a request to the data team, and it will provide structured data as needed, including accounting for (and even avoiding) data flagged for deletion. As confidence in the accuracy and integrity of our human annotation pipeline improves, we are also exploring the possibility of identifying personal data elements at scale so that they could be removed or tokenized in lieu of fully deleting the transcript. While difficult, this challenge suggests a way in which privacy-promoting, data-minimizing techniques could secure much more robust access to ML training data.

Takeaways

Integrating with privacy and data retention policies undoubtedly introduces challenges in ML pipelines, especially those powering the primary use cases of a customer-facing product/service. In our use case, working toward a more inclusive ASR model for Dialects, we first learned that even a little diversity in our training data makes the model more robust. In traditional ML practices, we tend to emphasize the size of the training data, but our results demonstrate that quality—and specifically, diversity—is irreplaceable. More importantly, we can get diversity without probing into users' privacy by using the model confidence measure.

Secondly, although these efforts at diversity were complicated by our commitment to honoring customer choice in their data usage, we discovered that with careful curation, we could engineer robustness and reproducibility into our ML pipeline, alongside efficiency gains, by standardizing dataset creation with a dedicated team. We believe abandoning the "trade-off" narrative ultimately improves our access to needed customer data by demonstrating we are willing to put in the effort to be good stewards. Efforts like our Diversity and Dialects initiatives likewise demonstrate to customers the value of wide participation in, and representation by, ML training sets.

2. Continuous ML Model Impacting Traffic

By Todd Phillips, Google

Background

Here's a story from an incident within Google from several years ago. For confidentiality, we obscure some of the details and don't say exactly which system was impacted, but the broad strokes of the story are still worth retelling.

The system in question included a continuous ML model that helped predict the likelihood of clicks on certain kinds of results in a search engine setting, continually updating on new data as it came in. Data came in from several sources, including web browsers and specialized apps on mobile devices. (See Chapter 10 for more background on continuous ML models.)

One day, an improvement was made to one of the apps that contributed a particularly large amount of traffic to the system. As part of the improvement, code was included that asked the app to issue the most recent query a second time after an app update was made, in order to make sure that the served results had the freshest state. This improvement was pushed out to all installations of the app as one instantaneous update. For the sake of ominous foreshadowing, we will call this mass instantaneous update *issue A*.

Problem and Resolution

What happened over the next day and a half was interesting. The moment each device in the world received the update, it reissued the most recent query. This resulted in a huge spike in traffic, but because the queries were being issued automatically, there were no additional user result clicks. This data with many more queries but no additional clicks was staged for retraining of the continuous ML model.

Because of an unrelated issue, which we will call *issue B*, the original push was rolled back. Each device that got the original update now updated back to the previous version. This, of course, caused each one to follow the protocol of reissuing the most recent query yet again, causing a third round of duplicate queries and even more data that had no associated clicks.

At this point, the continuous ML model was now happily training on all the corrupted data. Because the corrupted data included a lot of traffic with no associated clicks, the model was getting a signal that the overall click-through rate in the world was now about half of what it had been just a few hours before. The resulting learning within the model soon led to changes in the served models and, not surprisingly, lower served predictions than normal. This soon set off numerous alerts, and the model ops folks started to notice a problem with no obvious root cause because the app owners and the model owners had no visibility into each other's systems—indeed, they were in completely different areas of a much larger organization.

Meanwhile, the unrelated issue B that had caused the rollback was fixed, and the app folks pushed the update again. This caused—you guessed it!—another round of updates on each mobile device with the app, and still yet another round of duplicate queries.

By this time, the ops folks had pushed the stop button on the continuous ML model training, and all model training was stopped. The most up-to-date version of the model was therefore one that has been impacted by the corrupted data.

Also by this time, word has gotten through the organization, and the root cause has been traced to the recent app push, but the specific cause of the changes in model predictions due to the app update was not readily apparent. The behavior that an app update caused a duplicate query was not widely known, and those who knew about it on the app side did not make the connection to the way that it could impact training data in a continuous ML model. Thus, it was assumed that the update may have contained another bug, and the decision was made to re-roll back the re-update of the app and observe for a few hours after that. And of course, this re-rollback created still yet one more round of duplicate queries and corrupted data.

Once the rollback was completed and several hours of system observations were done, there was enough information available to be sure that the problem was just in the way that the pushes were being done. The mitigation turned out to be simple: stop making updates to the app in terms of rollbacks and re-updates, and then let the continuous ML system roll forward and catch up to the new state of the world. The ML model eventually saw clean data with the appropriate click-through rates.

Takeaways

One thing we took away from this study is that the many attempted mitigating actions actually ended up doing as much harm as good, and in some ways extended the period of impact. In retrospect, if we had just allowed the model to roll through, the system likely would have recovered much more smoothly and gracefully than it did in the wake of all our attempts to fix things. Sometimes it pays to just hold tight and roll through.

3. Steel Inspection

By Ivan Zhou, Landing AI

Background

Manufacturers in many industries rely on visual inspections to detect critical defects during production of steel rolls. I am an ML engineer at Landing AI and wrote this case study to show some data-centric techniques we used to develop deep learning models for visual inspection tasks.

Recently, my team and I worked on a steel inspection project (Figure 15-1). The customer had been developing a visual inspection model for years and had it running in production. But their models achieved only 70% to 80% accuracy. I was able to rapidly prototype a new deep-learning model that achieved 93% accuracy at detecting defects in the project.

Figure 15-1. This case study focuses on detecting defects in steel rolls that may have occurred in production

The goal of this project was to accurately classify the defects among the customer's hot rolling dataset; *hot rolling* is a key stage in the production pipeline of steel rolls. The defects were spread across 38 classes, and many defect classes had only a couple hundred examples.

The customers had been working on this problem for almost 10 years, and the best performance their model was able to achieve was only 80% accuracy, which was not sufficient for the client's needs. Over the years, the customers had tried to bring on several other AI teams to improve the accuracy of their models. All attempted to improve the performance by architecting several state-of-the-art models, but ultimately none were able to make any improvements.

Problem and Resolution

I went onsite to work on this project. I hired three local interns to help me label those images. In the first week, I spent almost all of my time learning about defect classes, managing the labeling work, and reviewing their labels. I gave data to interns in small batches. Every time they finished, I would review their labels and share with them feedback if there was a labeling error.

We didn't label all data at once. We labeled 30% of the data per class in the first iteration, pinpointed all the ambiguities, addressed them, and then labeled the next 30%. So we had three iterations of labeling over two weeks. We focused on defects that might be introduced in the "roll" stage of the manufacturing pipeline, after the metal is cast but before it is finished. Defects can occur from a variety of physical conditions in the hot process, and are grouped by category. In the end, we labeled 18,000 images and threw away more than 3,000 that we thought were confusing (Figures 15-2 and 15-3).

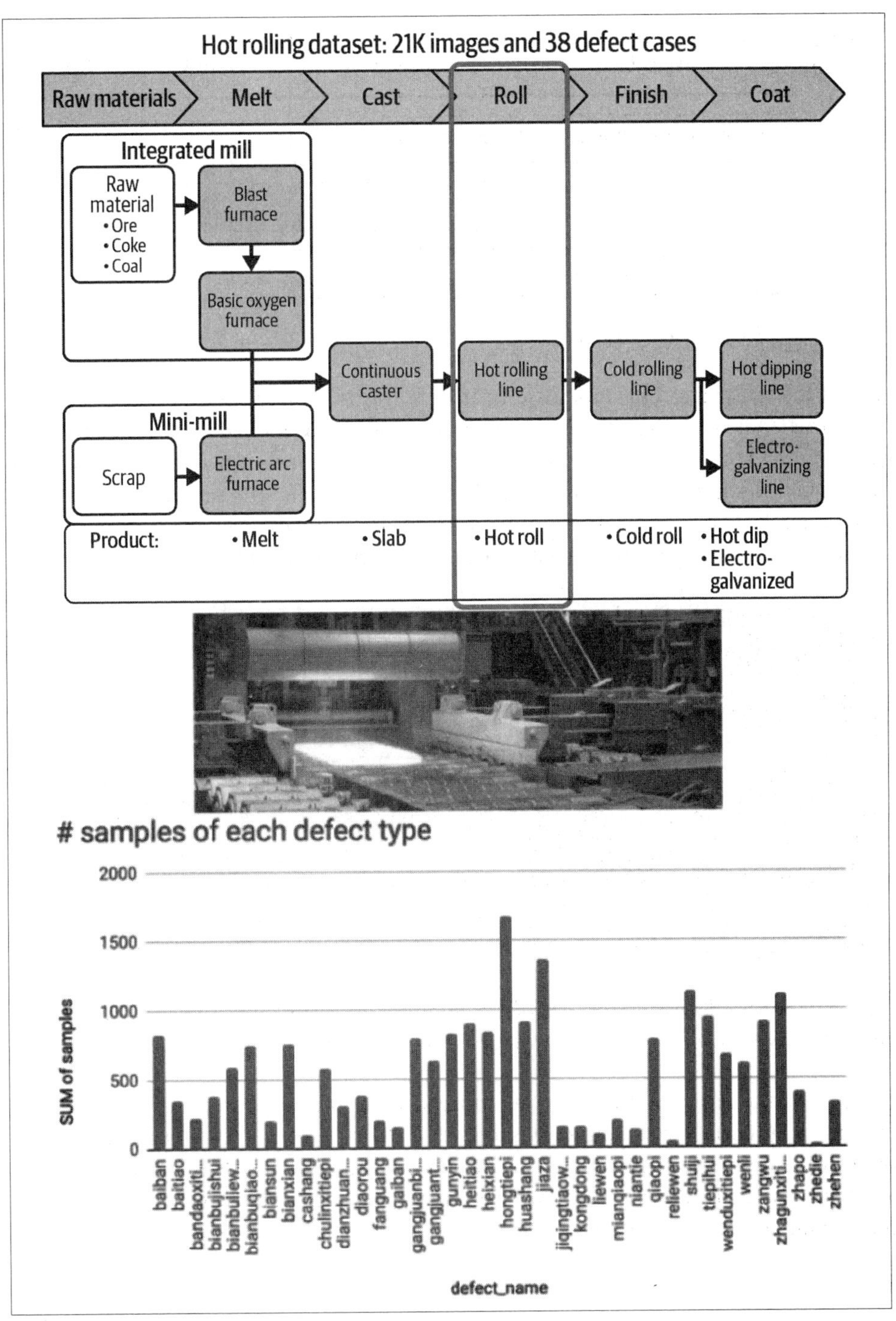

Figure 15-2. Details of the hot rolling setting

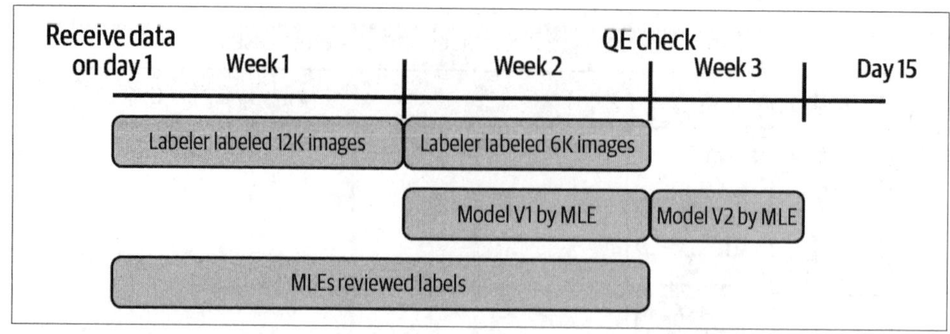

Figure 15-3. Timeline for data labeling, label review, and model training

One of the challenges that took us lots of time was to manage and update the defect consensus. Out of 38 defect classes, many pairs of classes looked very similar at first glance, so they easily confused the labelers. We had to constantly discuss ambiguous cases when disagreement occurred, and we had to update our defect definitions to maintain defect consensus among three labelers and ML engineers. For example, can you tell there are three distinctive defect classes from the nine images in Figure 15-4?

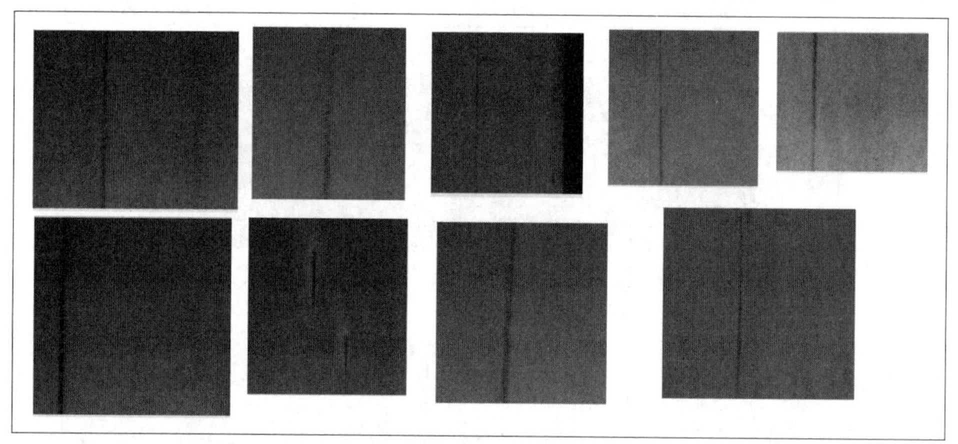

Figure 15-4. Visually identifying the distinctions among the three classes (black line, black stripe, and scratch) from nine images is far from trivial

So here are the answers. After we saw more samples of these three classes, we could continuously correct the boundaries between these three defect types and update their defect definitions (Figure 15-5). We spent lots of effort for labelers to maintain a defect consensus. For samples that were really hard to identify, we had to remove them from our training dataset.

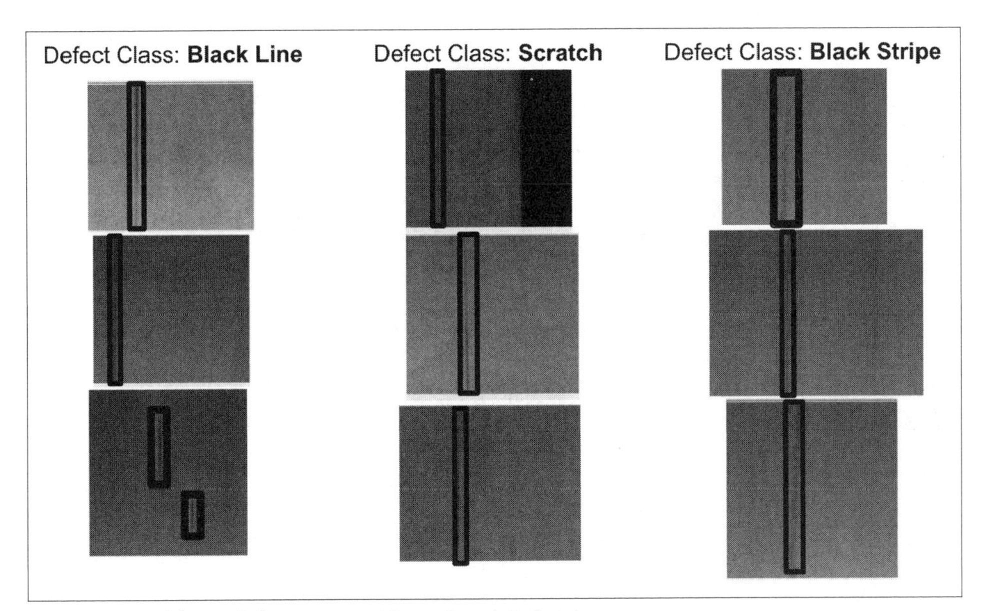

Figure 15-5. Three defect types with updated definitions

Besides the defect definition, it was also critical to establish labeling consensus. The labelers were not only expected to tell defect classes accurately, but since we were doing object detection, we also wanted their bounding boxes labeling to be tight and consistent.

For example, the samples shown in Figure 15-6 were from a defect class called *roller iron sheet*, which featured very dense holes or black dots. When labelers labeled the images, they were expected to draw tight bounding boxes around all areas with clear patterns of defects. If discontinuity occurred, they needed to annotate with separate boxes, like the third image (Figure 15-6). However, the fourth image was rejected during labeling reviewing, because the box was too wide and loosely covered a defective area. If we allowed this label to be added to our training set, it would mislead the model when calculating the losses, and we should avoid that.

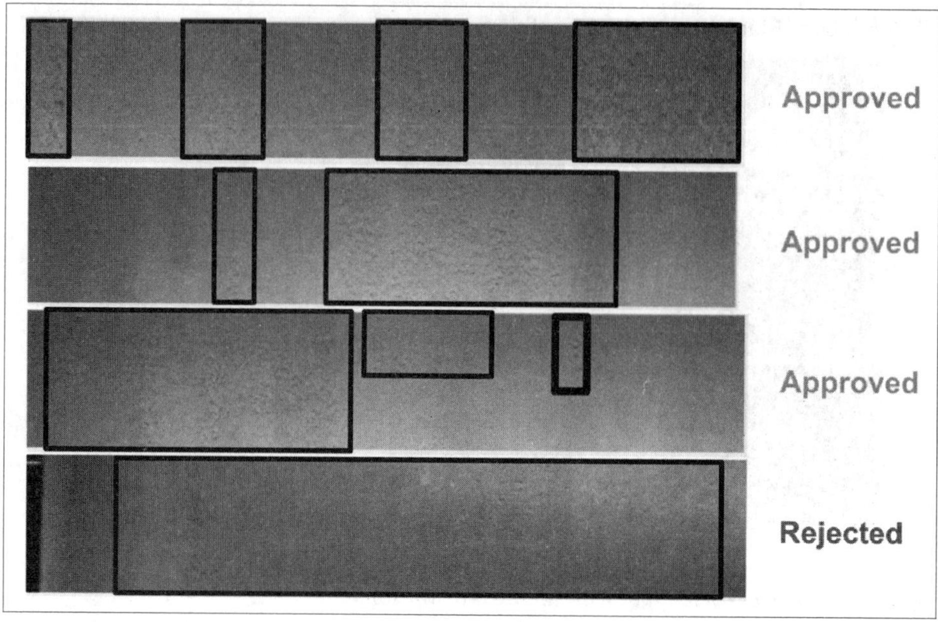

Figure 15-6. The third image shows a well-annotated discontinuity, while the fourth image was rejected during labeling review because the box too loosely covers a defective area

Takeaways

We spent less than 10% of our time doing model iterations. After each time we trained a model, I spent most of my time reviewing falsely predicted examples and identified root causes of errors. Then I took those insights back to further clean the dataset. After two iterations like this, we achieved 93% accuracy on the test set, or a 65% reduction in error rate. This far exceeded the baseline and the expectations that the customers had at that time and met their needs.

4. NLP MLOps: Profiling and Staging Load Test

By Cheng Chen, Dialpad

Background

Dialpad's AI team builds NLP applications to help users get more from their calls, including real-time transcript formatting, sentiment detection, action-item extraction, and more. Developing and deploying large NLP models systems is a challenge. Fitting them within the constraints of real-time, cost-effective performance makes that challenge significantly more complex.

In 2019, the large language model BERT achieved state-of-the-art NLP performance.[1] We planned to leverage it to provide more accurate NLP capabilities, including punctuation restoration and date, time, and currency detection. To reduce cloud cost, however, our real-time production environment has very limited resources assigned to it (GPU is not an option, and we have one CPU at most for many models). In addition, our hard limit on model inference is 50 ms per utterance. Meanwhile, the BERT base model has 12 layers of transformer blocks with 110 million parameters. We knew it would be challenging to optimize it to fit into our real-time environment, but we still overlooked one critical piece: the difficulty of obtaining an accurate estimate on how much faster the model has to be to meet our real-time demand.

Problem and Resolution

Our team needed to perform local profiling in order to benchmark various NLP models, which ran model inference over a large number of randomly sampled utterances and calculated average inference time per utterance.

Once the average inference speed met a fixed threshold, the packaged model would be handed over to our data engineering (DE) team, which would then do canary deployment in our Google Kubernetes Engine (GKE) cluster, and monitor a dashboard of various real-time metrics (Figure 15-7) with an ability to drill down into specific metrics (Figure 15-8).

This was how we gradually deployed a new BERT-based punctuator model (whose goal is to restore punctuation) into production, and this was where the confusion started. For a large language model based on BERT, often DE teams discovered that the latency or queue time bumped up significantly and they had to roll back the deployment. Clearly, the local profiling configuration was not aligned with the actual pattern occurring in the production system. This discrepancy may come from two sources: cluster compute resource allocation and/or traffic pattern. However, the reality was that scientists didn't have the right tools to properly benchmark model inference. This was resulting in time-consuming deployments and rollbacks with lots of wasted effort on repeated manual work. Add to that the anxiety about deploying a potentially underperforming model into the system, with the resulting system congestion and service outages. As a stopgap measure, applied scientists and engineers agreed to increase the compute resource, such as adding one more CPU for inference, but we clearly needed a better approach to benchmarking.

1 See the 2019 paper "BERT: Pre-training of Deep Bidirectional Transformers for Language Understanding" (*https://oreil.ly/WEh8t*) by Jacob Devlin et al.

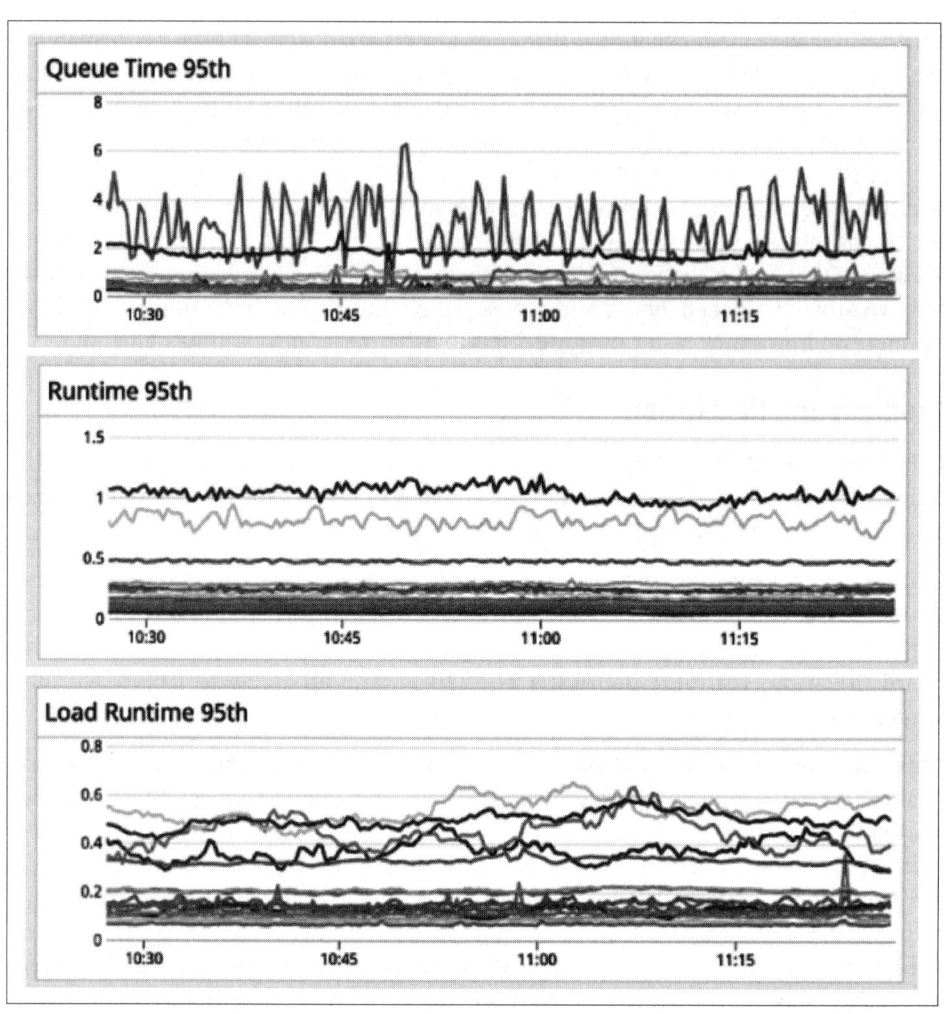

Figure 15-7. A dashboard monitoring real-time metrics

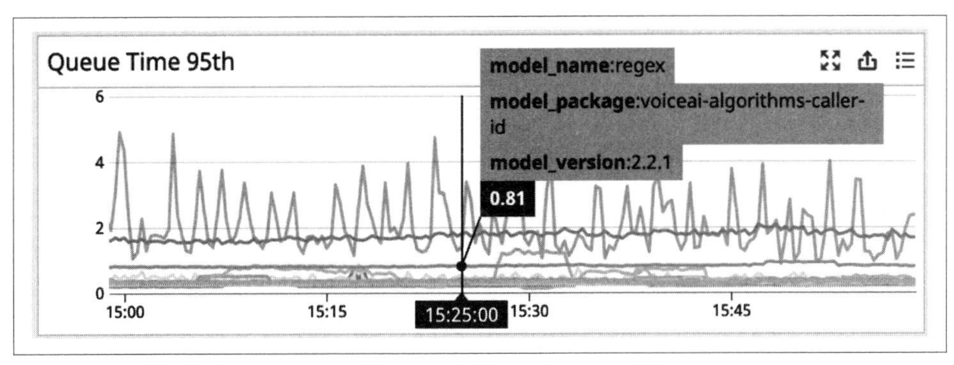

Figure 15-8. Drill-down to a specific metric for a given model

An improved process for benchmarking

What we needed was a way to allow NLP applied scientists to efficiently obtain benchmarking results that were close to production metrics. (Note that canary deployment was still required in production deployment.)

Apart from the production system, the DE team also maintained a staging environment where NLP models were deployed and integrated with the product interface (reference) prior to production deployment. Our QA team made test calls to test various call features, and applied scientists leveraged this environment to ensure that the model ran properly with the product UI. However, they had not used it to thoroughly benchmark large models.

The DE team proposed a comprehensive and self-serve load-test tool to help applied scientists benchmark model inference in the staging environment. When designing the load-test tool, we kept the following high-level points in mind:

- Load-test data should contain trigger phrases for the model.
- Load-test data should contain a healthy mix of utterance lengths. It is probably better to have longer utterances so as to give a better approximation of how the system will perform under stress.
- Load-test data should trigger model inference and not get short-circuited by optimizations/caches that would lead to misleadingly low runtime latencies.
- We use CircleCI workflows to control automatic deployments to staging.
- (Optional) Load-test data should have similar characteristics to data expected in production.

After the tool was developed, applied scientists had two options for performing a load test on staging:

- Audio-based (end-to-end) load test
  - This is a full end-to-end test.
  - This simulates calls on the system.
  - Data is sampled automatically from calls in the staging environment (e.g., QA calls) and tries to provide good coverage on certain features.
  - This audio can be customized so we can put NLP-specific audio datasets if needed.
- Text-based (model specific) load test
  - This targets only a single microservice (e.g., the punctuator or the sentiment model).
  - This allows us to pick the most difficult inputs to stress-test our models.

After scientists decided on the type of load test and had all the necessary data in place, they then deployed the changes to staging and started the load test.

Once the load test began, scientists could monitor the live dashboard for important metrics such as the Runtime 95th, as shown in Figure 15-9. That is the most significant value when evaluating inference speed. As a rule of thumb, anything below 1 second satisfies the requirement. Currently, most models tend to be clustered at or below 0.5 seconds.

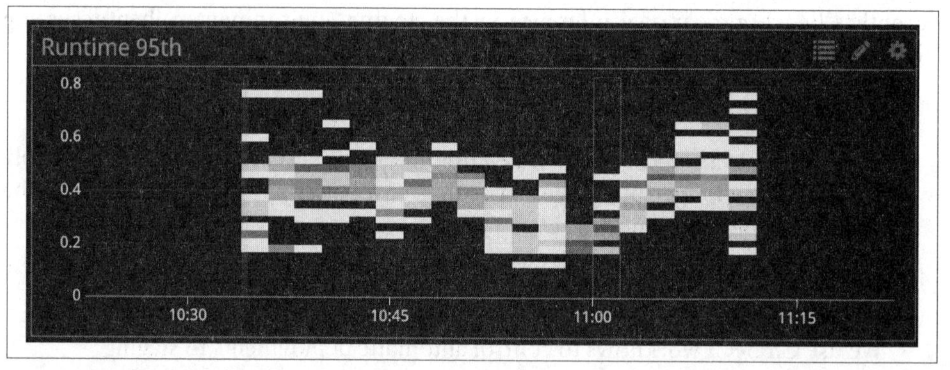

Figure 15-9. Runtime 95th percentile

With this tool in place, scientists could launch staging tests themselves, without asking for help from the DE team. The Datadog dashboard also provided a comprehensive breakdown of the runtime performance of each model so that applied scientists could monitor the metric numbers more closely. Therefore, the load-test tool significantly reduced communication overhead during our rapid development cycle.

Takeaways

Cramming state-of-the-art NLP performance into a resource-constrained real-time production environment requires very high confidence that benchmarks during testing will be borne out in production. When our testing methods started failing on the more resource-intensive BERT model, we reached into our staging environment to give our scientists a more representative environment to test against and made it self-serve so they could iterate rapidly. The automatic staging benchmarking step has since become a standard process in the model development process. Both teams are now relieved as applied scientists are able to obtain close-to-production estimates on model inference with great confidence.

5. Ad Click Prediction: Databases Versus Reality

By Daniel Papasian, Google

Background

Google's ad-targeting systems aim to help maximize the long-term value of shown ads, which includes minimizing the frequency of displaying unwanted ads to users. They do so in part by using models to predict the probability that a given ad will be clicked. When there are opportunities to show ads, an auction is conducted, and as part of this auction a server uses a model to predict the probability that certain ads will be clicked. These probabilities are one of several inputs to the auction, and if the model underperforms, both user and advertiser experience suffers. The act of displaying the ad results in us inserting a row to our database. The row in the database corresponds to an ad being shown to the user, with columns associated with the features used for model training. Additionally, a Boolean column represents whether the ad resulted in a click. This column is defaulted to `false` when the row is inserted.

If the ad is clicked, it generates a record in a click log. The click-logging team runs a process to remove fraudulent clicks, and publishes a feed of "clean" clicks for consumption by other internal users. This feed is used to issue an update to the already created row in the database to mark the ad impression as having resulted in a click.

The model was trained by looking at the rows in this database as examples, and using the click bit as the label. The rows in the data were the raw inputs to the model and the label for each event recording either "resulted in a click" or "was not clicked." If we were to never update the models, they would work well for some time, but eventually degrade in accuracy because of changing user behaviors and advertiser inventories. To improve the overall behavior, we automated the retraining and deployment of the click prediction model.

Before a retrain is pushed to production, we validate that we are improving the model's accuracy. We hold back a portion of the dataset from training and use it as a test set. Our training process handled this with Bernoulli sampling: for each ad shown in the 48 hours before training, there would be a 99% chance we'd train on it, and a 1% chance we would reserve it for our test set. This was implemented with a `test_set` bit on the row that we'd set to `true` 1% of the time. When the training system read the events table, it would filter out all rows where this was true. Newly trained models would be sent to this validation system, which generated inferences on recent events with the `test_set` bit. It compared these inferences to the observed labels to generate statistics about model accuracy. Models would be pushed to production only if the new models performed better than the old model over the recent test set events.

Problem and Resolution

One Monday, we came in and were greeted by an automatic alert: we were showing ads at a rate far below normal. We quickly realized that our mean predicted probability of a click was one-tenth of what it typically was. Whether we show an ad at all was gated in part on how probable we thought it was that the ad would be clicked. As such, the widespread underprediction of click probability explained the alert on the rate of ads being shown. But questions still remained: did our users' behavior change? Or had our model update somehow hurt us, despite our validation measures?

We queried the database for all rows in the 48 hours before the training cutoff. No matter how we aggregated it, we saw click rates that were astoundingly typical. The model was acting as if clicks were far more rare, but the data in our database didn't reflect that. But why didn't our validation system block the model from going to production? We tasked both our data science and production engineering teams to dig into the situation to understand what happened.

The data science team started by looking at the validation system: this was supposed to keep us from pushing out models that performed worse than the versions we were replacing. The validation system computed a loss metric by generating inferences over the test set. Lower loss was supposed to mean better models. The logs from Sunday's validation run indicated we processed the test events as expected, and that the loss statistic was lower for the new model than our old model. With a hunch, someone decided to rerun the validation system with the same pair of models across the test set. The test set was reread from the database, and the inferences were generated as expected. This time, the loss metric indicated the new model was worse than the old model—the opposite result from Sunday. What changed?

The production engineering teams checked a range of data from a large set of systems, trying to see whether any unexplained anomalies were in relevant systems. Curiously, a graph showed revenue of $0 for Wednesday through Sunday, and then a spike to very large amounts of revenue in the early hours of Monday. The graph was produced by a system that watched the feed of verified clicks.

When the production engineer and data scientist teams conferred with each other and shared their findings, they realized the model was underpredicting because of a failure of the infrastructure responsible for processing the raw click logs and distributing the clean click feed to consumers. The clean clicks arrived to the ML training system late—not until early Monday morning, after the model had last trained. Without any evidence to the contrary, the model believed that every ad shown in this period didn't result in a click. Every event was a true negative or a false negative, and that's all our test set contained as well. The model that we trained concluded that the ads were awful and no one would click on any of them, which was accurate given the data that we had when we trained the model. When the click-processing feed caught up, the validation data was relabeled so that ads that resulted in clicks were labeled that way. This explained why subsequent attempts at validation of the already trained model were still failing. The issue was resolved by retraining on the corrected data.

Takeaways

In retrospect, it's important to note that our model never directly predicted the probability of a click. Rather, it predicted the probability of a shown ad being marked as "was clicked" in our database at the time that training happens. Indeed, while we expect that our database is an accurate reflection of reality, bugs or failures may cause differences. In this case, a production failure in an upstream system caused the meaning of this label to diverge from what we wished it to reflect. Our models built using supervised learning techniques predict labels in our training set, and it's of critical importance that our labels reflect reality.

The teams collaborated to write a postmortem to analyze what happened and how to prevent it. This turned out to be a significant learning experience for our organization. We gathered the timeline: from the perspective of people working on the click prediction models, the problem wasn't detected until Monday. We later learned that the team that worked on the click logs noticed their pipeline was broken on Wednesday with a software release, and were aware of the problem the same day when their alerting indicated the feed wasn't being processed. They put in place mitigations to ensure that clicks would still be eventually billed, and figured they'd fix the rest of the data-feed processing first thing on Monday. They hadn't realized their system was a data dependency of an ML process downstream, and how that system was making assumptions about the completeness of data. We believe many ML pipelines make assumptions about the completeness of input data and correctness of provided labels without verifying these assumptions, and as such are at risk for similar problems.

We listed every potential cause of the outage we could muster; we knew we'd prioritize these based on effort to make an improvement and expected value of the improvement. The causes included the lack of integration testing leading to the click logs breaking, the training system's reliance on the click log processing being more reliable than agreed-upon service availability targets, and our assumption that the most recent events would be representative of all events.

Our follow-ups included establishing an availability target for the click log processing systems, expanding our validation system to check that the test set's positive ratio wasn't suspiciously low or high, and establishing a process for the click log team to communicate outages and pause training if serious problems with model health occurred.

6. Testing and Measuring Dependencies in ML Workflow

By Harsh Saini, Dialpad

Background

At Dialpad, we have a speech recognition and processing engine that has several ML dependencies. Audio comes in from our telephony backend and gets transcribed in real time through our proprietary ASR models, where formatting and readability improvements are also made. The output is then fed into our language-specific NLP models like NER and sentiment analysis. This pipeline can be simplified as a flowchart (Figure 15-10).

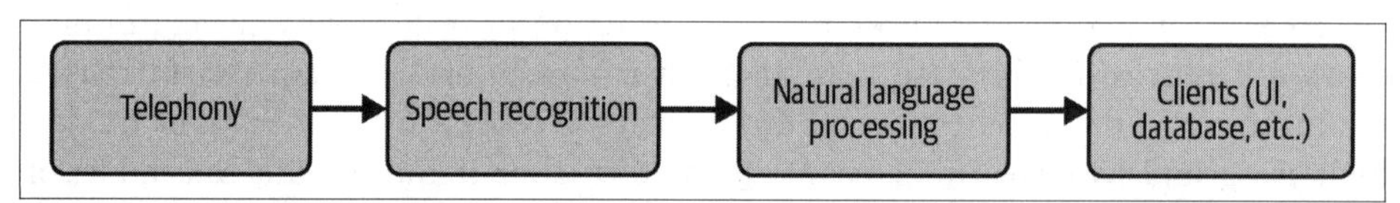

Figure 15-10. Flowchart of Dialpad speech-recognition and processing pipeline

However, in reality, the pipeline is not as straightforward as shown in this simplified diagram. Multiple speech-recognition models may be used, depending on the user's location and/or the product line they are using within Dialpad. For instance, a user from the UK using the call-center product line will be provided with a speech-recognition model fine-tuned on the UK dialect of English and trained on call-center domain-specific knowledge. Similarly, a user from the US using the sales product line will have their call transcribed using a speech-recognition model trained on the US English dialect and domain-specific knowledge for sales calls.

Additionally, for the NLP, several task-specific models run in parallel to perform tasks such as sentiment analysis, question detection, and action-item identification. With this in mind, the simplified flowchart can be extended to highlight the diversity of models in Dialpad's production ML pipeline.

Problem and Resolution

Figure 15-11 highlights some of the ML dependencies that exist for NLP task-specific models with regards to the upstream speech-recognition models. NLP model performance is sensitive to ASR model output artifacts. While most NLP models are not overly sensitive to minor changes to ASR model outputs, over time the data changes significantly enough, and NLP models experience a degradation in performance due to regression and data drift. A few common updates to ASR that result in a change in input data distribution for NLP models are as follows:

- Modifications in the vocabulary of the ASR system (e.g., the addition of the word *coronavirus*)
- Changes in the output of the ASR system (e.g., people are saying the same things, but we're getting better at accurately transcribing them)
- Topic drift, whereby people are actually talking about different things (e.g., suddenly everyone starts talking about elections in the US)

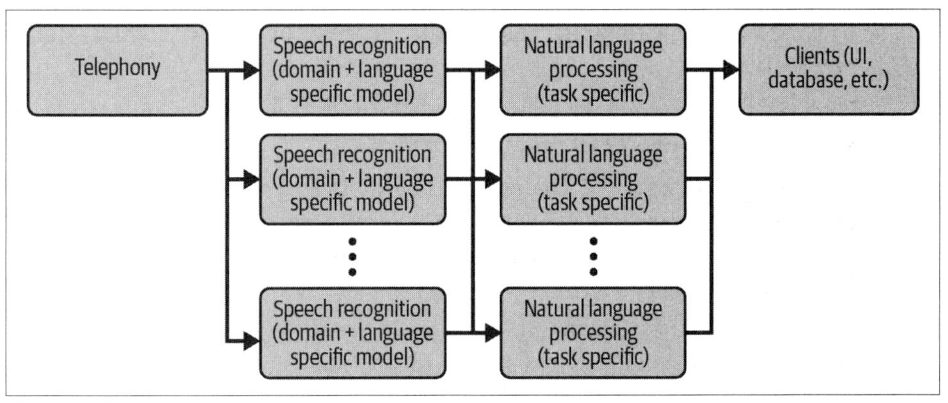

Figure 15-11. Some of the ML dependencies that exist for NLP task-specific models

To combat this phenomenon, the DE team at Dialpad, in collaboration with the data science team, built an offline testing pipeline that could measure NLP model performance for a given ASR model.

Building the regression-testing sandbox

Some of the key requirements for our regression testing and monitoring system were as follows:

- Ensure that monitoring of NLP model performance would happen automatically whenever newer ASR models are released.
- Simulate the behavior as observed in production by the models.
- Collect and report metrics submitted via the evaluation that can be viewed by stakeholders.
- Collect model inference artifacts and logs so as to assist in troubleshooting by scientists.
- Allow for ad hoc evaluation by data science teams when they wish to evaluate a model prerelease.
- Ensure that we could establish comparable baselines, since datasets for evaluation could be modified out of band.
- Be a scalable system so that multiple evaluations could occur simultaneously, and also not be bottlenecked as we increase either the dataset sizes or the number of models being tested.

Given these requirements, the following design decisions were made:

- Kubeflow Pipelines (KFP) was chosen as the platform to host the sandbox:
  - KFP allows users to write custom directed acyclic graphs (DAGs) called *pipelines*.
  - Each pipeline is sandboxed, and the platform as a whole can independently scale to the demands of all running pipelines.
  - The engineering teams at Dialpad are heavily invested in Kubernetes (*https://kubernetes.io*) and Argo Workflows (*https://argoproj.github.io/workflows*), which are the underlying technology powering KFP, so it seemed prudent to use this platform.
- The pipelines in KFP will build the correct infrastructure for evaluation by selecting the correct model deployment artifacts, given the evaluation criteria.
  - This will be done on the fly and will not be persisted to reduce cost.
  - The testbed will be decoupled from model versions and be aware of the order of dependencies only for correct orchestration.
- Outputs from every model will be persisted for 30 days for debugging purposes.
- Datasets for every task-specific NLP model would be versioned so as to track changes in evaluation data.

- Metrics will be collected for every combination of ASR model version, NLP model version, and dataset version.
  - This ensures that we can disambiguate among different dependencies correctly.
  - These metrics would then be visualized in a dashboard for observability.
- The input to the testing pipeline is raw audio recordings of conversations, since the idea was to capture whether an ASR model has changed in such a way that it alters the output enough that the downstream NLP model has varied performance.
  - Once audio samples were collected, they would be annotated to state whether they contain a specific NLP moment. For instance, a given audio snippet would be annotated by a human to verify whether it contained positive, negative, or neutral sentiment for the sentiment analysis task.
  - As you can see, this is an arduous task and is still one of the biggest bottlenecks for this project. It is extremely time-consuming to correctly slice, annotate, and store such samples for every NLP task (Figure 15-12).

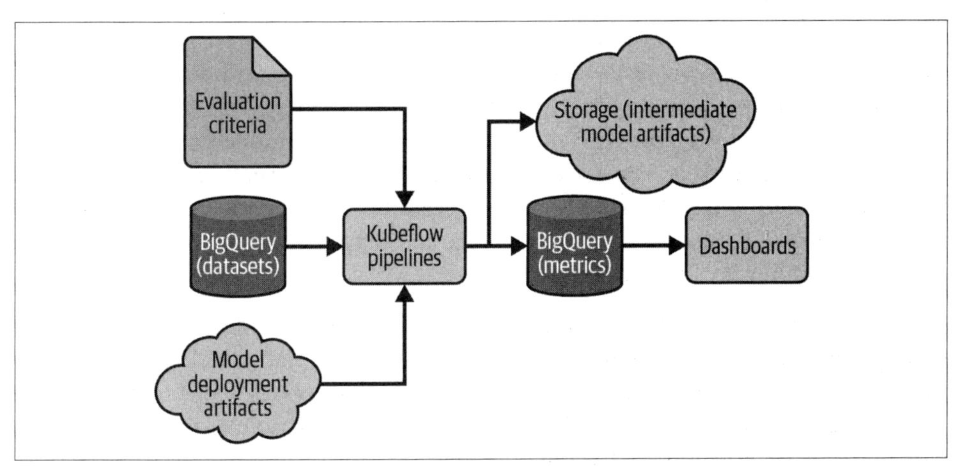

Figure 15-12. The regression-testing environment at a high level

And within KFP, a pipeline would simulate evaluation for a single combination of ASR model version, NLP model version, and dataset version. Since KFP allows us to run multiple pipelines in parallel, this would allow us to scale to all combinations of evaluation we would like to perform (Figure 15-13).

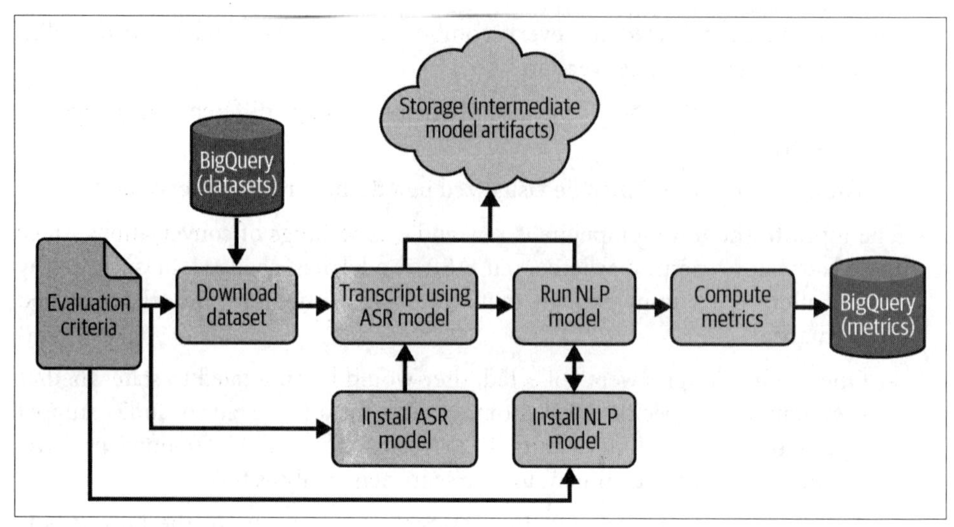

Figure 15-13. The architecture of the KFP pipeline as a DAG

Monitoring for regression

Once the pipelines were built on KFP, the next part of the project was to automatically perform regression tests whenever dependencies changed for NLP models. Luckily, at Dialpad we have mature CI/CD workflows managed by engineering, and they were updated to trigger KFP pipelines whenever ASR models were updated in the transcription service. The CI/CD workflow would send a signal to KFP with information about the ASR models, NLP models, etc., and the evaluation would then commence on KFP. Metrics would be stored, and Slack messages would be emitted containing a summary of the evaluation.

Once operational, this process captures performance evaluation data for all NLP task-specific models that have testing data available on the platform. For example, the F1-score of the NLP sentiment analysis model degraded by ~25% over the course of a year, as shown in Figure 15-14; the graph highlights the absolute difference from a baseline. This observation alerted the NLP team to investigate the issue and discover that accumulated data drift was the cause of the degradation. A new sentiment model was retrained using the latest ASR model output and released to production in just a few months.

Another tangential benefit of this process is that it allows for ad hoc evaluation of NLP models against different ASR models prior to production release. For instance, it is possible to measure the accuracy of a sentiment analysis model, prior to release, against an ASR model trained on new English dialects, such as Australian or New Zealand English.

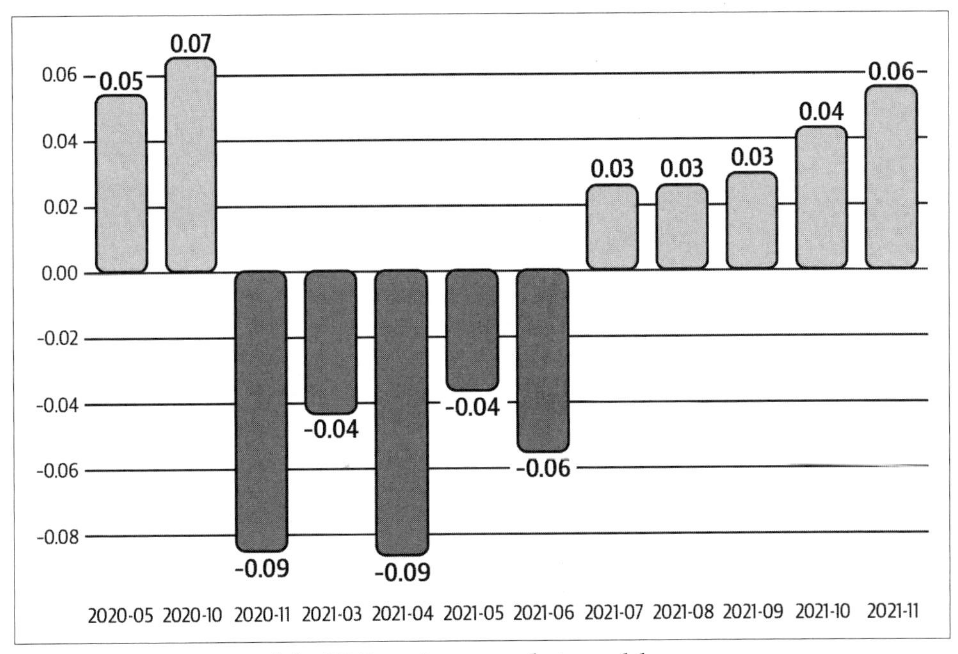

Figure 15-14. F1-score of the NLP sentiment analysis model

Takeaways

This ML regression-testing platform developed at Dialpad has provided data scientists and engineers with much improved visibility on the impact of new model releases on all dependent components in our production stack. Even with an incomplete knowledge of all the deployed production models, people are able to understand whether a proposed release is going to impact the stability and performance of other models in the production pipeline. This reduces the chances of a rollback and can provide an early indication if more work needs to be done to improve compatibility with existing components.

The testing platform is under active development. Other moving pieces are being addressed, one of which is keeping the sandbox orchestration in sync with production and allowing for other "live data" that only transiently lives during a call on production and is difficult to simulate in the regression-testing platform. Another feature being considered is how to provide automated alerting when a proposed release has significant impact on downstream models rather than the current human-in-the-loop approach.

Index

A

B

C

D

E

F

G

H

I

J

K

L

M

N

O

P

Q

R

S

T

U

About the Authors

Cathy Chen, CPCC, MA, specializes in coaching tech leaders to enable development of their own skills in leading teams. She has held the role of technical program manager, product manager, and engineering manager. She has led teams in large tech companies and startups launching product features, internal tools, and operating large systems. Cathy has a BS in electrical engineering from UC Berkeley and an MA in organizational psychology from Teachers College at Columbia University. Cathy lives with her partner in Pittsburgh, Pennsylvania, and works at Google in SRE.

Niall Richard Murphy has worked in internet infrastructure since the mid-1990s, specializing in large online services. He has worked with all of the major cloud providers from their Dublin, Ireland offices, and most recently at Microsoft, where he was global head of Azure site reliability engineering (SRE). His first exposure to machine learning came with managing the Ads ML teams in Google's Dublin office and working with Todd Underwood in Pittsburgh, though it has continued to fascinate him since. He is the instigator, coauthor, and editor of the two Google SRE books, and he is probably one of the few people in the world to hold degrees in computer science, mathematics, and poetry studies. He lives in Dublin with his wife and two children, and works on a startup involving ML in the SRE space.

Kranti Parisa is the vice president and head of product engineering at Dialpad. His teams build large-scale, cloud native, real-time business communications and collaboration software with industry-leading in-house AI/ML and telephony technology. Before Dialpad, he led teams responsible for search and personalization platforms, products, and services at Apple. Kranti was a cofounder, CTO, and technical advisor of multiple startups focusing on cloud computing, SaaS, and enterprise search. He has contributed to the Apache Lucene/Solr community and coauthored *Apache Solr Enterprise Search Server*. For his outstanding contributions to search and discovery, the US government has recognized him as a Person of Extraordinary Ability (EB1A).

D. Sculley is the CEO of Kaggle and GM of Third-Party ML Ecosystems at Google, and previously was a director of the Google Brain Team and the lead of some of Google's most critical production machine learning pipelines. He has focused on issues of technical debt in machine learning, along with robustness and reliability of models and pipelines, and has led teams applying machine learning to problems as diverse as ad click-through prediction and abuse prevention to protein design and scientific discovery. Additionally, he helped create Google's Machine Learning Crash Course, which has taught ML to millions of people worldwide.

Todd Underwood is a senior director at Google and leads machine learning SRE. He is also site lead for Google's Pittsburgh office. ML SRE teams build and scale internal and external ML services, and are critical to almost every significant product at Google. Before working at Google, Todd held a variety of roles at Renesys (in charge of operations, security, and peering for internet intelligence services), now part of Oracle Cloud, and before that he was CTO of Oso Grande, an independent internet service provider in New Mexico.

Colophon

The insect on the cover of *Reliable Machine Learning* is the honeypot ant (*Myrmecocystus mimicus*). Honeypot ants are found in southwest North America and parts of Mexico.

Similar to other ants, honeypot ant colonies consist of a variety of worker ants who scavenge food from flowers, fruit, and other insects. What is most notable about honeypot ants is how they store food. The repletes—one type of worker ant in the colony—grow large abdomens that they use to store the liquid they scavenge. During times when food supply is low, the repletes regurgitate liquid for the rest of the colony to eat. Repletes have a hard time moving around because of the size of their abdomen, so they are often found hanging from the roof of their nest.

Many of the animals on O'Reilly covers are endangered; all of them are important to the world.

The cover illustration is by Karen Montgomery, based on an antique line engraving from *Bibliothéque de Merveilles: Les Fourmis*. The cover fonts are Gilroy Semibold and Guardian Sans. The text font is Adobe Minion Pro; the heading font is Adobe Myriad Condensed; and the code font is Dalton Maag's Ubuntu Mono.

O’Reilly Media, Inc.介绍

O’Reilly以“分享创新知识、改变世界”为己任。40多年来我们一直向企业、个人提供成功必需之技能及思想，激励他们创新并做得更好。

O’Reilly业务的核心是独特的专家及创新者网络，他们通过我们分享知识。我们的在线学习（Online Learning）平台提供独家的直播培训、图书及视频，使客户更容易获取业务成功所需的专业知识。几十年来O’Reilly图书一直被视为学习开创未来之技术的权威资料。我们全年举办的诸多会议是活跃的技术聚会场所，来自各领域的专业人士在此建立联系，讨论最佳实践并发现可能影响技术行业未来的新趋势。

我们的客户渴望作出推动世界前进的创新，我们能祝您一臂之力。

业界评论

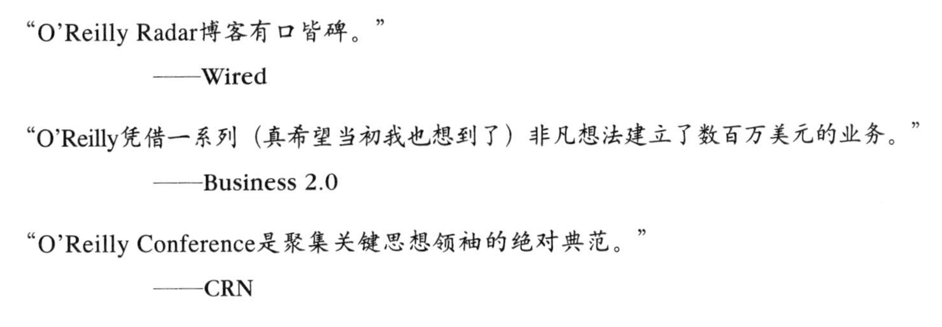

“O’Reilly Radar博客有口皆碑。”

——Wired

“O’Reilly凭借一系列（真希望当初我也想到了）非凡想法建立了数百万美元的业务。”

——Business 2.0

“O’Reilly Conference是聚集关键思想领袖的绝对典范。”

——CRN

“一本O’Reilly的书就代表一个有用、有前途、需要学习的主题。”

——Irish Times

“Tim是位特立独行的商人，他不光放眼于最长远、最广阔的视野并且切实地按照Yogi Berra的建议去做了：‘如果你在路上遇到岔路口，走小路（岔路）。’回顾过去Tim似乎每一次都选择了小路，而且有几次都是一闪即逝的机会，尽管大路也不错。”

——Linux Journal